Stem Cell Therapy for Neurodegenerative Disorders

This book comprehensively reviews the role of stem cell therapy against various neurodegenerative disorders. The initial chapters present different neurodegenerative disorders and discuss the application of stem cell therapy for Alzheimer's disease, Parkinson's disease, amyotrophic lateral sclerosis (ALS), and Huntington's disease. Aging and its impact on neurodegenerative disorders are also covered, addressing neuronal dysfunctions with age, neuromuscular dysfunctions like sarcopenia and frailty, and the potential role of stem cell interventions in aging. The subsequent section provides an overview of the different types of therapies, including cell replacement and neuroprotection. It also covers clinical trials of stem cell therapies for neurodegenerative disorders and discusses the advantages and disadvantages of various stem cell types and delivery methods. Further, the book explores the mechanisms of action, detailing how stem cells induce neural regeneration and repair. It also examines their role in reducing inflammation and oxidative stress, alongside potential risks and safety concerns associated with stem cell therapies. Toward the end, the book presents advanced delivery systems for stem cell therapies, the therapeutic potential of stem cell-derived exosomes and microvesicles in neurodegeneration, preclinical studies in animal models, and the role of organoids in neurodegenerative disorders. This book is useful for researchers and students of neuroscience, neurology, stem cells, pharmaceutical sciences, and biotechnology.

Key Features:

- Provides an in-depth review of stem cell therapy for various neurodegenerative disorders, including Alzheimer's disease, Parkinson's disease, ALS, and Huntington's disease
- Addresses aging-related neuronal and neuromuscular dysfunctions, such as sarcopenia and frailty
- Offers a detailed analysis of different types of stem cell-based therapies, including cell replacement and neuroprotection
- Covers clinical trials of stem cell therapies for neurodegenerative disorders, discussing the advantages and disadvantages of various stem cell types
- Explores the mechanisms by which stem cells promote neural regeneration and repair and reduce inflammation and oxidative stress
- Discusses cutting-edge delivery systems for stem cell therapies, enhancing the effectiveness and precision of treatments

This book is useful for researchers and students of neuroscience, neurology, stem cells, pharmaceutical sciences, and biotechnology.

Stem Cell Therapy for Neurodegenerative Disorders

Mechanisms, Applications, and Advances

Edited by
Masroor Anwar, Misbahuddin Rafeeq, and Mukhtiar Baig

CRC Press
Taylor & Francis Group
Boca Raton London New York

CRC Press is an imprint of the
Taylor & Francis Group, an **informa** business

Designed cover image: Shutterstock

First edition published 2027
by CRC Press
2385 NW Executive Center Drive, Suite 320, Boca Raton FL 33431

and by CRC Press
4 Park Square, Milton Park, Abingdon, Oxon, OX14 4RN

CRC Press is an imprint of Taylor & Francis Group, LLC

ISBN: 9781032949840 (hbk)
ISBN: 9781032949864 (pbk)
ISBN: 9781003582687 (ebk)

DOI: 10.1201/9781003582687

Typeset in Times
by Newgen Publishing UK

Contents

Preface

Neurodegenerative diseases constitute one of the most formidable biomedical challenges of the twenty-first century, encompassing a broad spectrum of disorders, including Alzheimer's disease, Parkinson's disease, Huntington's disease, amyotrophic lateral sclerosis, and multiple sclerosis that collectively affect millions worldwide. Despite distinct clinical phenotypes, these disorders share convergent pathogenic mechanisms involving aberrant protein aggregation, mitochondrial dysfunction, impaired autophagy, oxidative stress, and chronic neuroinflammation. This book presents these mechanisms while delineating the emerging regenerative and cell-based therapeutic strategies that aim to restore neuronal structure and function.

The first chapter establishes a molecular and pathophysiological foundation by focusing on the cellular cascades responsible for neurodegeneration. It discusses the interplay between proteostasis failure, mitochondrial impairment, and glial-mediated neuroinflammation, elucidating how genetic predisposition and environmental factors drive neuronal loss. The second chapter reviews this perspective by determining aging as the primary risk for neurodegenerative pathologies. It systematically analyzes the hallmarks of aging, focusing on genomic instability, telomere attrition, epigenetic drift, mitochondrial decline, and immunosenescence that collectively precipitate neuronal vulnerability and cognitive deterioration. By framing these processes within the context of neurobiology, the chapter emphasizes the need for geroprotective interventions that address the upstream biological determinants of neurodegeneration rather than its symptomatic outcomes.

The subsequent chapter on extracellular vesicles (EVs) delineates their dualistic role in neurodegeneration and repair. While EVs can propagate misfolded proteins such as amyloid-beta, tau, and α-synuclein, they also possess neuroprotective potential as endogenous nanocarriers capable of traversing the blood–brain barrier to deliver trophic and regulatory molecules. Stem cell-derived exosomes have emerged as a promising cell-free therapeutic platform for modulating neuroinflammation, promoting synaptic recovery, and facilitating neuroregeneration.

The following chapter focuses on mesenchymal stem cells (MSCs), highlighting their pleiotropic mechanisms of action encompassing neurotrophic support, immunomodulation, angiogenesis, and stimulation of endogenous progenitor cells. Extensive preclinical evidence and early clinical trials indicate that MSCs can attenuate neuroinflammation, enhance neuronal survival, and promote remyelination, thereby positioning them as a viable therapeutic modality for diverse neurodegenerative contexts. Complementarily, the chapter on immunomodulatory properties of stem cells elucidates how MSCs and neural stem cells orchestrate immune homeostasis through secretion of anti-inflammatory cytokines, induction of regulatory T-cell subsets, and reprogramming of microglia from pro-inflammatory to neuroprotective phenotypes. This immunological homeostasis is the crucial determinant of successful neural repair.

Progressing toward translational advancement, the chapter on nanoparticle-assisted stem cell therapies presents nanotechnology as a critical adjunct to regenerative medicine. The integration of nanoparticles with stem cells enhances cell survival, engraftment, and targeted delivery across the blood–brain barrier while permitting real-time imaging and controlled therapeutic release. This convergence of nanotechnology and cell therapy signifies a paradigm shift toward precision neuroregeneration.

The final chapter examines the role of induced pluripotent stem cells (iPSCs) as a transformative platform for disease modeling and personalized therapy. iPSC-derived neurons, astrocytes, and organoids faithfully recapitulate human neurodegenerative phenotypes, thereby offering unprecedented insight into disease mechanisms and individualized drug testing. The chapter also discusses advances in CRISPR/Cas9-mediated genome editing and three-dimensional brain organoid technologies, which together facilitate the development of next-generation autologous and allogeneic cell therapies.

This book reflects the convergence of molecular neuroscience, stem cell biology, immunology, and nanomedicine. By integrating fundamental mechanisms with translational innovations, the volume underscores a shift from symptomatic management toward mechanistically driven, regenerative interventions capable of modifying disease trajectories. It aims to serve as a comprehensive reference for students and researchers of neuroscience, biomedical sciences, nanotechnology, and stem cell biology.

About the Authors

Masroor Anwar is a Scientist D in the Department of Biophysics at the All India Institute of Medical Sciences, New Delhi. His research focuses on aging and age-related diseases, specifically skeletal muscle regeneration and muscle stem cell biology. He also has a keen interest in mesenchymal stem cell regeneration. Currently, Dr. Anwar is working on neurodegenerative diseases, including Alzheimer's and dementia. He has published several research articles in high-impact international journals and serves on the editorial boards of several journals.

Misbahuddin Rafeeq is an Associate Professor in the Department of Pharmacology at the Faculty of Medicine (Rabigh), King Abdulaziz University, Saudi Arabia. He completed his MBBS and MD in Pharmacology from Aligarh Muslim University, India. Dr. Rafeeq serves on the editorial boards of several other journals. He has received numerous awards and fellowships, including the Excellent Leadership Award from IDB, Saudi Arabia, and the Distinguished Achievement Award from KAU, Saudi Arabia. His research focuses on identifying novel genetic mutations in hereditary syndromes, exploring therapeutic targets through molecular docking, PK/PD, and toxicity studies, and studying the effects of natural compounds on animal models. Dr. Rafeeq has more than 65 high-impact publications and has secured funding for several large grants as a Principal/Co-Investigator.

Mukhtiar Baig is a Distinguished Professor of Clinical Biochemistry and Medical Educationist, currently serving as the Coordinator of the Assessment Unit at the Faculty of Medicine in Rabigh, King Abdulaziz University, Jeddah, KSA. He holds multiple advanced degrees, including an MBBS from Liaquat Medical College, an MPhil and PhD in Biochemistry from the University of Karachi, and an MHPE from Dow University of Health Sciences. Dr. Baig has been globally recognized, being included in Stanford University's list of the top 2% of scientists worldwide for 2022 and 2023. He has authored 160 international refereed publications and contributed to seven book chapters. Dr. Baig's research interests cover clinical biochemistry, identification of genetic mutations, and the effects of natural compounds on animal models.

1 Introduction to Neurodegenerative Disorders

1.1 INTRODUCTION

Neurodegenerative disorders are a heterogeneous group of diseases that are characterized by the progressive degeneration of the nervous system, primarily affecting neurons within specific regions of the brain and spinal cord. These disorders include Alzheimer's disease (AD), Parkinson's disease (PD), Huntington's disease (HD), amyotrophic lateral sclerosis (ALS), and multiple sclerosis (MS), among others. Each of these conditions is marked by distinct neuropathological features and clinical manifestations, yet they share common underlying mechanisms such as abnormal protein aggregation, impaired proteostasis, mitochondrial dysfunction, neuroinflammation, oxidative stress, and genetic susceptibility (Table 1.1).

At a cellular and molecular level, neurodegenerative diseases typically involve the accumulation and aggregation of misfolded proteins, which can form toxic oligomers and fibrillar deposits. For instance, AD is characterized by extracellular amyloid-beta plaques and intracellular neurofibrillary tangles (NFTs) composed of hyperphosphorylated tau proteins (Table 1.2; Guo et al. 2020). Similarly, PD involves aggregation of α-synuclein protein into Lewy bodies, leading to dopaminergic neuron loss in the substantia nigra pars compacta (Sulzer et al. 2017). In HD, expanded polyglutamine tracts in huntingtin protein trigger widespread neuronal dysfunction and death through transcriptional dysregulation and impaired proteostasis (Bates et al. 2015). ALS, a devastating motor neuron disorder, involves pathological aggregates of proteins like TDP-43 and mutant SOD1, which disrupt essential cellular processes and lead to degeneration of motor neuron (Ling et al. 2013).

Emerging evidence suggests the critical role of genetic and environmental factors as well as epigenetic modifications in modulating disease etiology and progression. For example, mutations in APP, PSEN1, and PSEN2 significantly contribute to familial AD, whereas common variants in genes like APOE strongly influence the risk for sporadic forms (Serrano-Pozo et al. 2011). Similarly, PD is associated with mutations in genes such as SNCA, LRRK2, and GBA1, which significantly influence α-synuclein pathology, lysosomal function, and mitochondrial functions (Table 1.3; Nalls et al. 2019). The identification of genetic modifiers in diseases such as Huntington's and ALS have further enhanced the understanding of disease mechanisms and provide potential therapeutic targets.

Neuroinflammation is increasingly recognized as a key pathogenic factor common to many neurodegenerative disorders. Activation of microglia and astrocytes contributes significantly to disease pathology through the release of pro-inflammatory cytokines and reactive oxygen species (ROSs), thereby exacerbating neuronal injury and dysfunction (Hanslik and Ulland 2020). Moreover, the complex interplay between neurons, glial cells, and immune cells highlights the noncell autonomous nature of neurodegeneration, underscoring the importance of integrated approaches in therapeutic strategies.

DOI: 10.1201/9781003582687-1

TABLE 1.1
Major Neurodegenerative Disorders and their Key Pathological Features

Disorder	Key Protein Aggregates	Primary Brain Region Affected	Major Clinical Symptoms
Alzheimer's disease	Amyloid-β, Tau	Hippocampus, cortex	Memory loss, cognitive decline
Parkinson's disease	α-Synuclein	Substantia nigra	Tremor, rigidity, bradykinesia
Huntington's disease	Mutant huntingtin (mHTT)	Striatum, cortex	Chorea, psychiatric symptoms
Amyotrophic lateral sclerosis	TDP-43, SOD1	Motor cortex, spinal cord	Muscle weakness, atrophy
Multiple sclerosis	Myelin-specific immune response	CNS white matter	Visual/motor disturbances

TABLE 1.2
Shared Molecular Mechanisms across Neurodegenerative Diseases

Mechanism	Neurodegenerative Disorders	Description
Protein aggregation	AD, PD, HD, ALS	Misfolded proteins form toxic aggregates
Mitochondrial dysfunction	All	Impaired ATP production and increased ROS
Neuroinflammation	All	Glial activation and cytokine release
Impaired autophagy	AD, PD, HD, ALS	Failure of cellular waste clearance
Genetic susceptibility	All	Inherited mutations and risk variants

TABLE 1.3
Key Genetic Mutations in Major Neurodegenerative Disorders

Disease	Gene	Mutation/Polymorphism	Functional Effect
Alzheimer's disease	APP, PSEN1/2, APOE4	Missense, ε4 allele	Aβ overproduction or impaired clearance
Parkinson's disease	SNCA, LRRK2, GBA1	Duplication, point mutations	α-Synuclein aggregation, lysosomal dysfunction
Huntington's disease	HTT	CAG repeat expansion	Polyglutamine toxicity
Amyotrophic lateral sclerosis	SOD1, C9orf72, TARDBP	Missense, repeat expansion	Protein/RNA toxicity
Multiple sclerosis	HLA-DRB1*15:01, IL2RA	Polymorphisms	Immune dysregulation

In recent years, advances in biomarkers and neuroimaging have led to early diagnosis, monitoring, and management of neurodegenerative diseases. Biomarkers such as neurofilament light chain (NfL), phosphorylated tau species, and α-synuclein oligomers have demonstrated potential in accurately predicting disease onset and progression and evaluating therapeutic interventions (Kang et al. 2022; Parnetti et al. 2019).

Stem cell-based therapies offer promising avenues for treating neurodegenerative diseases, harnessing their potential for neuronal regeneration, modulation of inflammation, and restoration of cellular homeostasis. The application of stem cells in therapeutic context continues to evolve rapidly, guided by advancements in understanding disease pathogenesis, stem cell biology, and regenerative medicine. This chapter provides an overview of the molecular pathology, clinical features, and emerging therapeutic strategies of different neurodegenerative disorders.

1.2 ALZHEIMER'S DISEASE

AD is neuropathologically defined by extracellular amyloid-β (Aβ) plaques and intracellular NFTs of hyperphosphorylated tau (Guo et al. 2020). Multiple molecular pathways are involved to generate these lesions: APP processing to toxic Aβ, aberrant tau phosphorylation and aggregation, genetic risk factors that modulate these processes, impaired protein clearance via endolysosomal and autophagic systems, mitochondrial/oxidative dysfunction, and chronic neuroinflammation (Guo et al. 2020; Hanslik and Ulland 2020). The following sections detail each of these mechanisms at the cellular and molecular level.

1.2.1 Amyloid-β Pathology

In AD, the amyloid precursor protein (APP) is sequentially cleaved by β-secretase (BACE1) and γ-secretase to generate Aβ peptides, whereas cleavage by an initial α-secretase (e.g. ADAM10) precludes Aβ formation (Tang 2020). In the nonamyloidogenic pathway, α-secretase yields soluble sAPPα and a C83 fragment; γ-secretase then produces a short nonpathogenic p3 peptide. In contrast, the amyloidogenic pathway involves β-secretase cleavage to release sAPPβ and the C99 fragment, which is trimmed by γ-secretase to Aβ peptides of various lengths (Figure 1.1; Guo et al. 2020). The predominant species are Aβ-40 and the more hydrophobic Aβ-42. The longer Aβ-42 species aggregates more readily and seeds pathology (Guo et al. 2020). The overproduction or impaired clearance of Aβ, especially elevated Aβ-42/Aβ-40 ratio, initiates plaque formation (Guo et al. 2020).

Once secreted, Aβ monomers misfold and coalesce into soluble oligomers, protofibrils, and eventually insoluble fibrils that deposit as plaques (Figure 1.1; Guo et al. 2020). Soluble Aβ oligomers,

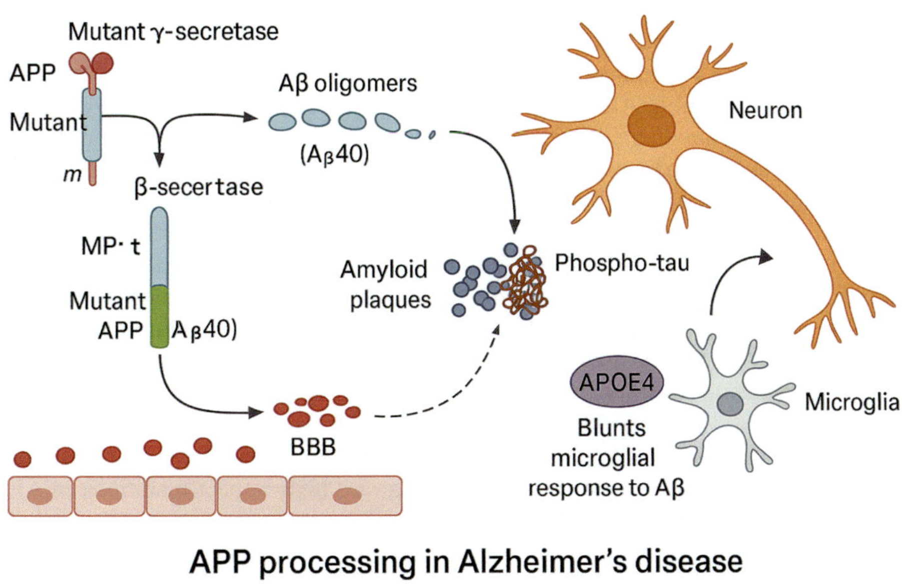

FIGURE 1.1 Amyloid Precursor Protein (APP) Processing in Alzheimer's Disease.

ranging from dimers and trimers up to dodecamers, are most neurotoxic species (Guo et al. 2020). These oligomers perturb neuronal function by forming pore-like structures in membranes, dysregulating calcium homeostasis, and eliciting excitotoxic glutamate release (Guo et al. 2020). Oligomeric Aβ also impairs synaptic plasticity and triggers oxidative stress and mitochondrial injury (Guo et al. 2020). Thus, aberrant APP processing produces aggregation-prone Aβ peptides that accumulate into toxic oligomers and plaques, driving synaptic dysfunction and neurodegeneration (Tang 2020; Guo et al. 2020).

1.2.2 Tau Pathology

Tau is a microtubule-associated protein that stabilizes axonal microtubules in neurons. In AD, tau becomes hyperphosphorylated at several serine/threonine residues (Guo et al. 2020). Kinases such as GSK3β, CDK5, and MAPKs are upregulated or dysregulated, while phosphatases like PP2A are downregulated, yielding hyperphosphorylated tau (Guo et al. 2020). Hyperphosphorylation causes tau to detach from microtubules and misfold. The detached tau self-assembles into paired helical filaments and NFTs within neurons (Guo et al. 2020). These tangles disrupt cytoskeletal integrity and impair neuronal transport. The regional distribution of tau pathology follows characteristic Braak stages and correlates with severity of dementia (Guo et al. 2020).

Compelling evidence indicates that tau aggregates spread in a prion-like manner through neural circuits. The misfolded tau can seed aggregation of normal tau in neighboring neurons, propagating trans-synaptically pathology (Guo et al. 2020). Pathological tau species are released into the extracellular space via exosomes or synaptic release and taken up by other neurons, acting as templates that induce further tau misfolding (Guo et al. 2020). Thus, tau pathology advances through connected brain regions. Hyper phosphorylated tau (p-tau) can also be detected in CSF early in AD and serves as a biomarker (Guo et al. 2020).

1.2.3 Role of Rare Mutations and Common Variants in AD

Rare autosomal mutations in *APP*, *PSEN1*, and *PSEN2* cause early-onset familial AD by altering Aβ production (Guo et al. 2020). APP duplications or point mutations increase total Aβ or shift γ-secretase cleavage toward Aβ-42. PSEN1/2 mutations impair γ-secretase function, often raising the Aβ-42/Aβ-40 ratio. Over 250 pathogenic PSEN1 mutations, and dozens of *APP* or *PSEN2* mutations, have been identified in familial AD (Guo et al. 2020). By contrast, the *APOE* gene that encodes apolipoprotein E strongly modulates sporadic AD risk: the ε4 allele is the single strongest risk factor for late-onset AD (Serrano-Pozo et al. 2011). APOE4 impairs Aβ clearance and enhances aggregation and also influences tau pathology and glial responses (Serrano-Pozo et al. 2011). Possession of one or two ε4 alleles multiplies AD risk several-fold, whereas ε2 is somewhat protective (Serrano-Pozo et al. 2011). Rare variants in the microglial receptor TREM2 also elevate AD risk (Hou et al. 2022). For example, the R47H variant increases risk ~2–3 times by reducing TREM2 function, which blunts microglial response to Aβ. In addition, genome-wide association studies (GWASs) have uncovered dozens of AD risk loci beyond APOE and the Mendelian genes (Kunkle et al. 2019). Many of these loci encode proteins involved in lipid metabolism, including CLU, ABCA7 as well as immune response, and endocytosis/endosomal trafficking (Kunkle et al. 2019). For instance, SORL1 and PICALM variants disrupt APP trafficking through the endosomal–lysosomal system (Knupp et al. 2021). Overall, genetic factors including rare high-impact mutations in APP/PSEN and common variants in APOE and GWAS loci converge on the pathways of Aβ metabolism, tau processing, lipid metabolism, and immune response (Kunkle et al. 2019; Serrano-Pozo et al. 2011; Hou et al. 2022).

1.2.4 Endosomal, Lysosomal, and Autophagic Pathways in AD Pathogenesis

Normal neurons rely on endosomes, lysosomes, and autophagy to recycle proteins and clear aggregates. In AD, this clearance machinery is impaired. A hallmark is enlarged early endosomes in neurons and accumulation of autophagic vacuoles loaded with APP and its fragments (Knupp et al. 2021). Many AD risk genes, e.g., *SORL1, BIN1, CD2AP, PICALM* regulate endosomal trafficking, and their dysfunction leads to mistargeting of APP and Aβ (Knupp et al. 2021). Familial *PSEN1/2* mutations also disrupt lysosomal acidification and protease function, thus exacerbating clearance defects (Knupp et al. 2021).

In AD, autophagic dysfunction is characterized by disrupted autophagosome–lysosome fusion and decreased enzymatic activity of lysosomal hydrolases, contributing to the accumulation of pathological substrates (Mançano et al. 2024). Aβ and tau aggregates themselves impair autophagy-lysosomal function, creating a vicious cycle (Mançano et al. 2024). For example, accumulation of Aβ peptides can inhibit cathepsins and block proteolytic flux (Mançano et al. 2024). Overall, failure of the endosomal–lysosomal and autophagic pathways impedes degradation of Aβ and tau, allowing toxic species to build up (Knupp et al. 2021; Mançano et al. 2024). This impairment in the proteostasis is considered a central hub of AD pathology (Knupp et al. 2021).

1.2.5 Role of Mitochondrial Dysfunction in AD Pathogenesis

In AD, neuronal mitochondria exhibit significant dysfunction, characterized by disrupted oxidative phosphorylation, reduced energy output, and increased generation of ROS, which contribute to neuronal injury and disease progression (Misrani and Tabassum 2021). Increased ROS inflict oxidative damage on lipids, proteins, and DNA in neurons. Aβ oligomers localize to mitochondria and disrupt electron transport, further raising ROS levels (Misrani and Tabassum 2021). Mitochondrial calcium overload via excitotoxicity and aberrant fission/fusion balance are also observed in AD models, contributing to energy failure and apoptosis (Misrani and Tabassum 2021). Notably, hyperphosphorylated tau impairs axonal transport of mitochondria, compounding neuronal energy deficits (Misrani and Tabassum 2021).

1.2.6 Microglial Inflammasome Activation and Genetic Modulators of Neuroinflammation in AD

Chronic activation of the innate immune system is a hallmark of AD, with microglia accumulating around amyloid plaques and NFTs, adopting a reactive phenotype and secreting pro-inflammatory cytokines and chemokines (Hanslik and Ulland 2020). Pattern-recognition receptors on microglia detect Aβ and tau aggregates, triggering pathways such as the NLRP3 inflammasome. Upon sensing Aβ or cell damage signals (DAMPs), the NLRP3 inflammasome assembles with ASC and caspase-1, leading to cleavage and secretion of IL-1β and IL-18 (Hanslik and Ulland 2020). These cytokines drive further inflammation and can induce neuronal dysfunction or pyroptotic death. Importantly, extracellular ASC specks released during inflammasome activation can bind Aβ and seed new plaques, creating a feed-forward cycle of pathology (Hanslik and Ulland 2020).

Genetic factors are further implicated in neuroinflammation in the pathogenesis of AD. The *APOE4* allele biases microglia toward a pro-inflammatory phenotype, while TREM2 loss-of-function variants impair microglial sensing and response to pathological hallmarks, such as amyloid-β and tau aggregates (Serrano-Pozo et al. 2011; Hou et al. 2022). Under normal conditions, TREM2 signaling facilitates the transition of microglia into a disease-associated microglial (DAM) phenotype, characterized by enhanced phagocytic activity and efficient clearance of cellular debris and amyloid-β. However, loss-of-function mutations in TREM2 hinder this phenotypic shift, retaining microglia in a homeostatic state and thereby compromising their ability to clear Aβ aggregates (Hou et al. 2022). Thus, in AD, the interplay of Aβ/tau with innate immunity mediated by microglia and inflammasomes exacerbates neurodegeneration (Hanslik and Ulland 2020; Hou et al. 2022).

1.2.7 Biomarkers for Monitoring AD Pathology

AD pathology can be monitored in vivo through molecular biomarkers. In cerebrospinal fluid (CSF), the salient biomarker profile includes decreased levels of $A\beta_{42}$ which is indicative of amyloid plaque sequestration and elevated concentrations of total tau (t-tau) and phosphorylated tau (p-tau), reflecting neuronal injury and tau pathology, respectively. (Kang et al. 2022). High CSF t-tau indicates neuronal injury, while specific p-tau species, e.g., p-tau181 and p-tau217 correlate closely with tau tangle burden (Guo et al. 2020; Kang et al. 2022). PET imaging allows visualization of plaques and tangles in vivo, Aβ-PET tracers (e.g. PiB, florbetapir) bind fibrillar amyloid, and tau-PET tracers (e.g. 18F-flortaucipir) reveal NFT distribution (Table 1.4; Kang et al. 2022). Neurodegeneration can also be monitored by CSF or blood neurofilament light, a general axonal damage marker (Kang et al. 2022).

1.3 HUNTINGTON'S DISEASE

HD is an autosomal dominant neurodegenerative disorder manifesting as progressive motor dysfunction, cognitive decline, and psychiatric disturbances. The disease is caused by an expanded CAG trinucleotide repeat in the HTT gene, leading to the production of mutant huntingtin (mHTT) protein with an elongated polyglutamine tract. This mutation results in toxic gain-of-function effects which contributes to neuronal dysfunction and death.

1.3.1 Genetic Basis and Molecular Pathogenesis of HD

HD is caused by a CAG trinucleotide repeat expansion in exon 1 of the huntingtin (HTT) gene, encoding a mHTT protein with an elongated polyglutamine (polyQ) tract. Normal alleles range from 6 to 35 repeats, while disease-associated alleles exceed 36, while fully penetrant alleles typically have more than 40 repeats (Figure 1.2; Bates et al. 2015). The mutant HTT protein is characterized by pathological gain-of-function properties that arise from both its altered structure and its proteolytic processing.

The full-length huntingtin protein (~350 kDa) is composed of multiple HEAT, i.e., Huntingtin, Elongation factor 3, PR65/A, TOR repeat motifs, forming a solenoid structure that facilitates protein–protein interactions and axonal transport (Saudou and Humbert 2016). The N-terminal region, containing the polyQ tract, is particularly vulnerable to proteolytic cleavage by caspases, notably caspase-3 and -6 and calpains. These cleavages yield toxic N-terminal fragments (~90–150 amino acids) that misfold and aggregate within the cytoplasm and nucleus (Figure 1.2; Gafni and Ellerby 2002). The exon 1 fragment alone is sufficient to recapitulate disease in transgenic mouse models (Mangiarini et al. 1996).

TABLE 1.4
Hallmark Biomarkers and Imaging Modalities

Disease	CSF Biomarkers	Blood Biomarkers	Imaging Tools
Alzheimer's disease	Aβ42, p-tau, t-tau	NfL, p-tau181	PET, MRI
Parkinson's disease	α-Synuclein	NfL	DaT-SPECT
Huntington's disease	mHTT	NfL	Structural MRI, fMRI
Amyotrophic lateral sclerosis	NfL, pTDP-43	Creatinine, IL-6	DTI, PET
Multiple sclerosis	OCBs, NfL, CHI3L1	KFLC	MRI, TSPO-PET

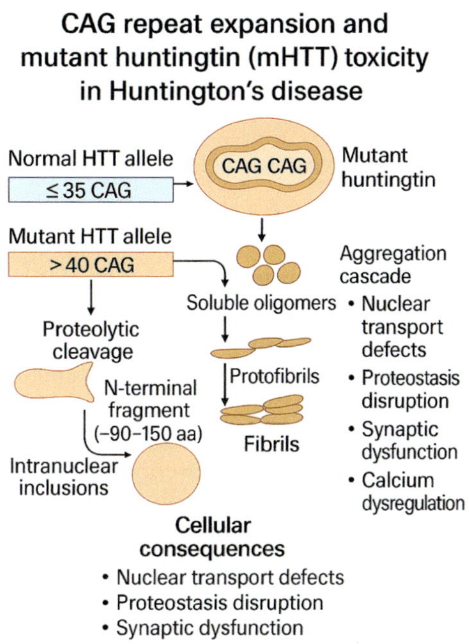

FIGURE 1.2 CAG Repeat Expansion and Mutant Huntingtin (mHTT) Toxicity in Huntington's Disease.

These N-terminal fragments assemble into soluble oligomers, fibrillar intermediates, and ultimately, intranuclear inclusions. Although inclusions were initially thought to be the primary toxic species, evidence now points to soluble oligomers and preamyloid aggregates as the most pathogenic forms (Arrasate et al. 2004). These toxic species interfere with nuclear transport, disrupt proteostasis, impair synaptic function, and dysregulate calcium signaling, collectively contributing to neuronal dysfunction and death (Twelvetrees et al. 2016).

1.3.2 RNA-Mediated Toxicity and Splicing Dysregulation

In addition to protein-mediated toxicity, recent studies have highlighted the contribution of mutant HTT RNA to cellular pathology. The expanded CAG tract in HTT mRNA forms stable hairpin structures that sequester RNA-binding proteins (RBPs), such as muscleblind-like (MBNL) proteins and nucleolin, disrupting normal splicing and RNA metabolism (Banez-Coronel et al. 2012). This mechanism mirrors the toxic RNA gain-of-function seen in other repeat expansion disorders, such as myotonic dystrophy.

The formation of RNA foci in both nuclear and cytoplasmic compartments is a consistent feature of HD neurons. These foci not only bind and inactivate splicing regulators but also interfere with nucleocytoplasmic transport by engaging components of the nuclear pore complex (Neueder and Bates 2014). Furthermore, antisense transcripts containing CUG repeats originating from bidirectional transcription of the HTT locus also form foci and may contribute to RNA-mediated toxicity (Banez-Coronel et al. 2012).

Aberrant splicing patterns have been detected in numerous HD-relevant transcripts, including those encoding synaptic proteins, mitochondrial enzymes, and components of the cytoskeleton (Lin et al. 2016). Such spliceopathies are further compounded by altered phosphorylation of splicing factors like SRSF6 and hnRNPs. These changes contribute to neuronal vulnerability, as precise splicing is essential for maintaining neuronal identity and function.

Emerging evidence also suggests that the toxic RNA species may undergo repeat-associated non-AUG (RAN) translation, producing additional dipeptide repeat proteins from expanded repeat

RNAs without canonical start codons (Cleary and Ranum 2017). Although RAN translation is better established in disorders like C9 or f72-ALS, recent studies have detected poly-Ala and poly-Ser RAN peptides in HD tissues.

1.3.3 Transcriptional Dysregulation and Epigenetic Alterations in HD

mHTT exerts profound and widespread effects on transcriptional regulation in neurons. One key mechanism involves mHTT's interaction with the REST/NRSF (Repressor Element-1 Silencing Transcription Factor) complex. In healthy neurons, wild-type HTT sequesters REST in the cytoplasm, allowing transcription of neuronal survival genes such as brain-derived neurotrophic factor (BDNF). However, in HD, mHTT fails to restrain REST, permitting its nuclear translocation and repression of REST target genes, including BDNF, SNAP25, and other synaptic effectors (Zuccato et al. 2003; Zuccato and Cattaneo 2009).

Additionally, mHTT interacts with various epigenetic regulators, notably CBP (CREB-binding protein), a histone acetyltransferase essential for transcriptional activation. mHTT sequesters CBP and inhibits its HAT activity, leading to global hypoacetylation of histone H3 and H4 tails and subsequent chromatin condensation (Steffan et al. 2000). The epigenetic repression is not limited to CBP activity. HD brains show aberrant recruitment of histone deacetylases (HDACs), particularly HDAC1, 3, and 4, which exacerbate chromatin compaction and silence neuronal genes critical for plasticity and energy homeostasis.

Genome-wide chromatin immunoprecipitation and ATAC-seq analyses have revealed enhancer inactivation and altered promoter usage in HD models. This includes disruption of superenhancers near synaptic genes and transcriptional downregulation of mitochondrial master regulators such as PGC-1α. Moreover, altered chromatin topology at the HTT locus itself that is driven by CAG repeat instability further contributes to epigenomic disarray. Therefore, transcriptional and epigenetic misregulation in HD is multifaceted, driven by both mHTT sequestration of coactivators and secondary chromatin remodeling, culminating in neuronal identity loss and synaptic breakdown.

1.3.4 Proteostasis Network Dysfunction in HD

The continuous production of misfolded mHTT overwhelms the neuronal proteostasis network, which comprises the ubiquitin–proteasome system (UPS), autophagy–lysosome pathway (ALP), molecular chaperones, and protein quality control compartments. Aggregated mHTT impairs proteasome activity by clogging the 20S catalytic core and altering the substrate recognition of the 19S regulatory cap, as demonstrated by fluorogenic peptide assays in HD neurons (Jana et al. 2001). This proteasomal inhibition leads to accumulation of polyubiquitylated substrates, further exacerbating proteotoxic stress.

Autophagy is likewise dysfunctional in HD, while autophagosome biogenesis is preserved or even upregulated, cargo recognition and lysosomal degradation are impaired. Martinez-Vicente et al. (2010) showed that mHTT interferes with the autophagy adaptor p62/SQSTM1, preventing selective loading of organelles and misfolded proteins into autophagosomes. Additionally, mHTT alters lysosomal pH and protease activity, delaying cargo degradation. Besides these deleterious effects, molecular chaperone systems are also compromised in HD, with notable downregulation of key chaperones such as HSP70 and HSP40 in affected brain regions. These heat shock proteins, which normally facilitate the refolding of misfolded or denatured proteins, exhibit reduced expression, thereby impairing proteostasis and contributing to the accumulation of toxic protein aggregates. Furthermore, mHTT disrupts BAG-family cochaperone function and disassembles chaperone complexes essential for triage decisions between refolding and degradation (Brehme et al. 2014). When chaperone and degradation pathways fail, mHTT is redirected to perinuclear inclusion bodies known as aggresomes, surrounded by vimentin cages and ER elements. Although aggresomes may initially buffer toxicity, they eventually become inert inclusions that occupy significant cellular volume and correlate with terminal neuronal pathology.

1.3.5 Mitochondrial Dysfunction and Oxidative Stress in HD

A salient aspect of HD is mitochondrial impairment as evidenced by early studies which have shown that the reduced oxidative phosphorylation efficiency in affected brain regions. Notably, deficiencies in respiratory chain complexes II and IV have been consistently observed in the striatum and cortex of HD patients and transgenic models, contributing to energy deficits and neuronal vulnerability (Gu et al. 1996; Browne et al. 1997). The subsequent research has revealed that mHTT directly associates with mitochondrial membranes, interacting with components such as the translocase of the outer membrane (TOM) complex, and disrupts mitochondrial import and membrane potential (Yano et al. 2014). These effects result in impaired oxidative phosphorylation, reduced ATP production, and increased proton leak.

A central feature of mitochondrial toxicity in HD is increased generation of ROSs. Damaged mitochondria release superoxide radicals and hydrogen peroxide, which overwhelm the cell's antioxidant defenses. Oxidative damage is evident in HD postmortem brains, with elevated 8-oxo-deoxyguanosine (8-oxo-dG), malondialdehyde (MDA), and protein carbonyls (Browne et al. 1999). mHTT also impedes the activity of peroxisome proliferator-activated receptor gamma coactivator 1-alpha (PGC-1α), a transcriptional regulator of mitochondrial biogenesis and antioxidant enzymes, via direct interference with the CREB/TAF4 complex (Cui et al. 2006).

Additionally, mHTT sensitizes mitochondria to calcium overload and enhances susceptibility to mitochondrial permeability transition pore (mPTP) opening. This leads to mitochondrial swelling, cytochrome C release, and caspase activation (Panov et al. 2002; Milakovic and Johnson 2005). The balance of mitochondrial fission and fusion is also disrupted. mHTT promotes pathological Drp1 activation and mitochondrial fragmentation, which contributes to synaptic failure and neuronal death (Shirendeb et al. 2012).

Together, these mechanisms establish a feed-forward loop wherein defective bioenergetics, redox imbalance, and calcium dysregulation propel the neuronal degeneration in HD.

1.3.6 Somatic CAG Repeat Instability as a Driver of HD Progression

While the inherited CAG repeat length in the HTT gene is the primary determinant of HD onset, recent evidence shows that somatic instability of the CAG tract, particularly expansion in the striatum, is a critical driver of disease progression. Repeat instability occurs postnatally and is influenced by DNA repair processes, especially those involving mismatch repair (MMR) and base excision repair pathways (Swami et al. 2009; Manley et al. 1999). Key enzymes include MSH3 and MLH1, which promote somatic expansion by stabilizing DNA loop-outs during replication and repair (Kovtun et al. 2007). The mouse models lacking Msh2 or Msh3 show significantly reduced somatic expansion and delayed HD phenotypes (Dragileva et al. 2009). Similarly, Fanconi anemia pathway members like FAN1 appear to limit expansion and act as disease-modifying factors. GWAS from the Genetic Modifiers of HD (GeM-HD) Consortium have confirmed that polymorphisms in *MSH3, FAN1, RRM2B*, and *MLH1* correlate with variation in age of onset among patients with identical inherited CAG lengths (Genetic Modifiers Consortium 2019).

Recent longitudinal studies using ultra-deep sequencing of peripheral blood and brain DNA have confirmed that longer somatic expansions in vulnerable regions correlate with faster clinical progression (Ciosi et al. 2021). These findings validate somatic instability as both a mechanistic driver and a biomarker of disease trajectory. Therapeutically, strategies aimed at modulating somatic instability either through small-molecule inhibition of DNA repair factors or antisense-based regulation are under investigation. Targeting these pathways may provide opportunity to delay onset and slow progression in HD, especially in premanifest gene carriers.

1.3.7 Glial Modulators of HD Pathogenesis

Although HD is genetically driven by neuronal expression of mHTT, however, glial cells, including astrocytes and microglia, play an active and independent role in pathogenesis. Astrocytes, which regulate synaptic neurotransmission and extracellular ion balance, express mHTT and display a reactive phenotype in HD brains. Notably, mHTT impairs expression and membrane localization of the astrocytic glutamate transporter GLT-1 (EAAT2), leading to elevated extracellular glutamate and excitotoxic stress on vulnerable medium spiny neurons (Faideau et al. 2010). Astrocytes in HD also exhibit disrupted potassium level and altered connexin expression, and mitochondrial abnormalities.

It has been shown that microglia are activated early in HD even in premanifest gene carriers. PET imaging using [11C]-PK11195 and other TSPO ligands demonstrate increased microglial activation in the striatum and cortex, preceding symptom onset (Pavese et al. 2006). Transcriptomic and proteomic profiling of microglia from HD models and patient tissue reveals upregulation of inflammatory mediators, including IL-1β, tumor necrosis factor-alpha (TNF-α), complement factors, and NF-κB signaling components (Crotti and Glass 2015). Further compounding neuroinflammation is NLRP3 inflammasome activation. In HD models, mHTT triggers mitochondrial ROS release and potassium efflux, which are the key signals for NLRP3 priming and activation (Saez-Atienzar and Masliah 2020). Once activated, inflammasome complexes cleave pro-caspase-1 to its active form, catalyzing IL-1β and IL-18 secretion, which exacerbates neuronal injury and fosters a prodegenerative environment. Together, astrocytic failure in glutamate and ion homeostasis and microglial-driven innate immune signaling act as amplifiers of mHTT toxicity, shifting HD from a cell autonomous to a network-based disease.

1.3.8 Fluid and Imaging Biomarkers in HD

Although HD is genetically confirmed by HTT genotyping, the need for biomarkers to track disease onset, progression, and therapeutic efficacy is important. Fluid and imaging biomarkers have gained prominence, particularly those reflecting neuronal injury and glial activity. Neurofilament light chain is elevated in both CSF and plasma in HD patients. Longitudinal data from TRACK-HD and HD-CSF cohorts show that plasma NfL levels rise years before clinical diagnosis and correlate with striatal atrophy, cognitive decline, and motor worsening (Byrne et al. 2017; Rodrigues et al. 2020). Mutant HTT protein itself is detectable in CSF using single molecule counting assays. CSF mHTT concentrations increase over time and correlate with disease burden, making it a promising pharmacodynamic biomarker in trials of HTT-decreasing therapies.

Also, neuroimaging biomarkers provide critical support for early diagnosis in HD. Structural MRI consistently demonstrates progressive atrophy of the caudate nucleus, putamen, and various cortical region, changes that are detectable up to 15 years prior to the clinical onset of motor symptoms (Tabrizi et al. 2009). Diffusion tensor imaging (DTI) identifies white matter degeneration, while resting-state fMRI reveals altered connectivity in sensorimotor and cognitive networks. PET imaging with TSPO tracers (e.g. [11C]-(R)-PK11195, [18F]DPA-714) highlights microglial activation, while novel synaptic density tracers (e.g. [11C]UCB-J targeting SV2A) are investigated for tracking synapse loss in HD. Emerging blood-based transcriptomic and exosomal RNA signatures also show promise in stratifying HD stage and therapeutic responsiveness.

1.4 PARKINSON'S DISEASE

PD is a progressive neurodegenerative disorder characterized by the selective loss of dopaminergic neurons in the substantia nigra pars compacta and the widespread presence of intracellular inclusions known as Lewy bodies. These inclusions are primarily composed of misfolded and aggregated α-synuclein protein, and they are central to PD pathology. The progressive neurodegeneration in PD affects not only motor pathways but also various nonmotor systems, contributing to its complex clinical presentation. The molecular and pathological mechanisms underlying PD are described below, with a focus on α-synuclein aggregation, genetic susceptibility, proteostasis dysfunction, mitochondrial impairment, neuroinflammation, along with a focus on emerging biomarkers.

1.4.1 Pathological Misfolding and Propagation of α-Synuclein in PD

Under physiological conditions, α-synuclein is a soluble, intrinsically disordered protein predominantly localized at presynaptic terminals, where it plays a role in synaptic vesicle trafficking and neurotransmitter release. However, under pathological conditions, α-synuclein undergoes misfolding and adopts β-sheet-rich structures that aggregate into toxic oligomers and fibrils (Sulzer et al. 2017). These aggregates accumulate into larger insoluble inclusions called Lewy bodies and Lewy neurites, hallmark lesions in both sporadic and familial PD (Figure 1.3; Shahmoradian et al. 2019). Oligomeric α-synuclein species are believed to be particularly neurotoxic, as they disrupt cellular membranes, impair mitochondrial function, and interfere with calcium homeostasis (Kalia and Lang 2015). Moreover, α-synuclein aggregates propagate in a prion-like fashion, transferring between neurons and acting as template for misfolding of native α-synuclein, thereby contributing to the progressive nature of PD (Braak et al. 2003; Chu and Kordower 2015).

1.4.2 Role of Genetic Factors in PD Pathogenesis

Genetic studies have identified both rare Mendelian mutations and common susceptibility variants linked to PD. The first mutations were discovered in the *SNCA* gene, which encodes α-synuclein, thereby establishing a direct connection between α-synuclein pathology and familial PD. These mutations lead to increased α-synuclein expression or altered protein conformation, enhancing its aggregation propensity (Polymeropoulos et al. 1997). Similarly, gene duplications or triplications of SNCA result in a dosage-dependent increase in α-synuclein levels and earlier onset of disease (Singleton et al. 2003). Another major genetic contributor is the *LRRK2* gene, encoding leucine-rich repeat kinase 2. The G2019S mutation is the most common cause of autosomal dominant PD and results in enhanced kinase activity, which affects vesicular trafficking, cytoskeletal dynamics, and autophagic pathways (Paisán-Ruíz et al. 2004; Bonet-Ponce and Cookson 2019).

Autosomal recessive forms of PD have been linked to mutations in PRKN (Parkin), PINK1, and DJ-1. Parkin is an E3 ubiquitin ligase involved in targeting damaged mitochondria for degradation via mitophagy, while PINK1 recruits Parkin to depolarized mitochondria, initiating mitophagic clearance (Pickrell and Youle 2015). Mutations in these genes impair mitochondrial quality control, leading to the accumulation of dysfunctional mitochondria and increased oxidative stress. DJ-1, a redox-sensitive chaperone that protects neurons against oxidative damage, and its loss-of-function mutations further impede cellular stress (Repici and Giorgini 2019).

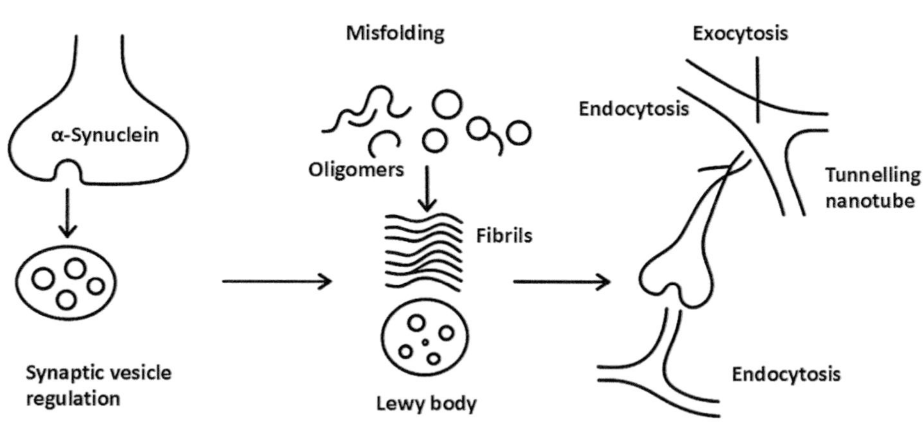

FIGURE 1.3 Pathological Misfolding and Propagation of α-Synuclein in Parkinson's Disease.

Variants in the GBA1 gene, which encodes the lysosomal enzyme glucocerebrosidase, represent the most common genetic risk factor for sporadic PD. GBA1 mutations reduce enzyme activity, leading to lysosomal dysfunction and accumulation of glucosylceramide, which promotes α-synuclein aggregation (Singleton et al. 2003). In addition to these high-impact mutations, GWASs have identified over 90 risk loci, implicating genes involved in lysosomal biology, endocytosis, and immune regulation (Nalls et al. 2019).

1.4.3 Disruption of Cellular Degradative Pathways in PD

A central feature of PD pathogenesis is the disruption of proteostasis, particularly the autophagy–lysosomal pathway (ALP) and the ubiquitin-proteasome system (UPS). α-Synuclein is normally degraded by both macroautophagy and chaperone-mediated autophagy (CMA). However, in PD, aggregated α-synuclein impairs lysosomal function and inhibits its own degradation, creating a toxic cycle (Vogiatzi et al. 2008). Mutations in *GBA1* and other lysosomal genes such as *ATP13A2* compromise lysosomal acidification and enzyme trafficking, reducing degradative capacity (Dehay et al. 2012). Furthermore, CMA is specifically blocked by mutant or aggregated α-synuclein binding to LAMP2A, a lysosomal membrane receptor essential for CMA substrate translocation (Cuervo et al. 2004). In addition, UPS dysfunction due to oxidative modification of proteasome subunits or overwhelmed capacity leads to accumulation of misfolded proteins and contributes to Lewy body formation (McNaught et al. 2001). Thus, the failure of cellular degradative systems permits the accumulation of α-synuclein and other neurotoxic proteins, ultimately leading to neuronal damage.

1.4.4 Mitochondrial Dysfunction and Impaired Mitophagy in PD

A hallmark of PD pathology is mitochondrial impairment, particularly the reduced activity of complex I in the substantia nigra, as consistently observed in postmortem studies. This deficit disrupts ATP production and promotes oxidative stress, contributing to dopaminergic neuronal degeneration (Schapira et al. 1990). This leads to decreased ATP production and increased generation of ROS, which damage mitochondrial DNA, lipids, and proteins. Dopaminergic neurons are particularly vulnerable due to their high metabolic demand and inherent ROS production from dopamine metabolism.

Genes linked to familial PD, such as PINK1 and Parkin, are essential for mitochondrial quality control. In healthy cells, PINK1 accumulates on the outer membrane of damaged mitochondria and recruits Parkin, which ubiquitinates outer membrane proteins, targeting the organelle for autophagic removal (Pickrell and Youle 2015). Mutations in these genes prevent mitophagy, allowing damaged mitochondria to persist and generate ROS. Moreover, aggregated α-synuclein impairs mitochondrial dynamics by disrupting fusion/fission balance and inhibiting axonal transport of mitochondria (Chinta and Andersen 2008). The result is an energy-deficient, oxidatively stressed neuron which are prone to degeneration.

1.4.5 Neuroinflammation and Innate Immune Activation in PD

Neuroinflammatory processes significantly contribute to PD progression. Activated microglia are found in the substantia nigra and other affected regions in PD, releasing pro-inflammatory cytokines such as TNF-α, IL-1β, and IL-6 (Hirsch and Hunot 2009). These cytokines induce nitric oxide production and exacerbate oxidative stress. α-Synuclein aggregates act as DAMPs, activating toll-like receptors (e.g. TLR2) on microglia and promoting inflammation. Importantly, α-synuclein also activates the NLRP3 inflammasome, a cytoplasmic complex that triggers caspase-1–mediated maturation of IL-1β. This contributes to a feedforward loop in which inflammation promotes further α-synuclein aggregation and neurodegeneration. It has been shown that astrocytes are also involved in the inflammatory milieu by releasing chemokines and undergoing reactive gliosis. Thus, chronic inflammation accelerates neuronal loss and impedes repair mechanisms in the PD brain.

1.4.6 Biomarkers for Diagnosis and Disease Progression in PD

The development of reliable biomarkers is crucial for early diagnosis and tracking of PD progression. α-Synuclein levels in CSF are invariably reduced in PD, but total α-synuclein assays lack diagnostic specificity (Parnetti et al. 2019; Table 1.4). More promising is seed amplification assays such as real-time quaking-induced conversion (RT-QuIC), which detects misfolded α-synuclein with high sensitivity and specificity (Fairfoul et al. 2016). Neurofilament light chain which is a marker of axonal damage is elevated in CSF and plasma of PD patients, particularly in those with rapid progression or cognitive decline. DaT-SPECT imaging using radioligands such as 123I-FP-CIT allows visualization of dopaminergic terminal loss and is widely used for differential diagnosis. Research is also focusing on plasma biomarkers such as phosphorylated α-synuclein, inflammatory cytokines, and exosomal cargo.

1.5 AMYOTROPHIC LATERAL SCLEROSIS

ALS is a progressive neurodegenerative disorder primarily affecting both upper motor neurons in the motor cortex and lower motor neurons in the brainstem and spinal cord. It manifests clinically as a combination of spasticity, muscle weakness, atrophy, and fasciculations, eventually culminating in respiratory failure. The average survival from symptom onset ranges from 3 to 5 years (Brown and Al-Chalabi 2017). The disease shows considerable clinical heterogeneity, including limb- or bulbar-onset forms, variable rates of progression, and phenotypic overlap with frontotemporal dementia (FTD). Globally, ALS has an annual incidence of approximately 2 per 100,000 and a lifetime risk of 1 in 400 (Logroscino et al. 2019). Though ALS is primarily sporadic (90–95%), approximately 5–10% of cases are familial (fALS) and inherited in a predominantly autosomal dominant fashion. Importantly, major breakthroughs in ALS genetics over the past decade have reframed the disease as a spectrum of genetically mediated disorders converging on common pathogenic cascades, such as protein misfolding, RNA toxicity, impaired nucleocytoplasmic transport, mitochondrial dysfunction, and noncell autonomous degeneration.

The discovery of diverse genetic mutations, including in *C9orf72*, *SOD1*, *TARDBP*, and *FUS*, has linked ALS mechanistically to other neurodegenerative diseases like FTD and has opened novel therapeutic avenues involving antisense oligonucleotides (ASOs), RNA interference, and gene correction technologies. Simultaneously, the field has seen significant advances in biomarkers, cellular modeling using patient-derived iPSCs, and high-throughput drug screening.

1.5.1 Genetic Basis of ALS

1.5.1.1 Genetic Determinants and Molecular Mechanisms of ALS

The earliest genetic discovery in ALS was the identification of mutations in *SOD1* (superoxide dismutase 1), an enzyme responsible for detoxifying superoxide radicals. Over 200 mutations have been described, accounting for ~20% of fALS and ~2% of all ALS cases (Rosen et al. 1993). Mutant SOD1 gains neurotoxicity not through loss of enzymatic activity but via misfolding and aggregation, aberrant metal binding, mitochondrial association, and oxidative damage (Pasinelli and Brown 2006). SOD1 aggregation disrupts axonal transport, proteostasis, and glial support systems.

The most frequent genetic cause of ALS is an intronic hexanucleotide (G4C2) repeat expansion in *C9orf72*, found in 30–40% of fALS and 7–10% of sporadic ALS cases (DeJesus-Hernandez et al. 2011). C9orf72-linked neurodegeneration is driven by a triad of interconnected pathogenic processes: (1) haploinsufficiency of C9orf72 protein, disrupting autophagy and endosomal trafficking;

(2) toxic gain-of-function at the RNA level, where expanded G4C2 repeats form nuclear foci that sequester essential RBPs; and (3) RAN translation of the expanded repeats, generating aggregation-prone dipeptide repeat proteins such as poly(GA), poly(GP), and poly(PR), which accumulate in neuronal and glial populations and contribute to cellular toxicity (Gendron and Petrucelli 2018).

TARDBP, encoding TDP-43, and *FUS* are RBPs that regulate splicing, mRNA transport, and stress granule dynamics. Mutations in both genes promote cytoplasmic mislocalization, aggregate formation, and disrupted RNA metabolism. TDP-43 inclusions are found in >95% of ALS cases, including those without *TARDBP* mutations (Ling et al. 2013; Mackenzie et al. 2010). These inclusions correlate with splicing defects in critical neuronal genes, such as *STMN2* and *UNC13A* (Melamed et al. 2019).

1.5.1.2 Genetic Insights into ALS

Besides the classical four genes, recent GWASs and rare variant analyses have uncovered several new ALS genes. *NEK1*, a kinase involved in DNA damage response and cytoskeletal organization, has been identified as a risk factor for both familial and sporadic ALS (Kenna et al. 2016). *TBK1*, involved in innate immunity and autophagy, is mutated in 1–2% of ALS cases and may contribute to impaired clearance of protein aggregates and damaged mitochondria (Freischmidt et al. 2015). Other genes such as *KIF5A*, motor protein, *CCNF*, which is involved in ubiquitin-mediated proteolysis, *CHCHD10* involved in mitochondrial cristae maintenance, and *ANXA11*, which is implicated in RNA granule stability suggest a broader convergence of ALS pathogenesis on organellar dynamics, proteostasis, and RNA processing (van Rheenen et al. 2021). Notably, many of these genetic risk factors overlap with genes implicated in FTD, AD, and myopathies, reflecting shared pathways in neurodegeneration. Further, somatic mutations, repeat instability, and epigenetic alterations, such as hypomethylation at the *C9orf72* locus, are being investigated as modulators of disease onset and progression. The cumulative burden of these factors, rather than a single mutation, likely determines phenotypic variability and therapeutic response.

1.5.2 Molecular and Cellular Pathogenesis

1.5.2.1 TDP-43 Proteinopathy as a Central Driver of ALS

TAR DNA-binding protein 43 (TDP-43), encoded by *TARDBP*, is a DNA/RBP with key roles in RNA splicing, transport, stability, and microRNA processing. In ~97% of ALS cases, TDP-43 mislocalizes from the nucleus to the cytoplasm, where it forms ubiquitinated, hyperphosphorylated aggregates (Neumann et al. 2006). This pathological redistribution results in both loss-of-function due to nuclear depletion and toxic gain-of-function effects in the cytoplasm. Nuclear depletion of TDP-43 disrupts alternative splicing of thousands of transcripts, notably those essential for axonal maintenance and neuromuscular junction function. One prominent target is *STMN2*, a microtubule regulator whose splicing is dysregulated when TDP-43 is lost from the nucleus, leading to axon regeneration defects (Klim et al. 2019). Similarly, cryptic exon inclusion in *UNC13A* has emerged as a biomarker and driver of TDP-43-mediated pathology.

In the cytoplasm, mislocalized TDP-43 aberrantly incorporates into stress granules, RNA–protein condensates that form under cellular stress. Prolonged incorporation leads to persistent granules that seed pathological aggregates. These granules interfere with normal mRNA translation and ribostasis (Ramaswami et al. 2013; Mann et al. 2019). Emerging evidence also implicates aberrant liquid–liquid phase separation (LLPS) of TDP-43 as a pathogenic mechanism, wherein physiological granules undergo pathological solidification into irreversible aggregates under chronic stress or mutation (Gasset-Rosa et al. 2019).

1.5.2.2 Role of Proteostasis Imbalance in ALS Pathogenesis

ALS is characterized by widespread disruption of proteostasis that is responsible for maintaining protein quality control. Mutant forms of SOD1, TDP-43, FUS, and C9orf72-DPR proteins tend to misfold and evade degradation, forming insoluble inclusions that overwhelm both the UPS and ALS. Proteasome inhibition has been observed in patient-derived motor neurons and mouse models, where accumulation of polyubiquitinated substrates indicates overwhelmed clearance machinery (Cheroni et al. 2022; Wang et al. 2020). The endoplasmic reticulum (ER) is particularly susceptible to stress in ALS, especially in SOD1-ALS and C9orf72-ALS. Misfolded proteins accumulate in the ER lumen, activating the unfolded protein response (UPR). Chronic UPR signaling, especially through the PERK-eIF2α-CHOP axis, leads to translational arrest, mitochondrial impairment, and ultimately apoptosis (Hetz and Saxena 2017). Postmortem spinal cords from ALS patients show upregulated expression of CHOP, ATF4, and XBP1s, indicating persistent ER stress. Autophagic flux is also impaired in ALS. In particular, *SQSTM1*, *OPTN*, and *TBK1*, all key autophagy mediators are mutated in ALS patients, disrupting autophagosome formation and clearance (Cirulli et al. 2015; Freischmidt et al. 2015). These mutations impair mitophagy and contribute to the accumulation of dysfunctional organelles.

1.5.2.3 Mitochondrial Dysfunction in ALS

Motor neurons exhibit heightened vulnerability to mitochondrial dysfunction owing to their extended axonal architecture, elevated metabolic demands, and reliance on precise calcium homeostasis. In ALS, mitochondrial abnormalities are both widespread and among the earliest detectable pathological aberrations. Electron microscopy of spinal cord neurons reveals swollen mitochondria with fragmented cristae and disorganized morphology. Both mutant SOD1 and TDP-43 associate with the outer mitochondrial membrane, impairing protein import via the TOM/TIM complex and disrupting respiratory chain activity (Magrané et al. 2014; Wang et al. 2013). Also, calcium homeostasis is perturbed, with mitochondrial calcium overload triggering mPTP opening and cytochrome C release (Tradewell et al. 2011). The resulting oxidative stress leads to excessive production of ROS, lipid peroxidation, and DNA damage. Mitochondrial fission and fusion dynamics are also disrupted; for example, SOD1-G93A mice exhibit abnormal expression of fission protein DRP1 and fusion protein MFN2, contributing to mitochondrial fragmentation. Moreover, suppressed mitophagy, a selective form of autophagy targeting damaged mitochondria, is a consistent finding in ALS, particularly in cases with mutations in *TBK1* and *OPTN* (Wong and Holzbaur 2014). These defects contribute to energy failure and synaptic degeneration in motor neurons.

1.5.2.4 Axonal Transport Defects

Efficient axonal transport is essential for motor neuron survival, given their extreme polarity and length. In ALS, both anterograde and retrograde transport are curtailed. TDP-43 pathology contributes to reduced trafficking of RNA granules, organelles, and synaptic vesicles. It also impairs local translation of essential mRNAs such as *Neurofilament-H* and *Stathmin-2* (Alami et al. 2014; Melamed et al. 2019).

Mutations in microtubule-associated proteins like *KIF5A*, *TUBA4A*, and *DCTN1* further disrupt cytoskeletal architecture and vesicle transport. For instance, ALS-linked *KIF5A* mutations cause defective cargo binding and axonal swellings. Retardation of axonal trafficking results in distal denervation, a prominent early feature of ALS pathology.

1.5.2.5 Glial Contributions to ALS Pathogenesis

ALS is a quintessential example of noncell-autonomous neurodegeneration. Both astrocytes and microglia contribute actively to disease progression. In SOD1-ALS models, astrocytes expressing mutant SOD1 exhibit reduced glutamate uptake due to loss of the transporter EAAT2/GLT-1, leading to excitotoxic injury of motor neurons (Rothstein et al. 1995; Nagai et al. 2007). These astrocytes also release nitric oxide, prostaglandins, and inflammatory cytokines. Microglia adopt an activated, disease-associated state early in ALS, characterized by upregulation of pro-inflammatory genes such as *Il1b*, *Tnf*, and *Ccl2* (Geloso et al. 2017). While early microglial activation may be neuroprotective, chronic activation promotes motor neuron degeneration via oxidative bursts, phagoptosis, and cytokine-mediated toxicity. Postmortem studies have shown reactive gliosis in the motor cortex and spinal cord of ALS patients, including increased expression of CD68, IBA1, and GFAP (Brettschneider et al. 2012). The extent of neuroinflammation correlates with clinical progression, and anti-inflammatory strategies are currently being tested in trials (McCauley and Baloh 2019).

1.5.3 Biomarkers

The development and validation of biomarkers in ALS have become essential for early diagnosis, patient stratification, monitoring disease progression, and assessing therapeutic efficacy (Table 1.4). Biomarkers are particularly valuable in ALS due to the clinical heterogeneity and absence of a single diagnostic test.

1.5.3.1 Fluid Biomarkers

Neurofilament light chain has emerged as a robust and reproducible biomarker for ALS. The elevated levels of NfL in CSF and plasma correlate with faster disease progression, reduced survival, and neuroimaging measures of white matter degeneration (Gaiani et al. 2017). Longitudinal studies demonstrate that plasma NfL levels remain stable over time in ALS, suggesting their utility in clinical trials as pharmacodynamic markers (Benatar et al. 2020). Another promising biomarker is phosphorylated TDP-43 (pTDP-43) in CSF, particularly relevant given the protein's central role in ALS pathogenesis. Although pTDP-43 assays are not yet standardized, several studies report increased levels in ALS patients versus controls (Steinacker et al. 2017). Additional fluid biomarkers under investigation include inflammatory cytokines metabolic products and extracellular vesicle signatures derived from glial and neuronal origin.

1.5.3.2 Imaging Biomarkers

Neuroimaging complements fluid biomarkers in diagnosing and tracking ALS. DTI reveals reduced fractional anisotropy in the corticospinal tract and corpus callosum, correlating with motor disability and functional scores (Menke et al. 2018). Magnetic resonance spectroscopy (MRS) demonstrates reduced N-acetylaspartate in motor cortex and spinal cord regions, indicative of neuronal loss. PET imaging with translocator protein (TSPO) ligands, such as [11C]-PK11195, provides a measure of microglial activation, and newer ligands show promise for better sensitivity and specificity. Functional MRI studies show disrupted connectivity in motor and prefrontal networks, which correlate with cognitive impairment in ALS-FTD cases.

1.5.3.3 Transcriptomic and Proteomic Biomarkers

Transcriptomic profiling of blood and CSF samples has identified dysregulated gene expression patterns predictive of ALS onset and progression. For example, reduced *STMN2* mRNA levels in peripheral blood may reflect TDP-43 pathology (Melamed et al. 2019). Further, proteomic studies using mass spectrometry and aptamer-based arrays have revealed altered expression of synaptic proteins, complement factors, and stress response elements in ALS fluids.

1.5.4 Therapeutics

1.5.4.1 Approved Treatments

Two disease-modifying drugs are currently approved for ALS. Riluzole, a glutamate release inhibitor, modestly prolongs survival by 2–3 months and reduces excitotoxicity via inhibition of voltage-gated sodium channels (Bensimon et al. 1994). Edaravone, an intravenous free radical scavenger, was approved based on trials showing slowed functional decline in early-stage ALS patients in Japan (Abe et al. 2017). Its efficacy in broader populations remains under debate (Table 1.5). Supportive

TABLE 1.5
Approved and Experimental Therapies for Neurodegenerative Diseases

Disease	FDA/EMA Approved Drugs	Mechanism of Action	Emerging Therapies
Alzheimer's disease	Donepezil, aducanumab	Cholinesterase inhibitor, anti-Aβ	Tau inhibitors, stem cells
Parkinson's disease	Levodopa, ropinirole	Dopamine replacement	Gene therapy, α-synuclein antibodies
Huntington's disease	Tetrabenazine	VMAT2 inhibition	ASOs, CRISPR, Neurotrophins
Amyotrophic lateral sclerosis	Riluzole, edaravone	Glutamate inhibitor, Antioxidant	ASOs, stem cell therapy
Multiple sclerosis	Ocrelizumab, fingolimod	B-cell depletion, S1P modulation	Remyelination agents, aHSCT

therapies, including noninvasive ventilation, gastrostomy, and multidisciplinary care, significantly improve quality of life and survival. However, they do not modify the underlying neurodegenerative process.

1.5.4.2 Antisense and RNA-Based Therapies

ASOs are designed to reduce expression of toxic transcripts or correct splicing errors. Tofersen, an ASO targeting *SOD1* mRNA, significantly lowers SOD1 protein levels in CSF and shows trends toward slowing disease progression in *SOD1*-ALS (Miller et al. 2022). Other ASOs are in development for *C9orf72*, *FUS*, and *ATXN2*. In parallel, small-interfering RNA (siRNA) and CRISPR-based systems are being tested in preclinical ALS models.

1.5.4.3 Small Molecules and Combination Therapies

AMX0035, a combination of sodium phenylbutyrate and taurursodiol, targets mitochondrial and ER stress. The CENTAUR trial showed modest slowing of ALS Functional Rating Scale-Revised (ALSFRS-R) decline, leading to FDA approval in 2022 (Paganoni et al. 2020). Other small molecules in clinical trials include masitinib (tyrosine kinase inhibitor), CNM-Au8 (nanocatalyst), and arimoclomol (HSP co-inducer).

1.5.4.4 Cell and Gene Therapy

Stem cell-based therapies, using mesenchymal stem cells, neural progenitors, or induced pluripotent stem cell (iPSC)-derived glia are under investigation. The rationale involves modulating inflammation and delivering trophic support. Intrathecal delivery of stem cells has been tested in phase I/II trials, although long-term efficacy remains unproven (Mazzini et al. 2019). Adeno-associated virus (AAV)-based gene therapy approaches, designed to deliver silencing construct, are increasingly being explored for neurodegenerative disorders. Nonetheless, obstacles such as precise targeting, potential immunogenicity, and the critical timing of administration remain major hurdles to clinical translation.

1.6 MULTIPLE SCLEROSIS

MS is a chronic, immune-mediated neurodegenerative disease of the central nervous system that is characterized by recurrent inflammation, demyelination, and subsequent axonal degeneration. The global prevalence of MS has increased steadily, now estimated at approximately 2.8 million people, with a notably higher incidence in Northern Europe and North America and a female-to-male ratio of approximately 3:1 (Walton et al. 2020). MS is clinically classified into distinct subtypes based on the temporal pattern of neurological deterioration, relapsing-remitting MS (RRMS), secondary progressive MS (SPMS), primary progressive MS (PPMS), and progressive relapsing MS (PRMS), with RRMS being the initial presentation in nearly 85% of patients. Over time, approximately 50% of untreated RRMS patients transition to SPMS, characterized by insidious neurological decline (Lublin et al. 2014). PPMS, affecting about 10–15% of cases, presents with a slow, progressive course from onset, typically without clear relapses or remissions.

Historically considered a demyelinating disorder, MS is now increasingly recognized as a disease involving both inflammatory and degenerative processes. Neurodegeneration, axonal transection, mitochondrial dysfunction, and cortical atrophy have been shown to occur early in disease progression, independent of focal demyelination (Trapp and Nave 2008; Lassmann 2018). A multidimensional model has emerged, implicating aberrant immune responses, mediated by T and B lymphocytes, combined with genetic susceptibility and environmental influences, including Epstein–Barr virus (EBV) exposure, smoking, and hypovitaminosis D.

1.6.1 Immunopathogenesis

1.6.1.1 T Cells and the Initiation of Autoimmunity

The traditional paradigm positions autoreactive CD4+ T cells, particularly Th1 and Th17 subsets as key initiators of MS. These cells are primed in the periphery by environmental antigens that mimic CNS components, traverse the BBB, and secrete cytokines such as interferon-gamma (IFN-γ), TNF-α, and IL-17A. Th17 cells, in particular, are adept at breaching the BBB by stimulating endothelial cells to express chemokines, e.g., CCL20 and matrix metalloproteinases (MMPs), which facilitate transmigration (Korn et al. 2009; Huppert et al. 2010). Once in the CNS, these T cells initiate a cascade of immune activation involving local antigen-presenting cells (APCs), microglia, and astrocytes. Notably, the persistence of CD8+ cytotoxic T cells in active lesions suggests that direct neurotoxicity via perforin and granzyme pathways contributes to axonal injury (Friese and Fugger 2005). While Tregs are responsible for dampening excessive inflammation, multiple studies have shown reduced suppressive function and altered phenotype in MS patients, contributing to the breakdown of immune tolerance (Viglietta et al. 2004; Durelli et al. 2009).

1.6.1.2 B Cell Contributions to the Pathogenesis of MS

B cells are now recognized as key mediators in the pathogenesis of MS, exerting their effects through multiple mechanisms, including antigen presentation to T cells, secretion of proinflammatory cytokines such as interleukin-6 (IL-6) and lymphotoxin-alpha (LTα), and production of autoreactive antibodies directed against central nervous system antigens. Clonally expanded, somatically hypermutated B cells have been found in both the CSF and parenchyma of MS patients (von Büdingen et al. 2012). These B cells very likely originate from peripheral germinal centers and subsequently migrate into the central nervous system, where they can form ectopic lymphoid-like structures within the meninges particularly in patients with SPMS. The presence of these meningeal follicles is associated with cortical demyelination and correlates with a more severe clinical course and poorer prognosis (Magliozzi et al. 2007). The therapeutic efficacy of B cell-depleting monoclonal antibodies strongly supports their pathogenic role. Also, anti-CD20 therapies reduce contrast-enhancing lesions and relapses by eliminating circulating and CNS-infiltrating memory B cells (Hauser et al. 2017).

1.6.1.3 Glial Contributions to Neuroinflammation in MS

Resident CNS glia, especially microglia and astrocytes, participate actively in the immune milieu. Microglia in early MS lesions express MHC-II, iNOS, and proinflammatory cytokines (e.g. IL-1β,

TNF-α), suggesting their transformation into an activated phenotype that can sustain inflammation and induce oligodendrocyte death (Lassmann et al. 2007). In MS, astrocytes transition from their homeostatic role to a reactive A1 phenotype within lesions. This phenotype is characterized by the upregulation of complement C3 and decreased production of neurotrophic factors, facilitating a proinflammatory environment that exacerbates neuronal and glial damage. Astrocyte-derived IL-6 and CXCL10 have been shown to enhance T cell recruitment and increasing the demyelination (Liddelow et al. 2017; Ponath et al. 2018).

1.6.2 Genetic and Environmental Risk Factors in MS

GWASs have identified over 230 loci associated with MS, the majority of which are linked to immune system function especially HLA-DRB1*15:01, which confers the highest genetic risk (International MS Genetics Consortium 2019). Other non-HLA genes include *IL2RA*, *IL7R*, *TYK2*, and *TNFRSF1A*, implicating regulatory pathways in T cell signaling and inflammation.

Among environmental factors, EBV has the strongest epidemiological association. A 2022 longitudinal study of U.S. military personnel revealed that seroconversion for EBV precedes MS onset by a median of 5 years and increases the risk by over 30-fold (Bjornevik et al. 2022). A proposed pathogenic mechanism involves molecular mimicry wherein EBV-encoded nuclear antigens share structural or sequence homology with central nervous system autoantigens, potentially triggering an autoimmune response. Also, low serum vitamin D levels are consistently associated with higher MS risk and activity, with experimental models suggesting that vitamin D modulates Treg function and IL-10 production (Ascherio and Munger 2016). Further, smoking and adolescent obesity also contribute to proinflammatory immune skewing, further elevating MS risk.

1.6.3 Molecular and Cellular Pathology

1.6.3.1 Demyelination Mechanisms

Demyelination in MS is a complex interplay of immune-driven injury, glial dysregulation, and metabolic compromise. Histologically, demyelinated plaques are categorized into active, chronic active (smoldering), and inactive lesions (Lassmann 2018). Active lesions contain dense infiltration of macrophages and microglia with phagocytosed myelin debris, whereas chronic lesions exhibit a hypocellular core surrounded by a rim of activated microglia (Absinta et al. 2019). Oligodendrocyte precursor cells (OPCs) normally differentiate into myelinating oligodendrocytes to restore myelin sheaths. However, in chronic MS lesions, remyelination is limited due to intrinsic blockade in OPC differentiation, myelin debris accumulation, or insufficient trophic support (Franklin and Ffrench-Constant 2017). Molecules like LINGO-1, Jagged1, and Notch1 inhibit OPC maturation and are upregulated in MS lesions (Mi et al. 2007).

1.6.3.2.1 Axonal Injury and Neurodegeneration

Axonal transection occurs even in early MS lesions and progresses independently of inflammation in progressive phases (Trapp et al. 1998; Mahad et al. 2015). Axons in demyelinated regions are metabolically stressed due to increased energy demand required for impulse propagation coupled with impaired axonal transport. Chronic neurodegeneration in MS is marked by mitochondrial DNA deletions, reduced expression of respiratory chain enzymes, and impaired calcium buffering. These changes are particularly pronounced in cortical and thalamic neurons, explaining cognitive dysfunction and fatigue in progressive MS (Campbell and Mahad 2018).

1.6.3.3 Blood–Brain Barrier Dysfunction in MS

The BBB functions as a selective interface between the peripheral circulation and the central nervous system, maintaining neural homeostasis. In MS, proinflammatory cytokines such as TNF-α and IFN-γ compromise BBB integrity by increasing endothelial permeability, while MMPs degrade components of the basement membrane, facilitating immune cell infiltration into the CNS. T cells crossing the BBB in MS express adhesion molecules such as VLA-4 and chemokine receptors like CCR6, which bind to their endothelial ligands, VCAM-1 and CCL20, respectively, facilitating their migration into the central nervous system. MRI with gadolinium contrast highlights areas of BBB compromise associated with acute inflammatory demyelination. However, in progressive MS, persistent BBB abnormalities have been observed even without contrast enhancement, indicating a low grade but continuous vascular pathology that may underlie chronic disease progression.

1.6.3.4 Mitochondrial Dysfunction and Oxidative Stress

Mitochondrial defects are central to MS neurodegeneration. Damaged mitochondria exhibit swollen morphology, reduced membrane potential, and impaired ATP production (Witte et al. 2009). Electron microscopy of MS lesions reveals loss of mitochondrial cristae and disrupted distribution along axons (Mahad et al. 2015). Oxidative stress in MS is driven by NADPH oxidase activity in microglia and astrocytes, releasing ROS that damage lipids, proteins, and DNA (Haider et al. 2011). Mitochondrial dysfunction also impairs calcium homeostasis and amplifies glutamate excitotoxicity, creating a feedforward loop of neuronal injury.

1.6.4 Biomarkers

1.6.4.1 CSF and Blood-Based Biomarkers

The oligoclonal bands (OCBs) are present in over 95% of MS patients and remain a key diagnostic criterion. However, they lack disease specificity and are often supplemented with other markers (Dobson and Giovannoni 2019). Further, the levels of neurofilament light chain in serum and CSF are strong indicators of axonal injury and disease activity. Elevated NfL predicts brain atrophy, disability progression, and response to therapy (Kuhle et al. 2019). The chitinase-3-like protein 1 (CHI3L1) is also linked to astrocyte activation, and its level correlate with lesion burden and progression risk (Comabella et al. 2010). One of the promising alternatives to OCBs is kappa free light chains (KFLC), which in CSF offers quicker and possibly more sensitive assessment of intrathecal immunoglobulin synthesis (Presslauer et al. 2016).

1.6.4.2 Advanced Imaging Biomarkers

In MRI, T2-weighted and FLAIR sequence detect white matter lesions, while T1 black holes indicate irreversible tissue damage. Cortical lesions are harder to visualize but contribute significantly to cognitive decline (Geurts et al. 2012). Besides, the DTI identifies microstructural damage in white matter and grey matter tracts, showing decreased fractional anisotropy in normal-appearing tissue. Also, in PET Imaging, the TSPO-PET visualizes microglial activation and is a potential tool for tracking neuroinflammation.

1.6.4.3 Transcriptomic and Proteomic Biomarkers in MS

High-throughput transcriptomic studies have identified differential gene expression patterns in T cells, monocytes, and CNS tissue. Genes related to IFN signaling, lipid metabolism, and oligodendrocyte function are frequently dysregulated (Corvol et al. 2018). Proteomic analyses of CSF have identified altered concentrations of neurogranin, contactin-1, and various complement components during relapse episodes and in progressive MS. These biomarkers hold potential for distinguishing disease stages and assessing therapeutic response, thereby enhancing personalized disease monitoring.

1.6.5 Therapeutics

1.6.5.1 Disease-Modifying Therapies

More than 20 disease-modifying therapies (DMTs) are now approved. These are classified by efficacy, route, and mechanism:

1) **Injectables:** Interferon-beta reduces relapse rates (~30%) and modulates T-cell trafficking. Glatiramer acetate induces Th2-type responses and promotes regulatory cells (Goodin 2014).
2) **Oral Therapies:**
   A) Fingolimod/Siponimod is a S1P receptor modulators sequester lymphocytes in lymph nodes.
   B) Dimethyl fumarate activates Nrf2, thus reducing oxidative stress and inflammation.
   C) Teriflunomid has been shown to inhibits dihydroorotate dehydrogenase and limiting T and B cell proliferation.
3) **Monoclonal Antibodies:**
   A) Natalizumab has been found to inhibit leukocyte adhesion (α4-integrin).
   B) Ocrelizumab has been shown to deplete B cell via anti-CD20 and is approved for RRMS and PPMS.
   C) Alemtuzumab depletes lymphocyte targeting CD52; however, it requires close monitoring for autoimmunity (Hauser et al. 2017).

1.6.5.2 Neuroprotective and Remyelinating Therapies

Clemastine fumarate showed efficacy in improving visual evoked potentials in optic neuritis, a proxy for remyelination (Green et al. 2017). However, long-term effects remain unclear. Further, high-dose biotin (MD1003) was tested in PPMS/SPMS; however, phase III results were mixed, showing minimal impact on progression (Tourbah et al. 2016). Other candidates like opicinumab (anti-LINGO-1) failed in clinical trials, underscoring the challenge of translating remyelination strategies.

1.6.5.3 Stem Cell and Gene Therapy

Autologous hematopoietic stem cell transplantation (aHSCT) has demonstrated significant efficacy in patients with treatment-refractory RRMS, achieving sustained clinical remission and disease stabilization. Notably, the MIST trial provided robust evidence for the superiority of aHSCT over conventional DMTs, showing greater reductions in both relapse rates and disability progression. (Burt et al. 2019). Mesenchymal stem cells possess notable immunomodulatory properties and provide trophic support to damaged neural tissue, positioning them as potential therapeutic candidates in MS. Early-phase clinical trials (Phase I/II) have established a favorable safety profile; however, conclusive evidence regarding their efficacy remains to be determined. In parallel, gene therapy strategies such as AAV vector-based delivery systems and CRISPR-mediated gene silencing are currently in preclinical development for MS. These approaches hold considerable promise for future targeted interventions aimed at modulating pathogenic pathways and promoting neuroprotection.

1.7 CONCLUSION

This chapter examines the molecular and cellular mechanisms underlying major neurodegenerative disorders, elucidating both their unique and convergent pathological features. AD is pathologically characterized by the extracellular accumulation of amyloid-β plaques, the intracellular formation of NFTs consisting of hyperphosphorylated tau protein, and the influence of genetic risk factors, most notably the APOE4 allele. The pathology of HD is driven by the expanded CAG trinucleotide repeat in the HTT gene, resulting in mHTT with toxic polyglutamine tracts, transcriptional dysregulation, and RNA-mediated toxicity. In PD, the aggregation of α-synuclein into Lewy bodies, along with mutations in genes such as *SNCA, LRRK2*, and *PINK1*, contribute to synaptic dysfunction and mitochondrial impairment. ALS is defined by the cytoplasmic mislocalization and aggregation of TDP-43 and mutant SOD1 proteins, with a significant genetic burden from C9orf72 hexanucleotide repeat expansions. MS is a chronic autoimmune demyelinating disease, driven by aberrant T and B cell responses, activation of microglia and astrocytes, and mitochondrial dysfunction within axons and oligodendrocytes. Also, the recent progress in the development of diagnostic and prognostic biomarkers, which are used for early detection and treatment, is also described. Finally, the therapeutic potential of stem cell-based interventions is outlined, focusing on their dual regenerative and immunomodulatory capacities in the context of neurodegeneration.

REFERENCES

Abe C, Inoue T, Inglis MA, Viar KE, Huang L, Ye H, Rosin DL, Stornetta RL, Okusa MD, Guyenet PG. C1 neurons mediate a stress-induced anti-inflammatory reflex in mice. *Nat Neurosci.* 2017;20(5):700–7. doi:10.1038/nn.4526

Absinta M, Sati P, Masuzzo F, Nair G, Sethi V, Kolb H, Ohayon J, Wu T, Cortese ICM, Reich DS. Association of chronic active multiple sclerosis lesions with disability in vivo. *JAMA Neurol.* 2019;76(12):1474–83. doi:10.1001/jamaneurol.2019.2399. Erratum in: *JAMA Neurol.* 2019 Dec 1;76(12):1520. doi: 10.1001/jamaneurol.2019.4331

Alami NH, Smith RB, Carrasco MA, Williams LA, Winborn CS, Han SSW, Kiskinis E, Winborn B, Freibaum BD, Kanagaraj A, Clare AJ, Badders NM, Bilican B, Chaum E, Chandran S, Shaw CE, Eggan KC, Maniatis T, Taylor JP. Axonal transport of TDP-43 mRNA granules is impaired by ALS-causing mutations. *Neuron.* 2014;81(3):536–43. doi:10.1016/j.neuron.2013.12.018

Arrasate M, Mitra S, Schweitzer ES, Segal MR, Finkbeiner S. Inclusion body formation reduces levels of mutant huntingtin and the risk of neuronal death. *Nature.* 2004 Oct 14;431(7010):805–10. doi: 10.1038/nature02998

Ascherio A, Munger KL. Epidemiology of multiple sclerosis: From risk factors to prevention-an update. *Semin Neurol.* 2016 Apr;36(2):103–14. doi: 10.1055/s-0036-1579693

Bañez-Coronel M, Porta S, Kagerbauer B, Mateu-Huertas E, Pantano L, Ferrer I, Guzmán M, Estivill X, Martí E. A pathogenic mechanism in Huntington's disease involves small CAG-repeated RNAs with neurotoxic activity. *PLoS Genet.* 2012;8(2):e1002481. doi:10.1371/journal.pgen.1002481

Bates GP, Dorsey R, Gusella JF, Hayden MR, Kay C, Leavitt BR, et al. Huntington disease. *Nat Rev Dis Primers.* 2015;1:15005. doi: 10.1038/nrdp.2015.5

Benatar M, Zhang L, Wang L, Granit V, Statland J, Barohn R, Swenson A, Ravits J, Jackson C, Burns TM, Trivedi J, Pioro EP, Caress J, Katz J, McCauley JL, Rademakers R, Malaspina A, Ostrow LW, Wuu J; CReATe Consortium. Validation of serum neurofilaments as prognostic and potential pharmacodynamic biomarkers for ALS. *Neurology.* 2020;95(1):e59–e69. doi:10.1212/WNL.0000000000009559

Bensimon G, Lacomblez L, Meininger V; ALS/Riluzole Study Group. A controlled trial of riluzole in amyotrophic lateral sclerosis. *N Engl J Med.* 1994;330(9):585–91. doi:10.1056/NEJM199403033300901

Bjornevik K, Cortese M, Healy BC, Kuhle J, Mina MJ, Leng Y, Elledge SJ, Niebuhr DW, Scher AI, Munger KL, Ascherio A. Longitudinal analysis reveals high prevalence of Epstein-Barr virus associated with multiple sclerosis. *Science.* 2022;375(6578):296–301. doi: 10.1126/science.abj8222

Bonet-Ponce L, Cookson MR. The role of Rab GTPases in the pathobiology of Parkinson' disease. *Curr Opin Cell Biol.* 2019;59:73–80. doi: 10.1016/j.ceb.2019.03.009

Braak H, Del Tredici K, Rüb U, de Vos RA, Jansen Steur EN, Braak E. Staging of brain pathology related to sporadic Parkinson's disease. *Neurobiol Aging.* 2003;24(2):197–211. doi:10.1016/s0197-4580(02)00065-9

Brehme M, Voisine C, Rolland T, Wachi S, Soper JH, Zhu Y, Orton K, Villella A, Garza D, Vidal M, Ge H, Morimoto RI. A chaperome subnetwork safeguards proteostasis in aging and neurodegenerative disease. *Cell Rep.* 2014 Nov 6;9(3):1135–50. doi: 10.1016/j.celrep.2014.09.042

Brettschneider J, Toledo JB, Van Deerlin VM, Elman L, McCluskey L, Lee VM, Trojanowski JQ. Microglial activation correlates with disease progression and upper motor neuron clinical symptoms in amyotrophic lateral sclerosis. *PLoS One.* 2012;7(6):e39216. doi:10.1371/journal.pone.0039216

Brown RH, Al-Chalabi A. Amyotrophic lateral sclerosis. *N Engl J Med.* 2017;377(2):162–72. doi:10.1056/NEJMra1603471

Browne SE, Bowling AC, MacGarvey U, Baik MJ, Berger SC, Muqit MM, Bird ED, Beal MF. Oxidative damage and metabolic dysfunction in Huntington's disease: selective vulnerability of the basal ganglia. *Ann Neurol*. 1997;41(5):646–53.

Burt RK, Balabanov R, Burman J, Sharrack B, Snowden JA, Oliveira MC, Fagius J, Rose J, Nelson F, Barreira AA, Carlson K, Han X, Moraes D, Morgan A, Quigley K, Yaung K, Buckley R, Alldredge C, Clendenan A, Calvario MA, Henry J, Jovanovic B, Helenowski IB. Effect of nonmyeloablative hematopoietic stem cell transplantation vs continued disease-modifying therapy on disease progression in patients with relapsing-remitting multiple sclerosis: A randomized clinical trial. *JAMA*. 2019;321(2):165–74. doi:10.1001/jama.2018.18743

Byrne S, Wade T, Hay P, Touyz S, Fairburn CG, Treasure J, Schmidt U, McIntosh V, Allen K, Fursland A, Crosby RD. A randomised controlled trial of three psychological treatments for anorexia nervosa. *Psychol Med*. 2017;47(16):2823–33. doi:10.1017/S0033291717001349

Campbell G, Mahad DJ. Mitochondrial dysfunction and axon degeneration in progressive multiple sclerosis. *FEBS Lett*. 2018;592(7):1113–21. doi:10.1002/1873-3468.13013

Cheroni C, Trattaro S, Caporale N, et al. Benchmarking brain organoid recapitulation of fetal corticogenesis. *Transl Psychiatry*. 2022;12:520. doi:10.1038/s41398-022-02279-0

Chinta SJ, Andersen JK. Redox imbalance in Parkinson's disease. *Biochim Biophys Acta*. 2008;1780(11):1362–7. doi:10.1016/j.bbagen.2008.02.005

Chu, Y, Kordower, JH. The prion hypothesis of Parkinson's disease. *Curr Neurol Neurosci Rep*. 2015;15:28. doi.org/10.1007/s11910-015-0549-x

Ciosi M, Cumming SA, Chatzi A, Larson E, Tottey W, Lomeikaite V, Hamilton G, Wheeler VC, Pinto RM, Kwak S, Morton AJ, Monckton DG. Approaches to sequence the HTT CAG repeat expansion and quantify repeat length variation. *J Huntingtons Dis*. 2021;10(1):53–74. doi:10.3233/JHD-200433

Cirulli ET, Lasseigne BN, Petrovski S, Sapp PC, Dion PA, Leblond CS, Couthouis J, Lu YF, Wang Q, Krueger BJ, Ren Z, Keebler J, Han Y, Levy SE, Boone BE, Wimbish JR, Waite LL, Jones AL, Carulli JP, Day-Williams AG, Staropoli JF, Xin WW, Chesi A, Raphael AR, McKenna-Yasek D, Cady J, Vianney de Jong JM, Kenna KP, Smith BN, Topp S, Miller J, Gkazi A; FALS Sequencing Consortium; Al-Chalabi A, van den Berg LH, Veldink J, Silani V, Ticozzi N, Shaw CE, Baloh RH, Appel S, Simpson E, Lagier-Tourenne C, Pulst SM, Gibson S, Trojanowski JQ, Elman L, McCluskey L, Grossman M, Shneider NA, Chung WK, Ravits JM, Glass JD, Sims KB, Van Deerlin VM, Maniatis T, Hayes SD, Ordureau A, Swarup S, Landers J, Baas F, Allen AS, Bedlack RS, Harper JW, Gitler AD, Rouleau GA, Brown R, Harms MB, Cooper GM, Harris T, Myers RM, Goldstein DB. Exome sequencing in amyotrophic lateral sclerosis identifies risk genes and pathways. *Science*. 2015;347(6229):1436–41. doi:10.1126/science.aaa3650

Cleary JD, Ranum LP. New developments in RAN translation: Insights from multiple diseases. *Curr Opin Genet Dev*. 2017;44:125–34. doi:10.1016/j.gde.2017.03.006

Comabella M, Fernández M, Martin R, Rivera-Vallvé S, Borrás E, Chiva C, Julià E, Rovira A, Cantó E, Alvarez-Cermeño JC, Villar LM, Tintoré M, Montalban X. Cerebrospinal fluid chitinase 3-like 1 levels are associated with conversion to multiple sclerosis. *Brain*. 2010;133(Pt 4):1082–93. doi:10.1093/brain/awq035

Corvol JC, Artaud F, Cormier-Dequaire F, Rascol O, Durif F, Derkinderen P, Marques AR, Bourdain F, Brandel JP, Pico F, Lacomblez L, Bonnet C, Brefel-Courbon C, Ory-Magne F, Grabli D, Klebe S, Mangone G, You H, Mesnage V, Lee PC, Brice A, Vidailhet M, Elbaz A. DIGPD study group. longitudinal analysis of impulse control disorders in Parkinson disease. *Neurology*. 2018;91(3):e189–e201. doi:10.1212/WNL.0000000000005816

Crotti A, Glass CK. The choreography of neuroinflammation in Huntington's disease. *Trends Immunol*. 2015;36(6):364–73. doi:10.1016/j.it.2015.04.007

Cuervo AM, Stefanis L, Fredenburg R, Lansbury PT, Sulzer D. Impaired degradation of mutant alpha-synuclein by chaperone-mediated autophagy. *Science*. 2004;305(5688):1292–5. doi:10.1126/science.1101738

Cui L, Jeong H, Borovecki F, Parkhurst CN, Tanese N, Krainc D. Transcriptional repression of PGC-1alpha by mutant huntingtin leads to mitochondrial dysfunction and neurodegeneration. *Cell*. 2006;127(1):59–69. doi:10.1016/j.cell.2006.09.015

Dehay B, Ramirez A, Martinez-Vicente M, Perier C, Canron MH, Doudnikoff E, Vital A, Vila M, Klein C, Bezard E. Loss of P-type ATPase ATP13A2/PARK9 function induces general lysosomal deficiency and leads to Parkinson disease neurodegeneration. *Proc Natl Acad Sci U S A*. 2012;109(24):9611–6. doi:10.1073/pnas.1112368109

DeJesus-Hernandez M, Mackenzie IR, Boeve BF, Boxer AL, Baker M, Rutherford NJ, Nicholson AM, Finch NA, Flynn H, Adamson J, Kouri N, Wojtas A, Sengdy P, Hsiung GY, Karydas A, Seeley WW, Josephs KA, Coppola G, Geschwind DH, Wszolek ZK, Feldman H, Knopman DS, Petersen RC, Miller BL, Dickson DW, Boylan KB, Graff-Radford NR, Rademakers R. Expanded GGGGCC hexanucleotide repeat in noncoding region of C9ORF72 causes chromosome 9p-linked FTD and ALS. *Neuron*. 2011;72(2):245–56. doi:10.1016/j.neuron.2011.09.011

Dobson R, Giovannoni G. Multiple sclerosis – a review. *Eur J Neurol*. 2019;26(1):27–40. doi: 10.1111/ene.13819

Dragileva E, Hendricks A, Teed A, Gillis T, Lopez ET, Friedberg EC, Kucherlapati R, Edelmann W, Lunetta KL, MacDonald ME, Wheeler VC. Intergenerational and striatal CAG repeat instability in Huntington's disease knock-in mice involve different DNA repair genes. *Neurobiol Dis*. 2009;33(1):37–47. doi:10.1016/j.nbd.2008.09.014

Durelli L, Conti L, Clerico M, Boselli D, Contessa G, Ripellino P, Ferrero B, Eid P, Novelli F. T-helper 17 cells expand in multiple sclerosis and are inhibited by interferon-beta. *Ann Neurol*. 2009;65(5):499–509. doi:10.1002/ana.21652

Faideau M, Kim J, Cormier K, Gilmore R, Welch M, Auregan G, Dufour N, Guillermier M, Brouillet E, Hantraye P, Déglon N, Ferrante RJ, Bonvento G. In vivo expression of polyglutamine-expanded huntingtin by mouse striatal astrocytes impairs glutamate transport: A correlation with Huntington's disease subjects. *Hum Mol Genet*. 2010;19(15):3053–67. doi:10.1093/hmg/ddq212

Fairfoul G, McGuire LI, Pal S, Ironside JW, Neumann J, Christie S, Joachim C, Esiri M, Evetts SG, Rolinski M, Baig F, Ruffmann C, Wade-Martins R, Hu MT, Parkkinen L, Green AJ. Alpha-synuclein RT-QuIC in the CSF of patients with alpha-synucleinopathies. *Ann Clin Transl Neurol*. 2016;3(10):812–18. doi:10.1002/acn3.338

Franklin, R., ffrench-Constant, C. Regenerating CNS myelin — from mechanisms to experimental medicines. *Nat Rev Neurosci*. 2017;18, 753–69.

Freischmidt A, Wieland T, Richter B, Ruf W, Schaeffer V, Müller K, Marroquin N, Nordin F, Hübers A, Weydt P, Pinto S, Press R, Millecamps S, Molko N, Bernard E, Desnuelle C, Soriani MH, Dorst J, Graf E, Nordström U, Feiler MS, Putz S, Boeckers TM, Meyer T, Winkler AS, Winkelman J, de Carvalho M, Thal DR, Otto M, Brännström T, Volk AE, Kursula P, Danzer KM, Lichtner P, Dikic I, Meitinger T, Ludolph AC, Strom TM, Andersen PM, Weishaupt JH. Haploinsufficiency of TBK1 causes familial ALS and fronto-temporal dementia. *Nat Neurosci*. 2015 May;18(5):631–6. doi:10.1038/nn.4000

Friese MA, Fugger L. Autoreactive CD8+ T cells in multiple sclerosis: a new target for therapy? *Brain*. 2005;128(Pt 8):1747–63. doi:10.1093/brain/awh578. Epub 2005 Jun 23. Erratum in: Brain. 2005 Sep;128(Pt 9):2215

Gafni J, Ellerby LM. Calpain activation in Huntington's disease. *J Neurosci*. 2002;22(12):4842–9. doi:10.1523/JNEUROSCI.22-12-04842.2002

Gaiani A, Martinelli I, Bello L, Querin G, Puthenparampil M, Ruggero S, Toffanin E, Cagnin A, Briani C, Pegoraro E, Sorarù G. Diagnostic and prognostic biomarkers in amyotrophic lateral sclerosis: neurofilament light chain levels in definite subtypes of disease. *JAMA Neurol*. 2017;74(5):525–32. doi:10.1001/jamaneurol.2016.5398

Gasset-Rosa F, Lu S, Yu H, Chen C, Melamed Z, Guo L, Shorter J, Da Cruz S, Cleveland DW. Cytoplasmic TDP-43 de-mixing independent of stress granules drives inhibition of nuclear import, loss of nuclear TDP-43, and cell death. *Neuron*. 2019;102(2):339–57.e7. doi:10.1016/j.neuron.2019.02.038

Geloso MC, Corvino V, Marchese E, Serrano A, Michetti F, D'Ambrosi N. The dual role of microglia in ALS: Mechanisms and therapeutic approaches. *Front Aging Neurosci*. 2017;9:242. doi:10.3389/fnagi.2017.00242

Gendron TF, Petrucelli L. Disease mechanisms of C9ORF72 repeat expansions. *Cold Spring Harb Perspect Med*. 2018;8(4):a024224. doi:10.1101/cshperspect.a024224

Genetic Modifiers of Huntington's Disease (GeM-HD) Consortium. 2019. CAG repeat not polyglutamine length determines timing of Huntington's disease onset. *Cell* 178 (4): 887–900.e14.

Geurts JJ, Calabrese M, Fisher E, Rudick RA. Measurement and clinical effect of grey matter pathology in multiple sclerosis. *Lancet Neurol*. 2012;11(12):1082–92. doi:10.1016/S1474-4422(12)70230-2

Goodin DS. The epidemiology of multiple sclerosis: insights to disease pathogenesis. *Handb Clin Neurol*. 2014;122:231–66. doi:10.1016/B978-0-444-52001-2.00010-8

Green AJ, Gelfand JM, Cree BA, Bevan C, Boscardin WJ, Mei F, Inman J, Arnow S, Devereux M, Abounasr A, Nobuta H, Zhu A, Friessen M, Gerona R, von Büdingen RC, Henry RG, Hauser SL, Chan JR. Clemastine fumarate as a remyelinating therapy for multiple sclerosis (ReBUILD): a randomised, controlled, double-blind, crossover trial. *The Lancet.* 2017;390(10111):2481–9.

Gu Y, Oyama F, Ihara Y. Tau is widely expressed in rat tissues. *J Neurochem.* 1996;67(3):1235–44. doi:10.1046/j.1471-4159.1996.67031235.x

Guo W, Dittlau KS, Van Den Bosch L. Axonal transport defects and neurodegeneration: Molecular mechanisms and therapeutic implications. *Semin Cell Dev Biol.* 2020;99:133–50. doi:10.1016/j.semcdb.2019.07.010

Haider L, Fischer MT, Frischer JM, Bauer J, Höftberger R, Botond G, Esterbauer H, Binder CJ, Witztum JL, Lassmann H. Oxidative damage in multiple sclerosis lesions. *Brain.* 2011;134(Pt 7):1914–24. doi:10.1093/brain/awr128

Hanslik KL, Ulland TK. The role of microglia and the Nlrp3 inflammasome in Alzheimer's disease. *Front Neurol.* 2020;11:570711. doi:10.3389/fneur.2020.570711

Hauser A, Attwood M, Rask-Andersen M, et al. Trends in GPCR drug discovery: New agents, targets and indications. *Nat Rev Drug Discov.* 2017;16:829–42. doi:10.1038/nrd.2017.178

Hetz C, Saxena S. ER stress and the unfolded protein response in neurodegeneration. *Nat Rev Neurol.* 2017;13(8):477–91. doi:10.1038/nrneurol.2017.99

Hirsch EC, Hunot S. Neuroinflammation in Parkinson's disease: a target for neuroprotection? *Lancet Neurol.* 2009;8(4):382–97. doi:10.1016/S1474-4422(09)70062-6

Hou J, Chen Y, Grajales-Reyes G, Colonna M. TREM2 dependent and independent functions of microglia in Alzheimer's disease. *Mol Neurodegener.* 2022 Dec 23;17(1):84. doi: 10.1186/s13024-022-00588-y.

Huppert J, Closhen D, Croxford A, White R, Kulig P, Pietrowski E, Bechmann I, Becher B, Luhmann HJ, Waisman A, Kuhlmann CR. Cellular mechanisms of IL-17-induced blood-brain barrier disruption. *FASEB J.* 2010;24(4):1023–34. doi:10.1096/fj.09-141978

International MS Genetics Consortium. Multiple sclerosis genomic map implicates peripheral immune cells and microglia in susceptibility. *Science.* 2019;365(6460): eaav7188. doi:10.1126/science.aav7188

Jana NR, Zemskov EA, Wang GH, Nukina N. Altered proteasomal function due to the expression of polyglutamine-expanded truncated N-terminal huntingtin induces apoptosis by caspase activation through mitochondrial cytochrome c release. *Hum Mol Genet.* 2001;10(10):1049–59.

Kalia LV, Lang AE. Parkinson's disease. *Lancet.* 2015;386(9996):896–912. doi:10.1016/S0140-6736(14)61393-3

Kang JE, Hansson O, Zetterberg H, Blennow K. Fluid biomarkers for Alzheimer's disease. *Lancet Neurol.* 2022;21(2):144–58. doi:10.1016/S1474-4422(21)00361-6

Kenna KP, van Doormaal PT, Dekker AM, Ticozzi N, Kenna BJ, Diekstra FP, van Rheenen W, van Eijk KR, Jones AR, Keagle P, Shatunov A, Sproviero W, Smith BN, van Es MA, Topp SD, Kenna A, Miller JW, Fallini C, Tiloca C, McLaughlin RL, Vance C, Troakes C, Colombrita C, Mora G, Calvo A, Verde F, Al-Sarraj S, King A, Calini D, de Belleroche J, Baas F, van der Kooi AJ, de Visser M, Ten Asbroek AL, Sapp PC, McKenna-Yasek D, Polak M, Asress S, Muñoz-Blanco JL, Strom TM, Meitinger T, Morrison KE; SLAGEN Consortium; Lauria G, Williams KL, Leigh PN, Nicholson GA, Blair IP, Leblond CS, Dion PA, Rouleau GA, Pall H, Shaw PJ, Turner MR, Talbot K, Taroni F, Boylan KB, Van Blitterswijk M, Rademakers R, Esteban-Pérez J, García-Redondo A, Van Damme P, Robberecht W, Chio A, Gellera C, Drepper C, Sendtner M, Ratti A, Glass JD, Mora JS, Basak NA, Hardiman O, Ludolph AC, Andersen PM, Weishaupt JH, Brown RH Jr, Al-Chalabi A, Silani V, Shaw CE, van den Berg LH, Veldink JH, Landers JE. NEK1 variants confer susceptibility to amyotrophic lateral sclerosis. *Nat Genet.* 2016;48(9):1037–42. doi:10.1038/ng.3626

Klim JR, Williams LA, Limone F, Guerra San Juan I, Davis-Dusenbery BN, Mordes DA, Burberry A, Steinbaugh MJ, Gamage KK, Kirchner R, Moccia R, Cassel SH, Chen K, Wainger BJ, Woolf CJ, Eggan K. ALS-implicated protein TDP-43 sustains levels of STMN2, a mediator of motor neuron growth and repair. *Nat Neurosci.* 2019;22(2):167–79. doi:10.1038/s41593-018-0300-4

Knupp A, Martinez RA, Jayadev S, Young JE. Retromer stabilizing compounds rescue enlarged endosome phenotypes in hiPSC-derived models of AD. *Alzheimer's Dement.* 2021;17:e055236. doi:10.1002/alz.055236

Korn T, Bettelli E, Oukka M, Kuchroo VK. IL-17 and Th17 cells. *Annu Rev Immunol.* 2009;27:485–517. doi:10.1146/annurev.immunol.021908.132710

Kovtun IV, Liu Y, Bjoras M, Klungland A, Wilson SH, McMurray CT. OGG1 initiates age-dependent CAG trinucleotide expansion in somatic cells. *Nature*. 2007;447(7143):447–52. doi:10.1038/nature05778

Kuhle J, Kropshofer H, Haering DA, Kundu U, Meinert R, Barro C, Dahlke F, Tomic D, Leppert D, Kappos L. Blood neurofilament light chain as a biomarker of MS disease activity and treatment response. *Neurology*. 2019;92(10):e1007–e1015. doi:10.1212/WNL.0000000000007032

Kunkle BW, Grenier-Boley B, Sims R, et al. Genetic meta-analysis of diagnosed Alzheimer's disease identifies new risk loci and implicates Aβ, tau, immunity and lipid processing. *Nat Genet*. 2019;51:414–30.

Lassmann H. Multiple sclerosis pathology. *Cold Spring Harb Perspect Med*. 2018;8(3):a028936. doi:10.1101/cshperspect.a028936

Lassmann H, Brück W, Lucchinetti CF. The immunopathology of multiple sclerosis: An overview. *Brain Pathol*. 2007;17(2):210–8. doi:10.1111/j.1750-3639.2007.00064.x

Logroscino G, Piccininni M. Amyotrophic lateral sclerosis descriptive epidemiology: The origin of geographic difference. *Neuroepidemiology*. 2019;52(1–2):93–103. doi:10.1159/000493386

Liddelow SA, Guttenplan KA, Clarke LE, Bennett FC, Bohlen CJ, Schirmer L, Bennett ML, Münch AE, Chung WS, Peterson TC, Wilton DK, Frouin A, Napier BA, Panicker N, Kumar M, Buckwalter MS, Rowitch DH, Dawson VL, Dawson TM, Stevens B, Barres BA. Neurotoxic reactive astrocytes are induced by activated microglia. *Nature*. 2017;541(7638):481–87. doi:10.1038/nature21029

Lin L, Park JW, Ramachandran S, Zhang Y, Tseng YT, Shen S, Waldvogel HJ, Curtis MA, Faull RL, Troncoso JC, Pletnikova O, Ross CA, Davidson BL, Xing Y. Transcriptome sequencing reveals aberrant alternative splicing in Huntington's disease. *Hum Mol Genet*. 2016;25(16):3454–3466. doi:10.1093/hmg/ddw187

Ling SC, Polymenidou M, Cleveland DW. Converging mechanisms in ALS and FTD: Disrupted RNA and protein homeostasis. *Neuron*. 2013;79(3):416–38. doi:10.1016/j.neuron.2013.07.033

Lublin FD, Reingold SC, Cohen JA, Cutter GR, Sørensen PS, Thompson AJ, Wolinsky JS, Balcer LJ, Banwell B, Barkhof F, Bebo B Jr, Calabresi PA, Clanet M, Comi G, Fox RJ, Freedman MS, Goodman AD, Inglese M, Kappos L, Kieseier BC, Lincoln JA, Lubetzki C, Miller AE, Montalban X, O'Connor PW, Petkau J, Pozzilli C, Rudick RA, Sormani MP, Stüve O, Waubant E, Polman CH. Defining the clinical course of multiple sclerosis: The 2013 revisions. *Neurology*. 2014;83(3):278–86. doi:10.1212/WNL.0000000000000560

Mackenzie IR, Rademakers R, Neumann M. TDP-43 and FUS in amyotrophic lateral sclerosis and frontotemporal dementia. *Lancet Neurol*. 2010;9(10):995–1007. doi:10.1016/S1474-4422(10)70195-2

Magliozzi R, Howell O, Vora A, Serafini B, Nicholas R, Puopolo M, Reynolds R, Aloisi F. Meningeal B-cell follicles in secondary progressive multiple sclerosis associate with early onset of disease and severe cortical pathology. *Brain*. 2007;130(Pt 4):1089–1104. doi:10.1093/brain/awm038

Magrané J, Cortez C, Gan WB, Manfredi G. Abnormal mitochondrial transport and morphology are common pathological denominators in SOD1 and TDP43 ALS mouse models. *Hum Mol Genet*. 2014;23(6):1413–24. doi:10.1093/hmg/ddt528

Mahad DH, Trapp BD, Lassmann H. Pathological mechanisms in progressive multiple sclerosis. *Lancet Neurol*. 2015;14(2):183–93.

Mançano ASF, Pina JG, Froes BR and Sciani JM. Autophagy-lysosomal pathway impairment and cathepsin dysregulation in Alzheimer's disease. *Front Mol Biosci*. 2024;11:1490275.

Mangiarini L, Sathasivam K, Seller M, Cozens B, Harper A, Hetherington C, Lawton M, Trottier Y, Lehrach H, Davies SW, Bates GP. Exon 1 of the HD gene with an expanded CAG repeat is sufficient to cause a progressive neurological phenotype in transgenic mice. *Cell*. 1996;87(3):493–506. doi:10.1016/s0092-8674(00)81369-0

Manley K, Shirley TL, Flaherty L, Messer A. Msh2 deficiency prevents in vivo somatic instability of the CAG repeat in Huntington disease transgenic mice. *Nat Genet*. 1999;23(4):471–3. doi:10.1038/70598

Mann JR, Gleixner AM, Mauna JC, Gomes E, DeChellis-Marks MR, Needham PG, Copley KE, Hurtle B, Portz B, Pyles NJ, Guo L, Calder CB, Wills ZP, Pandey UB, Kofler JK, Brodsky JL, Thathiah A, Shorter J, Donnelly CJ. RNA binding antagonizes neurotoxic phase transitions of TDP-43. *Neuron*. 2019;102(2):321–38.e8. doi:10.1016/j.neuron.2019.01.048

Martinez-Vicente M, Talloczy Z, Wong E, Tang G, Koga H, Kaushik S, de Vries R, Arias E, Harris S, Sulzer D, Cuervo AM. Cargo recognition failure is responsible for inefficient autophagy in Huntington's disease. *Nat Neurosci*. 2010;13(5):567–76. doi:10.1038/nn.2528

Mazzini L, Gelati M, Profico DC, Sorarù G, Ferrari D, Copetti M, Muzi G, Ricciolini C, Carletti S, Giorgi C, Spera C, Frondizi D, Masiero S, Stecco A, Cisari C, Bersano E, De Marchi F, Sarnelli MF, Querin G, Cantello R, Petruzzelli F, Maglione A, Zalfa C, Binda E, Visioli A, Trombetta D, Torres B, Bernardini L,

Gaiani A, Massara M, Paolucci S, Boulis NM, Vescovi AL; ALS-NSCs trial study group. Results from phase I clinical trial with intraspinal injection of neural stem cells in amyotrophic lateral sclerosis: A long-term outcome. *Stem Cells Transl Med.* 2019;8(9):887–97. doi:10.1002/sctm.18-0154

McCauley ME, Baloh RH. Inflammation in ALS/FTD pathogenesis. *Acta Neuropathol.* 2019;137(5):715–30. doi: 10.1007/s00401-018-1933-9

McNaught KS, Olanow CW, Halliwell B, Isacson O, Jenner P. Failure of the ubiquitin-proteasome system in Parkinson's disease. *Nat Rev Neurosci.* 2001;2(8):589–94. doi:10.1038/35086067

Melamed Z, López-Erauskin J, Baughn MW, Zhang O, Drenner K, Sun Y, Freyermuth F, McMahon MA, Beccari MS, Artates JW, Ohkubo T, Rodriguez M, Lin N, Wu D, Bennett CF, Rigo F, Da Cruz S, Ravits J, Lagier-Tourenne C, Cleveland DW. Premature polyadenylation-mediated loss of stathmin-2 is a hallmark of TDP-43-dependent neurodegeneration. *Nat Neurosci.* 2019;22(2):180–90. doi:10.1038/s41593-018-0293-z

Menke LA; DDD study; Gardeitchik T, Hammond P, Heimdal KR, Houge G, Hufnagel SB, Ji J, Johansson S, Kant SG, Kinning E, Leon EL, Newbury-Ecob R, Paolacci S, Pfundt R, Ragge NK, Rinne T, Ruivenkamp C, Saitta SC, Sun Y, Tartaglia M, Terhal PA, van Essen AJ, Vigeland MD, Xiao B, Hennekam RC. Further delineation of an entity caused by CREBBP and EP300 mutations but not resembling Rubinstein-Taybi syndrome. *Am J Med Genet A.* 2018;176(4):862–76. doi:10.1002/ajmg.a.38626

Mi S, Hu B, Hahm K, Luo Y, Kam Hui ES, Yuan Q, Wong WM, Wang L, Su H, Chu TH, Guo J, Zhang W, So KF, Pepinsky B, Shao Z, Graff C, Garber E, Jung V, Wu EX, Wu W. LINGO-1 antagonist promotes spinal cord remyelination and axonal integrity in MOG-induced experimental autoimmune encephalomyelitis. *Nat Med.* 2007;13(10):1228–33. doi:10.1038/nm1664

Milakovic T, Johnson GV. Mitochondrial respiration and ATP production are significantly impaired in striatal cells expressing mutant huntingtin. *J Biol Chem.* 2005;280(35):30773–82. doi:10.1074/jbc.M504749200

Miller TM, Andersen PM, Bradley JG, Forshew SJ, Monaco MS, et al. Trial of antisense oligonucleotide tofersen for SOD1 ALS. *N Eng J Med.* 2022;387(12):1099–110. https://doi.org/10.1056/NEJMoa2204705

Misrani A, Tabassum S, Yang L. Mitochondrial dysfunction and oxidative stress in Alzheimer's disease. *Front Aging Neurosci.* 2021;13:617588. doi:10.3389/fnagi.2021.617588

Nagai M, Re DB, Nagata T, Chalazonitis A, Jessell TM, Wichterle H, Przedborski S. Astrocytes expressing ALS-linked mutated SOD1 release factors selectively toxic to motor neurons. *Nat Neurosci.* 2007;10(5):615–22. doi:10.1038/nn1876

Nalls MA, Blauwendraat C, Vallerga CL, Heilbron K, Bandres-Ciga S, Chang D, Tan M, Kia DA, Noyce AJ, Xue A, Bras J, Young E, von Coelln R, Simón-Sánchez J, Schulte C, Sharma M, Krohn L, Pihlstrøm L, Siitonen A, Iwaki H, Leonard H, Faghri F, Gibbs JR, Hernandez DG, Scholz SW, Botia JA, Martinez M, Corvol JC, Lesage S, Jankovic J, Shulman LM, Sutherland M, Tienari P, Majamaa K, Toft M, Andreassen OA, Bangale T, Brice A, Yang J, Gan-Or Z, Gasser T, Heutink P, Shulman JM, Wood NW, Hinds DA, Hardy JA, Morris HR, Gratten J, Visscher PM, Graham RR, Singleton AB; 23andMe Research Team; System Genomics of Parkinson's Disease Consortium; International Parkinson's Disease Genomics Consortium. Identification of novel risk loci, causal insights, and heritable risk for Parkinson's disease: A meta-analysis of genome-wide association studies. *Lancet Neurol.* 2019;18(12):1091–1102. doi:10.1016/S1474-4422(19)30320-5

Neueder A, Bates GP. A common gene expression signature in Huntington's disease patient brain regions. *BMC Med Genomics.* 2014;7:60. doi:10.1186/s12920-014-0060-2

Neumann M, Sampathu DM, Kwong LK, Truax AC, Micsenyi MC, Chou TT, Bruce J, Schuck T, Grossman M, Clark CM, McCluskey LF, Miller BL, Masliah E, Mackenzie IR, Feldman H, Feiden W, Kretzschmar HA, Trojanowski JQ, Lee VM. Ubiquitinated TDP-43 in frontotemporal lobar degeneration and amyotrophic lateral sclerosis. *Science.* 2006;314(5796):130–3. doi:10.1126/science.1134108

Paganoni S, Macklin EA, Hendrix S, Berry JD, Elliott MA, Maiser S, Karam C, Caress JB, Owegi MA, Quick A, Wymer J, Goutman SA, Heitzman D, Heiman-Patterson T, Jackson CE, Quinn C, Rothstein JD, Kasarskis EJ, Katz J, Jenkins L, Ladha S, Miller TM, Scelsa SN, Vu TH, Fournier CN, Glass JD, Johnson KM, Swenson A, Goyal NA, Pattee GL, Andres PL, Babu S, Chase M, Dagostino D, Dickson SP, Ellison N, Hall M, Hendrix K, Kittle G, McGovern M, Ostrow J, Pothier L, Randall R, Shefner JM, Sherman AV, Tustison E, Vigneswaran P, Walker J, Yu H, Chan J, Wittes J, Cohen J, Klee J, Leslie K, Tanzi RE, Gilbert W, Yeramian PD, Schoenfeld D, Cudkowicz ME. Trial of sodium phenylbutyrate-taurursodiol for amyotrophic lateral sclerosis. *N Engl J Med.* 2020;383(10):919–30. doi:10.1056/NEJMoa1916945

Paisán-Ruíz C, Jain S, Evans EW, Gilks WP, Simón J, van der Brug M, López de Munain A, Aparicio S, Gil AM, Khan N, Johnson J, Martinez JR, Nicholl D, Martí Carrera I, Pena AS, de Silva R, Lees A, Martí-Massó JF, Pérez-Tur J, Wood NW, Singleton AB. Cloning of the gene containing mutations that cause PARK8-linked Parkinson's disease. *Neuron.* 2004;44(4):595–600. doi:10.1016/j.neuron.2004.10.023

Panov AV, Gutekunst CA, Leavitt BR, Hayden MR, Burke JR, Strittmatter WJ, Greenamyre JT. Early mitochondrial calcium defects in Huntington's disease are a direct effect of polyglutamines. *Nat Neurosci.* 2002;5(8):731–6. doi:10.1038/nn884

Parnetti L, Gaetani L, Eusebi P, Paciotti S, Hansson O, El-Agnaf O, Mollenhauer B, Blennow K, Calabresi P. CSF and blood biomarkers for Parkinson's disease. *Lancet Neurol.* 2019;18(6):573–86. doi:10.1016/S1474-4422(19)30024-9

Pasinelli P, Brown RH. Molecular biology of amyotrophic lateral sclerosis: insights from genetics. *Nat Rev Neurosci.* 2006;7(9):710–23. doi:10.1038/nrn1971

Pavese N, Gerhard A, Tai YF, Ho AK, Turkheimer F, Barker RA, Brooks DJ, Piccini P. Microglial activation correlates with severity in Huntington disease: A clinical and PET study. *Neurology.* 2006;66(11):1638–43. doi:10.1212/01.wnl.0000222734.56412.17

Pickrell AM, Youle RJ. The roles of PINK1, parkin, and mitochondrial fidelity in Parkinson's disease. *Neuron.* 2015;85(2):257–73. doi:10.1016/j.neuron.2014.12.007

Polymeropoulos MH, Lavedan C, Leroy E, Ide SE, Dehejia A, Dutra A, Pike B, Root H, Rubenstein J, Boyer R, Stenroos ES, Chandrasekharappa S, Athanassiadou A, Papapetropoulos T, Johnson WG, Lazzarini AM, Duvoisin RC, Di Iorio G, Golbe LI, Nussbaum RL. Mutation in the alpha-synuclein gene identified in families with Parkinson's disease. *Science.* 1997;276(5321):2045–7. doi:10.1126/science.276.5321.2045

Ponath G, Park C, Pitt D. The role of astrocytes in multiple sclerosis. *Front Immunol.* 2018;9:217. doi:10.3389/fimmu.2018.00217

Presslauer S, Milosavljevic D, Huebl W, Aboulenein-Djamshidian F, Krugluger W, Deisenhammer F, Senel M, Tumani H, Hegen H. Validation of kappa free light chains as a diagnostic biomarker in multiple sclerosis and clinically isolated syndrome: A multicenter study. *Mult Scler.* 2016;22(4):502–10. doi:10.1177/1352458515594044

Ramaswami M, Taylor JP, Parker R. Altered ribostasis: RNA-protein granules in degenerative disorders. *Cell.* 2013;154(4):727–36. doi:10.1016/j.cell.2013.07.038

Repici M, Giorgini F. DJ-1 in Parkinson's disease: Clinical insights and therapeutic perspectives. *J Clin Med.* 2019;8(9):1377. doi:10.3390/jcm8091377

Rodrigues FB, Byrne LM, Tortelli R, Johnson EB, Wijeratne PA, Arridge M, De Vita E, Ghazaleh N, Houghton R, Furby H, Alexander DC, Tabrizi SJ, Schobel S, Scahill RI, Heslegrave A, Zetterberg H, Wild EJ. Mutant huntingtin and neurofilament light have distinct longitudinal dynamics in Huntington's disease. *Sci Transl Med.* 2020 Dec 16;12(574):eabc2888. doi: 10.1126/scitranslmed.abc2888

Rosen DR, Siddique T, Patterson D, Figlewicz DA, Sapp P, Hentati A, Donaldson D, Goto J, O'Regan JP, Deng HX, et al. Mutations in Cu/Zn superoxide dismutase gene are associated with familial amyotrophic lateral sclerosis. *Nature.* 1993;362(6415):59–62. doi:10.1038/362059a0. Erratum in: Nature. 1993 Jul 22;364(6435):362. doi: 10.1038/364362c0

Rothstein JD, Van Kammen M, Levey AI, Martin LJ, Kuncl RW. Selective loss of glial glutamate transporter GLT-1 in amyotrophic lateral sclerosis. *Ann Neurol.* 1995;38(1):73–84. doi: 10.1002/ana.410380114

Saez-Atienzar S, Masliah E. Cellular senescence and Alzheimer disease: The egg and the chicken scenario. *Nat Rev Neurosci.* 2020;21(8):433–44. doi:10.1038/s41583-020-0325-z. Epub 2020 Jun 29. Erratum in: *Nat Rev Neurosci.* 2020 Oct;21(10):587. doi: 10.1038/s41583-020-0366-3

Saudou F, Humbert S. The biology of huntingtin. *Neuron.* 2016;89(5):910–26. doi:10.1016/j.neuron.2016.02.003

Schapira AH, Cooper JM, Dexter D, Clark JB, Jenner P, Marsden CD. Mitochondrial complex I deficiency in Parkinson's disease. *J Neurochem.* 1990;54(3):823–7. doi:10.1111/j.1471-4159.1990.tb02325.x

Serrano-Pozo A, Frosch MP, Masliah E, Hyman BT. Neuropathological alterations in Alzheimer disease. *Cold Spring Harb Perspect Med.* 2011;1(1):a006189. doi:10.1101/cshperspect.a006189

Shahmoradian SH, Lewis AJ, Genoud C, et al. Lewy pathology in Parkinson's disease consists of crowded organelles and lipid membranes. *Nat Neurosci.* 2019;22:1099–1109. doi:10.1038/s41593-019-0423-2

Shirendeb UP, Calkins MJ, Manczak M, Anekonda V, Dufour B, McBride JL, Mao P, Reddy PH. Mutant huntingtin's interaction with mitochondrial protein Drp1 impairs mitochondrial biogenesis and causes defective axonal transport and synaptic degeneration in Huntington's disease. *Hum Mol Genet.* 2012;21(2):406–20. doi:10.1093/hmg/ddr475

Singleton AB, Farrer M, Johnson J, et al. α-Synuclein locus triplication causes Parkinson's disease. *Science*. 2003;302(5646):841. doi:10.1126/science.1090278

Steffan JS, Kazantsev A, Spasic-Boskovic O, Greenwald M, Zhu YZ, Gohler H, Wanker EE, Bates GP, Housman DE, Thompson LM. The Huntington's disease protein interacts with p53 and CREB-binding protein and represses transcription. *Proc Natl Acad Sci U S A*. 2000;97(12):6763–8. doi:10.1073/pnas.100110097

Steinacker P, Semler E, Anderl-Straub S, Diehl-Schmid J, Schroeter ML, Uttner I, Foerstl H, Landwehrmeyer B, von Arnim CA, Kassubek J, Oeckl P, Huppertz HJ, Fassbender K, Fliessbach K, Prudlo J, Roßmeier C, Kornhuber J, Schneider A, Volk AE, Lauer M, Danek A, Ludolph AC, Otto M; FTLDc Study Group. Neurofilament as a blood marker for diagnosis and monitoring of primary progressive aphasias. *Neurology*. 2017;88(10):961–69. doi:10.1212/WNL.0000000000003688

Sulzer D, Alcalay RN, Garretti F, Cote L, Kanter E, Agin-Liebes J, Liong C, McMurtrey C, Hildebrand WH, Mao X, Dawson VL, Dawson TM, Oseroff C, Pham J, Sidney J, Dillon MB, Carpenter C, Weiskopf D, Phillips E, Mallal S, Peters B, Frazier A, Lindestam Arlehamn CS, Sette A. T cells from patients with Parkinson's disease recognize α-synuclein peptides. *Nature*. 2017;546(7660):656–61. doi:10.1038/nature22815. Epub 2017 Jun 21. *Erratum in: Nature*. 2017 Sep 13;549(7671):292. doi:10.1038/nature23896

Swami M, Hendricks AE, Gillis T, Massood T, Mysore J, Myers RH, Wheeler VC. Somatic expansion of the Huntington's disease CAG repeat in the brain is associated with an earlier age of disease onset. *Hum Mol Genet*. 2009;18(16):3039–47. doi:10.1093/hmg/ddp242

Tabrizi SJ, Langbehn DR, Leavitt BR, Roos RA, Durr A, Craufurd D, Kennard C, Hicks SL, Fox NC, Scahill RI, Borowsky B, Tobin AJ, Rosas HD, Johnson H, Reilmann R, Landwehrmeyer B, Stout JC; TRACK-HD investigators. Biological and clinical manifestations of Huntington's disease in the longitudinal TRACK-HD study: Cross-sectional analysis of baseline data. *Lancet Neurol*. 2009;8(9):791–801. doi:10.1016/S1474-4422(09)70170-X

Tang Y. The role of microglia in Alzheimer's disease: From homeostasis to neurodegeneration. *Front Aging Neurosci*. 2020;12:233. doi:10.3389/fnagi.2020.00233

Tourbah A, Lebrun-Frenay C, Edan G, Clanet M, Papeix C, Vukusic S, De Sèze J, Debouverie M, Gout O, Clavelou P, Defer G, Laplaud DA, Moreau T, Labauge P, Brochet B, Sedel F, Pelletier J; MS-SPI study group. MD1003 (high-dose biotin) for the treatment of progressive multiple sclerosis: A randomised, double-blind, placebo-controlled study. *Mult Scler*. 2016;22(13):1719–31. doi:10.1177/1352458516667568

Tradewell ML, Cooper LA, Minotti S, Durham HD. Calcium dysregulation, mitochondrial pathology and protein aggregation in a culture model of amyotrophic lateral sclerosis: Mechanistic relationship and differential sensitivity to intervention. *Neurobiol. Dis*. 2011;42:265–75. doi:10.1016/j.nbd.2011.01.016

Trapp BD, Nave KA. Multiple sclerosis: An immune or neurodegenerative disorder? *Annu Rev Neurosci*. 2008;31:247–69. doi:10.1146/annurev.neuro.30.051606.094313

Trapp BD, Peterson J, Ransohoff RM, Rudick R, Mörk S, Bö L. Axonal transection in the lesions of multiple sclerosis. *N Engl J Med*. 1998;338(5):278–85. doi:10.1056/NEJM199801293380502

Twelvetrees AE, Pernigo S, Sanger A, Guedes-Dias P, Schiavo G, Steiner RA, Dodding MP, Holzbaur EL. The dynamic localization of cytoplasmic dynein in neurons is driven by kinesin-1. *Neuron*. 2016;90(5):1000–15. doi:10.1016/j.neuron.2016.04.046

Viglietta V, Baecher-Allan C, Weiner HL, Hafler DA. Loss of functional suppression by CD4+CD25+ regulatory T cells in patients with multiple sclerosis. *J Exp Med*. 2004;199(7):971–9. doi:10.1084/jem.20031579

Vogiatzi T, Xilouri M, Vekrellis K, Stefanis L. Wild type alpha-synuclein is degraded by chaperone-mediated autophagy and macroautophagy in neuronal cells. *J Biol Chem*. 2008;283(35):23542–56. doi:10.1074/jbc.M801992200

von Büdingen HC, Kuo TC, Sirota M, van Belle CJ, Apeltsin L , Glanville J, Cree BA, Gourraud PA, Schwartzburg A, Huerta G, Telman D, Sundar PD, Casey T, Cox DR, Hauser SL. B cell exchange across the blood-brain barrier in multiple sclerosis. *J Clin Invest*. 2012;122(12):4533–43. doi:10.1172/JCI63842

Walton C, King R, Rechtman L, Kaye W, Leray E, Marrie RA, Robertson N, La Rocca N, Uitdehaag B, van der Mei I, Wallin M, Helme A, Angood Napier C, Rijke N, Baneke P. Rising prevalence of multiple sclerosis worldwide: Insights from the Atlas of MS, third edition. *Mult Scler*. 2020;26(14):1816–21. doi:10.1177/1352458520970841

Wang C, Pan R, Wan X, Tan Y, Xu L, McIntyre RS, Choo FN, Tran B, Ho R, Sharma VK, Ho C. A longitudinal study on the mental health of general population during the COVID-19 epidemic in China. *Brain Behav Immun*. 2020;87:40–48. doi:10.1016/j.bbi.2020.04.028

Wang P, Deng J, Dong J, Liu J, Bigio EH, Mesulam M, Wang T, Sun L, Wang L, Lee AY, McGee WA, Chen X, Fushimi K, Zhu L, Wu JY. TDP-43 induces mitochondrial damage and activates the mitochondrial unfolded protein response. *PLoS Genet.* 2019 May 17;15(5):e1007947. doi: 10.1371/journal.pgen.1007947

Witte AV, Fobker M, Gellner R, Knecht S, Flöel A. Caloric restriction improves memory in elderly humans. *Proc Natl Acad Sci USA*. 2009;106(4):1255–60. doi:10.1073/pnas.0808587106

Wong YC, Holzbaur EL. Optineurin is an autophagy receptor for damaged mitochondria in parkin-mediated mitophagy that is disrupted by an ALS-linked mutation. *Proc Natl Acad Sci U S A*. 2014;111(42):E4439–48. doi:10.1073/pnas.1405752111

Yano H, Baranov SV, Baranova OV, Kim J, Pan Y, Yablonska S, Carlisle DL, Ferrante RJ, Kim AH, Friedlander RM. Inhibition of mitochondrial protein import by mutant huntingtin. *Nat Neurosci*. 2014;17(6):822–31. doi:10.1038/nn.3721

Zuccato C, Cattaneo E. Brain-derived neurotrophic factor in neurodegenerative diseases. *Nat Rev Neurol*. 2009;5(6):311–22. doi:10.1038/nrneurol.2009.54

Zuccato C, Tartari M, Crotti A, Goffredo D, Valenza M, Conti L, Cataudella T, Leavitt BR, Hayden MR, Timmusk T, Rigamonti D, Cattaneo E. Huntingtin interacts with REST/NRSF to modulate the transcription of NRSE-controlled neuronal genes. *Nat Genet.* 2003;35(1):76–83. doi:10.1038/ng1219

2 Aging and Neurodegenerative Diseases

2.1 INTRODUCTION

Aging is universally recognized as the predominant risk factor for the development of neurodegenerative disorders, including Alzheimer's disease (AD), Parkinson's disease (PD), amyotrophic lateral sclerosis (ALS), and Huntington's disease (HD) (Hou et al. 2019; Wyss-Coray, 2016). Although these conditions differ in clinical manifestations and specific pathogenic mechanisms, they share underlying age-related vulnerabilities that accelerate their onset and progression. As the global population ages, the burden of neurodegenerative diseases is increasing, posing significant societal, economic, and healthcare challenges. Understanding how aging contributes to neural decline is thus central to developing effective strategies for prevention, treatment, and regenerative interventions.

At a cellular and molecular level, aging is characterized by a series of well-defined biological processes often referred to as the hallmarks of aging. These include genomic instability, telomere attrition, epigenetic alterations, loss of proteostasis, mitochondrial dysfunction, cellular senescence, deregulated nutrient sensing, stem cell exhaustion, and altered intercellular communication (Figure 2.1; López-Otín et al. 2013). Each of these hallmarks plays a critical role in diminishing neuronal health and resilience, rendering the aging brain more susceptible to neurodegenerative changes (Hou et al. 2019; Gonzales et al. 2022).

One of the earliest detectable changes in the aging brain is mitochondrial dysfunction (Table 2.1). Neurons are metabolically active and rely heavily on mitochondria for ATP production and calcium buffering. With age, mitochondrial DNA (mtDNA) damage accumulates, oxidative phosphorylation becomes less efficient, and reactive oxygen species (ROSs) levels rise, resulting in oxidative stress that damages proteins, lipids, and nucleic acids (Lautrup et al. 2019). Studies have shown that mitochondrial deficits precede and potentially drive pathological features in AD, PD, and ALS (Wang et al. 2019).

Parallel to mitochondrial decline is the progressive loss of proteostasis, the capacity of cells to maintain the integrity and functionality of their proteome (Figure 2.2). Aging compromises chaperone activity, ubiquitin-proteasome function, and autophagic degradation pathways, leading to the accumulation of misfolded and aggregated proteins, hallmark features of neurodegenerative diseases such as β-amyloid plaques in AD and α-synuclein inclusions in PD (Azam et al. 2021; Mir et al. 2023). Disruption in proteostasis not only impairs neuronal function but also triggers toxic cascades that promote cell death.

Neuroinflammation is another key mediator that links aging to neurodegeneration (Figure 2.2). Age-related alterations in the immune system, immunosenescence, result in a chronic, low-grade inflammatory state known as inflammaging (Mayne et al. 2020). In the brain, microglia and astrocytes become dysregulated with age, adopting a pro-inflammatory phenotype that contributes to synaptic

DOI: 10.1201/9781003582687-2

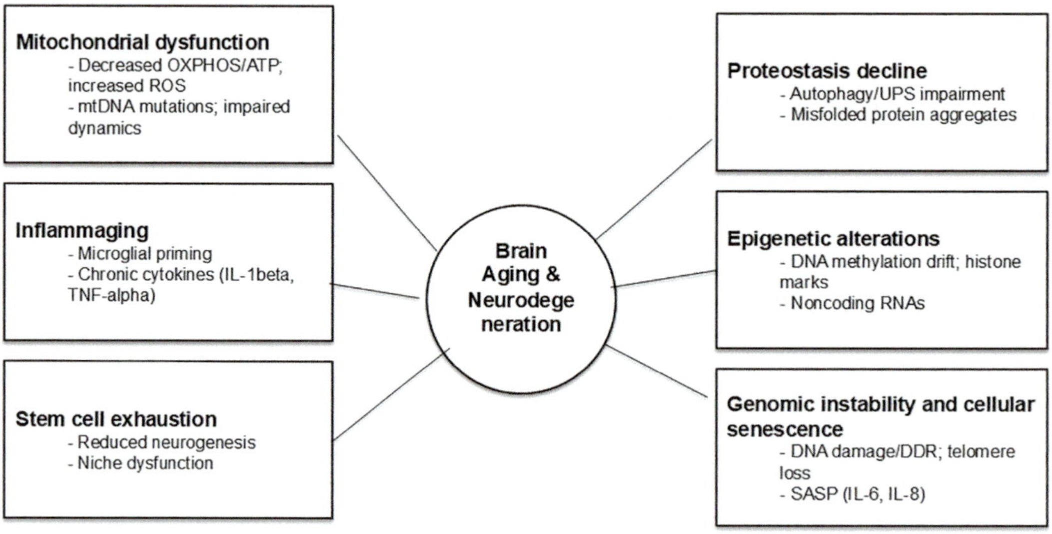

FIGURE 2.1 Hallmarks of Brain Aging and Neurodegeneration.

TABLE 2.1
Mitochondrial Alterations in Aging and Neurodegeneration

Feature	Effects on Aging	Implication in Neurodegenerative Diseases
Reduced ATP production	Energy deficit in neurons	Contributes to neuronal death in AD, PD
Increased ROS generation	Oxidative damage to proteins, lipids, DNA	Accelerates pathology in AD, PD, ALS
mtDNA mutations	Impaired mitochondrial function	Linked to early onset PD and ALS
Altered mitochondrial dynamics	Imbalance in fission/fusion	Associated with synaptic failure in AD

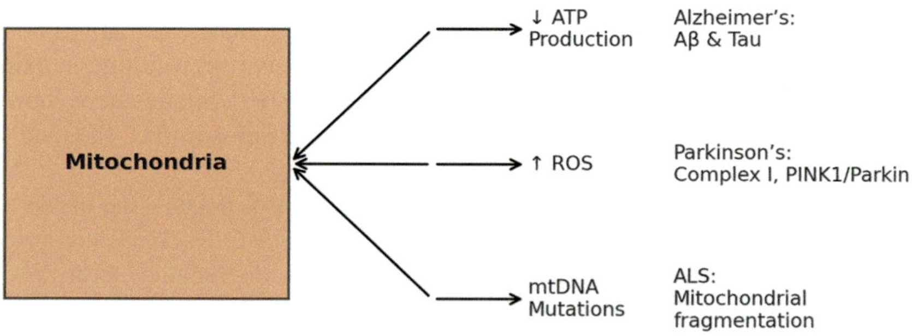

FIGURE 2.2 Mitochondrial Dysfunction in Aging Neurons.

dysfunction, blood–brain barrier breakdown, and neuronal loss (Wyss-Coray, 2016; Gonzales et al. 2022). Notably, longitudinal studies have indicated that elevated systemic inflammation predicts accelerated cognitive decline and an increased risk of dementia (Walker et al. 2019).

Epigenetic changes, including alterations in DNA methylation, histone modification, and non-coding RNA expression, accumulate over time and affect gene expression patterns critical for neuronal survival, synaptic plasticity, and repair mechanisms (Mir et al. 2023). These epigenetic drifts are increasingly recognized as key drivers of age-related cognitive decline and neurodegeneration. Evidence suggests that correcting epigenetic dysregulation may confer therapeutic benefits, with the potential to reprogram aged neurons and restore neural network functionality (Horvath and Raj 2018).

Another important factor is stem cell exhaustion. Neurogenesis, the generation of new neurons from neural stem cells (NSCs), significantly declines with age, particularly in the hippocampus brain region that is critical for learning and memory. The depletion of the NSC pool and the reduced capacity of aged progenitor cells to proliferate and differentiate contribute to impaired brain plasticity and decreased repair capacity following injury or disease. Consequently, age-related declines in neurogenesis are thought to be an early contributor to cognitive deficits and neurodegeneration.

At a systems biology level, aging also induces alterations in systemic factors that impact brain health. Circulating molecules in the blood, such as pro-inflammatory cytokines and metabolic byproducts, influence neuroinflammation, synaptic integrity, and vascular health (Wyss-Coray, 2016). Studies comparing heterochronic parabiosis connecting the circulatory systems of young and old mice have shown that young blood can rejuvenate aged brains, while old blood accelerates brain aging, highlighting the significant role of systemic aging in brain health (Castellano et al. 2017).

Importantly, while chronological age is nonmodifiable, biological aging processes represent promising therapeutic targets. Emerging evidence suggests that interventions aimed at enhancing mitochondrial function (Fang et al. 2019), clearing senescent cells (Baker et al. 2016), modulating immune responses (Nayak D et al. 2014), and restoring proteostasis (Kaushik and Cuervo 2015) could delay the onset or slow the progression of neurodegenerative diseases. Moreover, advances in stem cell technology, including induced pluripotent stem cells (iPSCs) and NSC transplantation, hold promise for restoring lost neural function in aging and neurodegenerative contexts (Plemel et al. 2020).

2.2 BIOLOGICAL MECHANISMS LINKING AGING TO NEURODEGENERATIVE DISORDERS

As outlined in the introduction, advancing age is the greatest risk factor for neurodegenerative disorders such as AD, PD, and ALS (Niccoli and Partridge 2012; Hou et al. 2019). In this section, each of these aging-related mechanisms are examined in depth, explaining their molecular mechanisms and illustrating how they link aging to neurodegenerative diseases.

2.2.1 Mitochondrial Dysfunction

One of the central features of aging cells is a decline in mitochondrial function. Mitochondria are powerhouses of cells, and their efficient operation is essential for high-energy demanding organs like the brain. Mitochondrial dysfunction in aging is characterized by a reduction in ATP production, diminished electron transport chain activity, increased generation of reactive oxygen species, and accumulation of mtDNA mutations (Table 2.1; Lin and Beal 2006; Waseem et al. 2021). Over time, oxidative damage to mitochondrial components and mtDNA leads to a self-perpetuating cycle of dysfunction. Impaired mitochondria produce excess ROS, which in turn further damages mitochondria and other cellular constituents (Lin and Beal 2006). In the aging brain, this manifests as chronic oxidative stress and an overall energy deficit that leaves neurons vulnerable to degeneration (Figure 2.2; Hou et al. 2019).

Mitochondrial dysfunction is a widely recognized hallmark of neurodegenerative disorders. Postmortem and experimental studies in Alzheimer's disease have shown deficits in mitochondrial enzymes, e.g., cytochrome oxidase and altered mitochondrial dynamics in affected neurons (Lin and Beal 2006). Amyloid-β (Aβ) and phosphorylated tau, which is the pathological proteins of

AD, can impair mitochondrial function. The Aβ aggregates have been found to be associated with mitochondria, where they may disrupt respiratory chain complexes and enhance ROS production, while tau pathology in neurons correlates with abnormal mitochondrial transport and fission–fusion balance (Waseem et al. 2021). In Parkinson's disease, dopaminergic neurons of the substantia nigra show reduced activity of Complex I of the respiratory chain and often contain mtDNA deletions acquired with age (Lin and Beal 2006). Several familial PD genes, e.g., *PINK1*, *Parkin*, *DJ-1* encode proteins that safeguard mitochondrial hemostasis, and mutations in these genes lead to defective mitophagy mitochondrial autophagy and accumulation of dysfunctional mitochondria, contributing to neurodegeneration (Exner et al. 2012; Hou et al. 2019). Likewise, in ALS and Huntington's disease, mitochondrial abnormalities such as swollen, fragmented mitochondria and impaired electron transport have been observed, linking energy failure to motor neuron death and striatal neuron death, respectively (Lin and Beal 2006; Yin et al. 2016). Thus, age-related mitochondrial decline directly converges with disease-specific pathways, e.g., increased protein aggregation or disruption of calcium and ROS hemostasis in stressed neurons, accelerating neurodegenerative disorder processes.

Therapeutic strategies targeting mitochondrial dysfunction are an active area of research. Antioxidants and redox modulators have been tested to mitigate oxidative damage in the aging brain. Notably, mitochondria-targeted antioxidants such as MitoQ and SS-31 (elamipretide) can concentrate within mitochondria and neutralize RO, and these agents have shown promise in preclinical models of neurodegeneration by improving cellular energy metabolism and reducing neurotoxic protein aggregates (Reddy et al. 2017; Yin et al. 2016). Other approaches include promoting mitochondrial biogenesis and dynamics. For example, activation of peroxisome proliferator-activated receptor gamma coactivator-1 alpha (PGC-1α), which is a central regulator of mitochondrial biogenesis, or targeted modulation of mitochondrial fission and fusion machinery restores mitochondrial network integrity and bioenergetic capacity in aged neurons. There is also interest in enhancing mitophagy to remove damaged mitochondria, small molecules that activate the PINK1-Parkin pathway or that mimic caloric restriction, e.g., via AMPK activation, helps in maintaining bioenergetically potent mitochondria in aging cells (Waseem et al. 2021). While clinical trials of some compounds like coenzyme Q10 or certain antioxidant mixtures have yielded mixed results in AD or PD, the continued refinement of mitochondrial-targeted therapies holds potential. Crucially preserving mitochondrial function in the elderly might not only delay aging-related physiological decline but also increases resistance of the brain to neurodegenerative pathology. Indeed, mitochondrial dysfunction provides a permissive environment for initiation and propagation of other deleterious processes such as protein misfolding and inflammation, underscoring its central role in linking aging to neurodegeneration.

2.2.2 Loss of Proteostasis and Protein Aggregation in Brain Aging and Neurodegeneration

Aging is characterized by a progressive decline in proteostasis, wherein the cellular mechanisms responsible for maintaining protein folding, stability, and degradation become increasingly compromised. Neurons being long-lived and postmitotic are especially dependent on robust proteostasis to regulate damaged or misfolded proteins (Labbadia and Morimoto 2015). Under young, healthy conditions, cells rely on an integrated proteostasis mechanism, including molecular chaperones, the ubiquitin-proteasome system (UPS), and autophagy-lysosome pathways to fold nascent proteins correctly, refold or degrade misfolded proteins, and clear protein aggregates. With advancing age, however, this quality control system becomes less efficient (Hipp et al. 2014; Höhn et al. 2020). Heat shock proteins (HSPs) and other chaperones are often downregulated or functionally compromised in aged tissues, diminishing the cell's ability to stabilize unfolded proteins. Similarly, the UPS activity declines, in part due to oxidative modifications of proteasome subunits and reduced expression of proteasome assembly factors in aging cells (Höhn et al. 2020). Autophagic efficiency also decreases with age, lysosomal hydrolase levels drop, and autophagosome formation or clearance is impaired, leading to accumulation of dysfunctional organelles and protein aggregates that would normally be removed (Table 2.2; Labbadia and Morimoto 2015). The cumulative effect of these age-related alterations is a disruption of proteostasis, wherein oxidatively damaged, glycated, or misfolded proteins accumulate more rapidly than they can be refolded or eliminated through degradation pathways (Figure 2.3).

Proteostasis failure has a direct and profound link to neurodegenerative disorders, which are often characterized by the accumulation of toxic protein species. Classic examples include extracellular Aβ plaques and intracellular tau neurofibrillary tangles in Alzheimer's disease, α-synuclein-rich Lewy bodies in Parkinson's disease, TDP-43 or SOD1 aggregates in ALS, and huntingtin protein aggregates in Huntington's disease. Aging appears to set the stage for these pathogenic proteins to misfold and aggregate. In an elderly brain, as the proteostatic mechanisms falter, even normal proteins are more prone to misfolding. Once misfolded, they may escape timely clearance and begin to aggregate (Hipp et al. 2014). For instance, age-related impairment of autophagy in the brain can lead to incomplete digestion of Aβ or α-synuclein, promoting the formation of oligomers and fibrils that are neurotoxic. Postmortem studies of AD brains reveal an abundance of autophagic vacuoles and lysosomal storage material, indicating a breakdown in the autophagy-lysosomal clearance pathway (Nixon 2013). Likewise, experimental inhibition of proteasomes or autophagy in neuron cultures produces inclusion bodies reminiscent of those in neurodegenerative diseases, highlighting the importance of these systems for preventing proteotoxicity (Höhn et al. 2020). Aging

TABLE 2.2
Proteostasis Network Impairments in Aging Brain

Component	Age-Related Changes	Pathological Impact
Chaperones (HSPs)	Decreased expression/activity	Increased protein misfolding
Ubiquitin-proteasome system	Reduced proteasome activity	Accumulation of damaged proteins
Autophagy	Impaired lysosomal clearance	Protein aggregates in AD, PD
ER unfolded protein response	Chronic activation	Synaptic dysfunction in prion disease

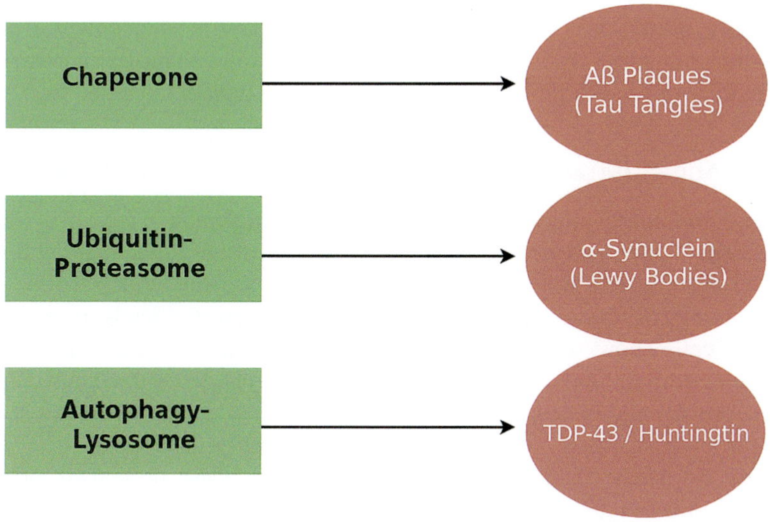

FIGURE 2.3 Proteostasis Dysregulation and Protein Aggregation.

also exacerbates the production of advanced glycation end-products (AGEs) and other modified proteins that tend to aggregate. Indeed, AGEs have been identified as a hallmark of AD brains, where they can cross-link with Aβ and tau, further stabilizing aggregates (Höhn et al. 2020).

There are specific molecular links known between proteostatic decline and neurodegeneration. One example is the unfolded protein response (UPR) in the endoplasmic reticulum and with age, the chronic ER stress from accumulating misfolded proteins can trigger UPR signaling that initially attempts to restore folding capacity but eventually leads to pro-apoptotic pathways. In AD and prion disease models, overactivation of the PERK arm of the UPR which reduces general protein synthesis has been implicated in synaptic dysfunction and memory deficits, and reducing this stress through PERK inhibition can reverse cognitive impairment in mouse models (Ma et al. 2013). In Parkinson's disease, mutations in genes like *LRRK2* and *GBA* are believed to impair lysosomal function and autophagy, directly linking genetic risk factors to proteostasis. Furthermore, the protein aggregates themselves can act in a prion-like manner to disrupt the normal folding of native proteins and propagate pathology from cell to cell, a phenomenon observed for misfolded tau, α-synuclein, and others. (Hipp et al. 2014). An aging brain exhibits diminished capacity to counteract seeded protein misfolding and propagation, owing to a decline in its proteostasis defense mechanisms.

Given the central role of proteostasis loss in neurodegeneration, enhancing protein quality control has become a prominent therapeutic goal. One therapeutic strategy involves enhancing chaperone expression or activity. Pharmacological chaperone inducers, e.g., arimoclomol, which amplifies HSP expression have been explored to help refold misfolded proteins in diseases like ALS and have reached clinical trials though with mixed outcomes (Kalmar et al. 2014). Another strategy is to enhance proteasome or autophagy function to clear aggregates more efficiently. For example, rapamycin and other mTOR inhibitors stimulate autophagy and have been shown to reduce the buildup of Aβ and tau in mouse models of AD (Caccamo et al. 2014). Small molecule enhancers of lysosomal activity e.g., agonists of TFEB, a transcription factor that upregulates lysosomal biogenesis, are also being investigated to determine if they can compensate for the age-related autophagy decline. In the case of PD, molecules that stabilize α-synuclein in a nonaggregating conformation or that promote degradation such as ambroxol, which increases lysosomal exocytosis, or modulators of the chaperone-mediated autophagy pathway are being tested. Importantly, lifestyle interventions known to preserve proteostasis, such as caloric restriction, exercise, or diets rich in polyphenols appear to induce cellular stress responses that keep the proteostasis network more robust with age (Labbadia and Morimoto 2015). Overall, maintaining proteostasis in the aging brain could reduce the burden of misfolded proteins and thereby delay the onset or progression of neurodegenerative pathology. Notably, proteostasis is interlinked with other aging mechanisms, for instance, excessive ROS from mitochondrial dysfunction damage proteins and enhance proteostasis, and chronic inflammation oxidize or modify proteins, tipping the balance toward aggregation.

2.2.3 Immune Aging, Inflammaging, and Neuroinflammation in Neurodegeneration

The aging process profoundly affects the immune system, a phenomenon termed immunosenescence while simultaneously fostering a pro-inflammatory milieu known as inflammaging (Franceschi et al. 2007; Costantini et al. 2018). Immunosenescence involves the functional decline of both innate and adaptive immunity with age, e.g., the thymus degeneration that leads to fewer naïve T cells, and peripheral T cells show reduced proliferative capacity and diversity; B-cell production in bone marrow decreases; and innate immune cells like macrophages and natural killer cells exhibit altered cytokine profiles and phagocytic ability (Costantini et al. 2018). Paradoxically, even as specific immune responses weaken, elderly individuals commonly have elevated levels of systemic pro-inflammatory cytokines (IL-6, TNF-α, IL-1β) and acute phase proteins, reflecting a state of chronic low-grade inflammation (Franceschi et al. 2007; Table 2.3). This inflammaging is thought to result from lifelong antigenic burden, accumulation of senescent cells with a pro-inflammatory

TABLE 2.3
Immunosenescence and Inflammaging in Neurodegenerative Disorders

Immune Changes	Aging Effect	Neurodegenerative Impact
Thymic involution	Reduced naïve T cells	Diminished adaptive immune response
Elevated cytokines (IL-6, TNF-α)	Systemic inflammation	Linked to faster cognitive decline
Microglial priming	Hyperreactive to stimuli	Pro-inflammatory state in AD
Astrocytic activation	Sustained reactivity	Contributes to BBB breakdown and neuronal loss

secretory phenotype, changes in the gut microbiome, and cell-intrinsic dysregulation of inflammatory pathways. In the central nervous system, aging causes microglia, the resident immune cells of the brain to adopt an activated state characterized by hypertrophic morphology, upregulation of inflammatory mediators, and heightened reactivity to stimuli (Heneka et al. 2015). Astrocytes in aged brains may also become reactive, contributing to an inflammatory environment. Concurrently, the blood–brain barrier becomes more permeable with age, allowing peripheral inflammatory factors and immune cells to infiltrate the brain more readily (Costantini et al. 2018). Together, these age-related immune alterations set the stage for chronic neuroinflammation in response to triggers and a diminished ability to resolve inflammation.

Chronic neuroinflammation is now recognized as a key driver of progression of neurodegenerative disease (Figure 2.4; Heneka et al. 2015). The pathologies of AD, PD, ALS, and others all include an inflammatory component that is exacerbated by aging. In Alzheimer's disease, for instance, insoluble Aβ plaques and tau tangles activate microglia and astrocytes, while acute activation of glia may help to clear Aβ initially. In aged brains, the response often becomes dysregulated and persistent and microglia surrounding plaques can enter a pro-inflammatory state, secreting cytokines like IL-1β, IL-6, TNF-α, and chemokines that damage neurons and synapses (Heneka et al. 2015). Aging microglia also exhibit impaired phagocytic capacity, hence reducing the clearance of Aβ aggregates, which allows plaques to accumulate perpetuating a vicious cycle of neuroinflammation and amyloid plaque deposition. Genome-wide association studies in AD have identified risk variants in several immune-related genes, e.g., *TREM2*, *CR1*, *CLU*, underscoring that innate immune pathways are centrally involved in AD pathogenesis (Heneka et al. 2015). In aged brains, TREM2-deficient microglia exhibit impaired capacity to contain amyloid deposits, leading to the accumulation of unrestrained plaques that aggravate neuroinflammation and thus neurotoxic damage. Similarly, in Parkinson's disease, there is evidence of active microgliosis in the substantia nigra even in early stages, and pro-inflammatory cytokines are elevated in the cerebrospinal fluid of patients (Bartels and Leenders 2007). The α-synuclein aggregates characteristic of PD can directly trigger the NLRP3 inflammasome within microglia, causing caspase-1 activation and release of IL-1β (Franceschi et al. 2007; Costantini et al. 2018). Indeed, a recent study demonstrated that aged glial cells with senescence exhibit impaired autophagy and accumulate α-synuclein, exacerbating further inflammation and neuronal damage (Hong et al. 2024). In amyotrophic lateral sclerosis and frontotemporal dementia (FTD), aberrant accumulation of misfolded proteins such as SOD1 and TDP-43 induces microglial and astroglial activation, thereby accelerating motor neuron degeneration. Aging likely lowers the threshold for these pathological proteins to elicit deleterious neuroimmune responses, exacerbating disease progression. Furthermore, systemic inflammation, e.g., infections or chronic inflammatory conditions common in the elderly can worsen neurodegenerative disease symptoms, indicating crosstalk between peripheral inflammaging and central neuroinflammation (Costantini et al. 2018).

The convergence of aging-related immune dysregulation with neurodegeneration suggests that anti-inflammatory or immunomodulatory interventions could be beneficial. Epidemiological

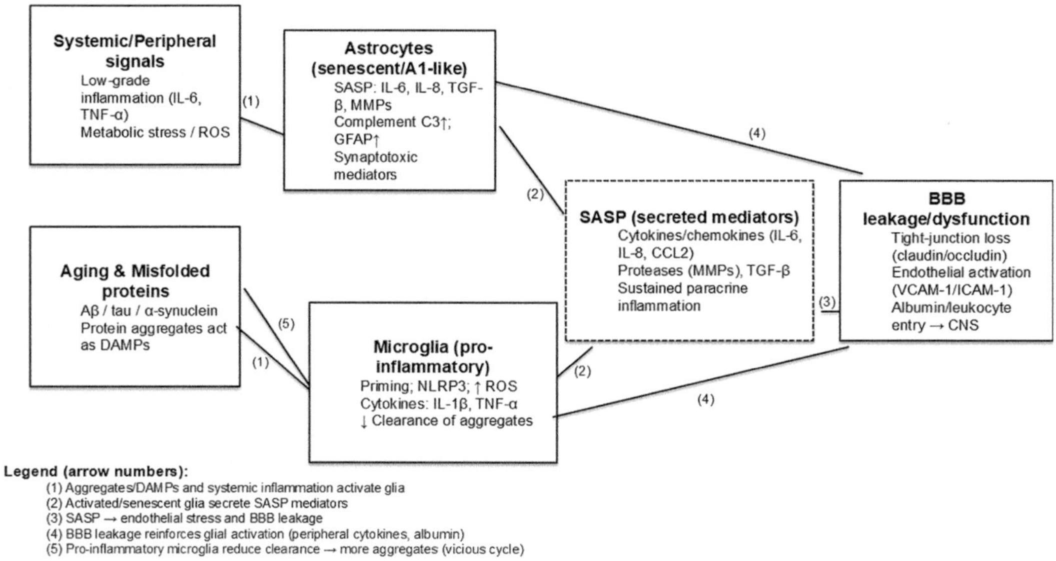

FIGURE 2.4 Neuroinflammation and Senescent Glial Cells.

observations that long-term use of nonsteroidal anti-inflammatory drugs (NSAIDs) was associated with lower incidence of AD promoted interest in targeting neuroinflammation (Heneka et al. 2015). Although NSAID trials in established AD did not yield clear benefits, researchers are developing more specific approaches. One strategy aims to inhibit the inflammasome and pro-inflammatory cytokine signaling in the aged brain. For example, pharmacological inhibitors of NLRP3 or neutralizing antibodies against IL-1β and TNF-α are being evaluated in preclinical models of AD and PD, where they have been shown to reduce neuronal loss and pathology (Heneka et al. 2015). Another promising strategy involves modulating microglial activation states. The temporary depletion of microglia using colony-stimulating factor-1 receptor (CSF1R) inhibitors, followed by repopulation, has been shown in AD mouse models to restore a more homeostatic, juvenile microglial phenotype and reduce amyloid pathology. Enhancing innate immune clearance is another approach, the stimulation of receptors such as TREM2 can promote microglial phagocytosis of pathological debris, while immunotherapies targeting Aβ or α-synuclein depend on functional microglia for effective clearance of aggregated protein. In the context of aging, combining such therapies with treatments that rejuvenate immune function could be crucial. On the systemic level, interventions such as exercise and dietary optimization have been shown to lower chronic inflammation and might indirectly mitigate neuroinflammation. There is even exploratory research into young blood plasma transfusions or parabiosis, which in mice have rejuvenated aspects of the aged immune system and reduced brain inflammation (Yousef et al. 2019). Ultimately, maintaining a balanced immune milieu in the aging body and brain, inhibiting deleterious inflammation while preserving pathogen defense and repair mechanisms might delay the onset of neurodegenerative changes or slow their progression. Notably, immune aging intersects with other hallmarks, for instance, senescent cells secrete inflammatory factors, and misfolded proteins can act as signals to the immune system. Thus, therapies like senolytics that remove senescent, pro-inflammatory cells could also ameliorate inflammaging and neuroinflammation concurrently.

2.2.4 Epigenetic Alterations in Brain Aging and Neurodegeneration

Epigenetic alterations represent a key hallmark of ageing, mediating the interaction between environment and temporal genomic regulation by modifying patterns of gene expression that are associated with the onset and progression of neurodegenerative diseases. During aging, the epigenetic landscape of cells undergoes widespread shifts often described as epigenetic drift. In general, aged cells exhibit aberrant DNA methylation patterns, certain gene promoters that are typically unmethylated under physiological conditions may acquire aberrant DNA methylation with age, while other regions become hypomethylated, potentially activating transposable elements or deleterious genes (Latorre et al. 2023). With aging, there is a decline in histone protein levels and shifts in histone modification patterns, for example, global histone acetylation and methylation patterns change, reflecting impaired chromatin state regulation (Latorre et al. 2023). These molecular changes are influenced by both intrinsic aging processes like cumulative DNA damage or changes in metabolite levels such as NAD^+ that affect chromatin-modifying enzymes and extrinsic factors like, toxins, stress, and that leads to dysregulation of gene expression. In the aging brain, genes involved in synaptic plasticity, stress response, and repair mechanisms are epigenetically repressed, whereas genes promoting cell senescence or inflammation are aberrantly activated (Gräff et al. 2012; Table 2.4). Indeed, an epigenetic clock based on DNA methylation has been developed that correlates with chronological age, and interestingly, studies have found that in neurodegenerative conditions like Alzheimer's, the epigenetic clock in certain tissues can appear accelerated (Levine et al. 2015). This suggests that cells in those individuals are older than expected epigenetically, possibly contributing to disease susceptibility.

Epigenetic changes exert a dual influence in neurodegeneration. The age-related epigenetic drift primes the genome for pathological gene expression, creating a permissive environment for disease initiation, while other epigenetic modifications are induced by disease pathology and subsequently intensify its progression. In Alzheimer's disease, postmortem analyses have revealed altered DNA methylation at numerous gene loci in neurons and glia. For example, promoters of genes involved in learning and memory such as brain-derived neurotrophic factor (BDNF) show increased DNA methylation in AD, potentially contributing to their reduced expression and the cognitive dysfunction that is observed in disease (Nagata et al. 2015). Additionally, repetitive DNA elements that are normally silenced through DNA methylation undergo demethylation during aging and in Alzheimer's disease, resulting in their aberrant activation and contributing to genomic instability. Histone modification patterns are also disrupted in Alzheimer's disease, with patient brains frequently exhibiting reduced histone acetylation at neuronal gene loci, a change partly attributed to the overexpression or hyperactivity of specific histone deacetylases (HDACs) (Gräff et al. 2012). HDAC2 has been found upregulated in models of neurodegeneration and in AD brains, where it binds to synaptic plasticity gene promoters and represses them, forming an epigenetic blockade to learning-associated gene expression (Gräff et al. 2012). In Parkinson's disease, there is emerging evidence of epigenetic dysregulation in dopaminergic neurons. For instance, environmental toxins linked to PD may induce persistent epigenetic changes such as α-synuclein gene promoter demethylation, which could elevate α-synuclein levels. In LRRK2-mutant models of PD, alterations in microRNA expression profiles and histone modification patterns regulating immune-related genes have been documented, indicating that epigenetic dysregulation contributes to the neuroinflammatory component of PD pathology (Latorre et al. 2023).

TABLE 2.4
Age-Associated Epigenetic Changes and Their Role in Neurodegeneration

Epigenetic Feature	Change with Age	Disease Consequence
DNA methylation	Global hypomethylation, promoter hypermethylation	Altered gene expression in AD
Histone modifications	Loss of acetylation/methylation	Repression of memory-related genes
Noncoding RNAs	Dysregulation of miRNAs	Perturbation of synaptic plasticity
Chromatin remodeling	Reduced chromatin accessibility	Impaired neural repair mechanisms

Meanwhile, in ALS/FTD, studies have noted aberrant DNA methylation in the brains and blood of patients, including at loci like *C9orf72* where a pathogenic repeat expansion is also associated with DNA hypermethylation as a possible compensatory mechanism and genes regulating neural identity. Broadly, epigenetically mediated transcriptional repression of neuroprotective pathways, coupled with aberrant activation of deleterious gene networks, represents a potential mechanism by which aging drives cellular vulnerability and promotes neurodegeneration.

Crucially, unlike fixed genetic mutations, epigenetic modifications are potentially reversible, making them attractive to therapeutic targets. One major area of interest is using epigenetic drugs to restore gene expression profile in the aging or diseased brain. HDAC inhibitors are a prime example. These compounds enhance chromatin accessibility by inhibiting the removal of acetyl groups from histones, thereby facilitating transcriptional activation. In mouse models of Alzheimer's disease, treatment with class I HDAC inhibitors such as vorinostat or more selective HDAC2 inhibitors has been shown to restore open chromatin configurations at memory-associated gene loci and improve cognitive performance, despite the presence of amyloid-β or tau pathology (Gräff et al. 2012). Similarly, HDAC inhibition can induce neurotrophic factors and reduce neuronal death in models of Huntington's disease, and some HDAC inhibitors have progressed to clinical trials for neurodegenerative conditions. Conversely, in age-related conditions characterized by aberrant activation of deleterious genes, pharmacological activation of DNA methyltransferases (DNMTs) or stimulation of histone methyltransferases may help re-establish transcriptional silencing of transposable elements and pro-inflammatory genes, although such epigenetic interventions remain largely at the preclinical stage.

2.2.4.1 Modulation of Sirtuins

NAD^+-dependent deacetylases that link cellular metabolism to chromatin states is another strategy. SIRT1, for instance, plays a crucial role in maintaining genomic stability and exerts neuroprotective effects; however, its activity diminishes with advancing age due to a decline in cellular NAD^+ levels. Augmenting SIRT1 either pharmacologically, e.g., with SIRT1 activators like resveratrol or NAD^+ precursors or via gene therapy has shown protective effects in models of AD and ALS, likely by enhancing DNA repair, reducing protein aggregation, and activating antioxidant genes (Herskovits and Guarente 2014). Furthermore, epigenetic alterations hold promise as biomarkers for the early detection of neurodegenerative disease risk, offering insight into preclinical molecular changes that precede overt pathology. For instance, certain DNA methylation patterns in blood or brain may predict cognitive decline, enabling early interventions that could rejuvenate the epigenome. Dietary and lifestyle interventions also have epigenetic effects, nutrients like folate or polyphenols can influence DNA/histone methylation, and exercise has been shown to promote a neuroprotective DNA methylation profile in the brain, which correlates with better cognitive aging (Latorre et al. 2023). By elucidating and ultimately modulating the epigenetic alterations that accrue with age, the goal is to preserve gene expression profiles conducive to neuronal health and synaptic plasticity, thereby enhancing resilience against neurodegenerative processes. It is worth noting that epigenetic dysregulation can interplay with other mechanisms such as DNA damage. DNA damage can recruit chromatin-modifying enzymes, and when such damage becomes persistent, it may result in the stable silencing of neighboring genes. Moreover, inflammatory signals can trigger epigenetic alterations that entrench cells in a chronically pro-inflammatory state.

2.2.5 Genomic Instability and Cellular Senescence in Brain Aging

Over the course of a lifetime, cells accumulate various forms of DNA damage from single-strand nicks and base lesions to more deleterious double-strand breaks and telomere shortening. Genomic instability, the increased frequency of unrepaired DNA damage and mutations, is a cardinal feature of aging (López-Otín et al. 2013). The brain, although largely composed of non-dividing neurons, is not immune to this process. Neurons have high metabolic rates and produce significant ROS as byproducts, which can attack DNA. Moreover, several neurons persist for decades, undergoing repeated cycles of activity-dependent DNA breakage and repair, including transient double-strand breaks associated with learning-induced gene activation, which, over time, may cumulatively introduce genomic errors. (Madabhushi et al. 2014). In aging, DNA repair pathways such as nucleotide excision repair, base excision repair, and double-strand break repair by homologous recombination or nonhomologous end joining become less efficient, leading to the persistence of DNA lesions (Maynard et al. 2015). For instance, proteins involved in recognizing and fixing DNA damage like ATM, ATR, PARP1, BRCA1, Ku70/80 show age-related functional decline or altered expression in the brain. The result is that aged neurons and glial cells often harbor a higher load of oxidative base modifications e.g., 8-oxo-guanine, abasic sites, and even somatic mutations in their genomes compared to younger cells (Croteau et al. 2015). Whole-genome sequencing of single human neurons has revealed that neurons of older individuals accumulate hundreds of somatic single-nucleotide variants and small indels, some of which may disrupt gene function (Lodato et al. 2018). In parallel, telomere attrition, progressive shortening of chromosomes ends with each cell division limits the replicative potential of dividing cells. While most neurons are postmitotic and thus spared telomere-driven division limits, NSCs and glial progenitors do divide and can experience telomere shortening with age, potentially impacting brain regenerative capacity (López-Otín et al. 2013). The aging brain experiences an accumulating burden of genomic lesions and instability that precipitate cellular dysfunction, i.e., neurons with DNA damage may aberrantly express genes or activate stress-response pathways, while glial progenitors with critically shortened telomeres may lose their proliferative capacity.

One major consequence of genomic instability in proliferative cells is the induction of cellular senescence. Senescence is a state of stable cell cycle arrest accompanied by a distinctive phenotype enlarged cell size, active DNA damage signaling e.g., p53 and p21CIP1, or p16INK4a upregulation, and a pro-inflammatory secretory profile known as the SASP (senescence-associated secretory phenotype) (Campisi 2013). Senescent cells accumulate in many tissues with age, and far from being inert, they actively secrete cytokines like IL-6, IL-8, proteases, and growth factors that can disrupt tissue microenvironments. Historically, neurons were thought not to undergo classical senescence due to their nondividing nature. However, recent research indicates that postmitotic cells can enter a senescence-like state characterized by markers such as DNA damage foci and SASP production without proliferation (Bussian et al. 2018; Baker and Petersen 2018). In the aging brain, glial cells are more clearly prone to senescence. Astrocytes and microglia exposed to chronic stressors, including oxidative stress, protein aggregates, and telomere dysfunction can become senescent and develop a SASP profile (Baker and Petersen 2018). Oligodendrocyte precursor cells in aged individuals also show senescent features, which may impair remyelination. Senescent glia has been identified in postmortem brains of patients with neurodegenerative diseases, marked by increased p16INK4a and senescence-associated β-galactosidase activity, especially in surrounding areas of pathology (Baker and Petersen 2018). The SASP factors released from senescent cells can accelerate neurodegeneration by promoting local inflammation, degrading the extracellular matrix, and damaging neighboring neurons through the secretion of proteases and reactive oxygen species.

There is mounting evidence that genomic instability and cellular senescence actively contribute to neurodegenerative disease progression. In progeroid syndromes caused by DNA repair deficiencies, such as ataxia-telangiectasia or Werner syndrome, patients often exhibit early-onset neurodegeneration, signifying a causal link between defective DNA maintenance and neuronal loss (Maynard et al. 2015). In the general aging population, brain from Alzheimer's patients show more DNA double-strand breaks and oxidative DNA lesions than age-matched controls, particularly in regions like the hippocampus (Lu et al. 2004). These DNA lesions can lead to neuronal dysfunction, for example, accumulating DNA breaks in neurons can trigger chronic activation of ATM and p53, which in turn may drive synaptic depression and eventual cell death. Tau protein pathology in AD has been linked to DNA damage as well as tau tangles sequester DNA repair proteins, compounding genomic instability in affected neurons. In Parkinson's disease, dopaminergic neurons are exposed to dopamine oxidation products that damage DNA, and their decline has been associated with reduced expression of DNA repair genes. Recent studies have shown that senescent astrocytes in a tauopathy mouse model secrete inflammatory mediators that worsen tau aggregation and neuronal death. Strikingly, the targeted elimination of these senescent glial cells prevented neurodegeneration and cognitive decline in the mice (Bussian et al. 2018). Similarly, in a model of synucleinopathy, senescent microglia were found to have impaired autophagic clearance of α-synuclein, leading to its buildup and enhanced propagation of pathology (Hong et al. 2024). These findings illustrate a direct mechanism by which senescence links aging to neurodegeneration. Senescent cells not only fail to perform their normal support functions but actively secrete factors that impact the neuronal environment. Furthermore, the chronic SASP-driven inflammation can reduce the threshold for other processes like protein aggregation or mitochondrial dysfunction to cause damage, thereby synergizing with the other hallmarks of aging.

Addressing genomic instability and cellular senescence as therapeutic targets is a cutting-edge strategy in neurodegeneration research. Enhancing DNA repair efficiency in the aging brain represents a promising avenue for maintaining genomic integrity and neuronal viability. Compounds that elevate NAD^+ levels, such as nicotinamide riboside, activate sirtuins, and PARPs involved in DNA damage responses, potentially improve repair and protect neurons from age-related DNA damage (Croteau et al. 2015). Preclinical studies have shown that enhancing expression of certain repair enzymes like OGG1 for base excision repair or components of the nonhomologous end-joining machinery in neurons can reduce DNA damage and improve cell survival under stress. However, the more dramatic intervention is the clearance of senescent cells using senolytics, that selectively induce apoptosis in senescent cells. In animal models, senolytic treatment has yielded exciting results for neurodegeneration. For example, treating tau transgenic mice with a combination of dasatinib plus quercetin cleared senescent astrocytes and microglia from the brain and resulted in reduced tau pathology and improved memory function (Bussian et al. 2018). This suggests that eliminating even a subset of the inflammatory, senescent cells can tilt the balance back toward neural health. Another study targeting senescent microglia in a Parkinson's model showed that removing these cells decreased neuroinflammation and protected dopaminergic neurons (Chinta et al. 2018). Besides elimination of senescent cells, interventions are being explored to suppress the SASP. For instance, inhibitors of IRAK4 or JAK/STAT, which are downstream of SASP cytokine signaling, decreases the inflammatory output of senescent cells without necessarily eliminating them (Baker and Petersen 2018). Although still in early stages, several senolytic and senomorphic agents have progressed to clinical trials for non-neurological age-related disorders, fostering the prospect that, contingent on safety validation and effective brain delivery, these could be repurposed for the treatment of neurodegenerative diseases. Therapeutic targeting of senescence requires careful modulation, since transient senescence following acute injury supports tissue repair and regeneration, while chronic persistence of senescent cells exacerbates inflammation and accelerates age-related decline.

Ultimately, the most effective strategy lies in preventing excessive genomic damage by minimizing exposure to environmental genotoxins, mitigating ROS-induced DNA lesions through antioxidants, and reinforcing genomic surveillance via gene therapy. These approaches enhance key DNA repair pathways that assist in preserving genomic integrity in the aging brain.

2.2.6 Stem Cell Exhaustion and Decline of Neurogenesis in Brain Aging

The adult brain harbors populations of stem and progenitor cells. These can generate new neurons and glial cells throughout life, primarily in the hippocampal dentate gyrus via NSCs in the subgranular zone and the olfactory bulb via progenitors from the subventricular zone. This process of adult neurogenesis and glial cell replacement is crucial for learning, memory, and repair after injury. With aging, however, the activity of NSCs declines markedly, a phenomenon known as stem cell exhaustion (López-Otín et al. 2013; Nicaise et al. 2020). In older individuals and animal models, there is a reduction in the number of NSCs and neural progenitors in the neurogenic niches. Those that remain often exhibit lower proliferative capacity and a bias in their differentiation outcomes as the aging NSCs tend to produce more astrocytes through a phenomenon of gliogenesis bias at the expense of generating new functional neurons (Encinas et al. 2011). This shift is partly driven by cell-intrinsic changes like shortening of telomere, accumulation of DNA damage, and epigenetic alterations in the NSCs and partly by changes in the stem cell niche environment such as increased inflammatory signals, reduced growth factors, and altered extracellular matrix in the aging niche (Nicaise et al. 2020). For example, aged hippocampal stem cells show higher expression of cell cycle inhibitors (p16INK4a, p21CIP1) consistent with entry into a quiescent or senescent state, and they accumulate metabolites that signal a switch away from neurogenesis (Encinas et al. 2011). In the aged hippocampus, the surrounding niche exhibits elevated levels of BMP and Wnt pathway inhibitors, which bias NSCs toward astrocytic differentiation or maintain them in a quiescent state. Furthermore, factors secreted by senescent cells or chronically activated microglia in the aging brain like IL-1β or HMGB1 have been shown to directly inhibit neurogenesis (Kalamakis et al. 2019). With advancing age, the brain exhibits a progressive decline in regenerative potential, marked by reduced neurogenesis, limited neuronal integration into functional circuits, and impaired replacement of lost or injured cells.

Stem cell exhaustion has clear implications in neurodegenerative disorders, where neuronal loss is a central feature. In conditions such as Alzheimer's disease, hippocampal neurogenesis appears to be further compromised beyond the decline observed in normal aging, likely due to the local neurotoxic environment and tau oligomers. Also, Aβ not only directly inhibit neurogenesis but also exacerbate stem cell–intrinsic deficits (Choi et al. 2008). Autopsy studies have invariably reported reduced numbers of proliferating progenitors in the dentate gyrus of AD patients, and those neurons that are generated often fail to survive or mature properly. The net effect is that the aged AD brain cannot effectively replenish neurons in regions that undergo degeneration like the entorhinal–hippocampal circuit, potentially accelerating cognitive decline. In Parkinson's disease, the principal neurodegeneration occurs in the substantia nigra of the midbrain, a region with minimal adult neurogenesis; consequently, considerable interest has focused on whether NSCs from other brain regions could be recruited or engineered to restore lost dopaminergic neurons. However, in the context of an aged PD brain, any endogenous regenerative attempt may be hampered by the paucity of active NSCs and the presence of inhibitory signals from chronic inflammation and α-synuclein pathology. Similarly, in demyelinating conditions and ALS, oligodendrocyte progenitor cells (OPCs) represent a glial stem cell pool capable of differentiating into myelinating oligodendrocytes. Aged OPCs show signs of exhaustion and senescence (p16INK4a upregulation) and are less efficient at migrating and repairing demyelinated lesions (Neumann et al. 2019). In progressive MS which is not a classic neurodegenerative disease like AD/PD, but shares features of chronic CNS aging, OPC senescence is thought to contribute to failure of remyelination (Nicaise et al. 2019). Similarly, in chronic neurodegenerative conditions, regenerative strategies aimed at remyelination, or glial replacement may be hindered by stem cell senescence and reduced proliferative capacity. Furthermore, the loss of neurogenesis in the aging hippocampus may itself contribute to cognitive aging and increased vulnerability to dementia (Frankland et al. 2013). Newly generated neurons contribute to processes such as pattern separation and mood regulation, and their decline may aggravate memory deficits and depressive symptoms commonly observed in the early stages of Alzheimer's disease.

To address stem cell exhaustion, multiple therapeutic strategies and experimental approaches are currently under investigation. One approach involves replenishing endogenous stem cells or restoring their niches. Evidence from heterochronic parabiosis where the circulatory systems of young and aged animals are joined demonstrated that factors present in young blood can partially restore the function of aged NSCs, thereby enhance neurogenesis and improving cognitive performance in older mice (Villeda et al. 2014). These factors include oxytocin, GDF11, and circulating microRNAs, which are being investigated for therapeutic delivery. Conversely, neutralizing proaging factors in old blood like certain chemokines or SASP factors might similarly promote stem cell activity. Modulating the signaling pathways that govern NSC quiescence and differentiation is another strategy; for instance, inhibition of TGF-β or BMP signaling in aged mice has been shown to reactivate NSCs and boost neurogenesis (Katsimpardi et al. 2016, Meyers and Fiona 2016). Also, exercise is a well-documented, nonpharmacological intervention that increases neurogenesis even in middle-aged animals, most likely by increasing BDNF and improving vascular support in the niche. In Parkinson's disease, transplantation of dopaminergic neurons derived from fetal tissue or pluripotent stem cells into the striatum of patients has shown some success in restoring motor function, though widespread adoption awaits improvements in graft survival and controlled growth. In disorders such as ALS and AD, where widespread neural networks are compromised, stem cell transplantation poses greater challenges; however, clinical trials employing mesenchymal stem cells or induced neural progenitors have shown potential to deliver trophic support or replace degenerated cells in a more localized fashion. For example, intrathecal infusion of mesenchymal stem cells in ALS patients hold promises of slowing disease progression, possibly by modulating the immune environment and providing growth factors (Bonab et al. 2012). Further, iPSC technology offers the possibility of creating patient-specific neural progenitors that could be transplanted, or even in vivo reprogramming approaches where glial cells in the brain are directly converted into neurons.

2.3 THERAPEUTIC IMPLICATIONS AND REGENERATIVE STRATEGIES

As discussed in preceding sections, the converging molecular mechanisms of aging, including mitochondrial dysfunction, loss of proteostasis, chronic inflammation, epigenetic drift, cellular senescence, and genomic instability play a pivotal role in driving neurodegenerative pathology. The current and emerging interventions designed to slow biological aging in the nervous system, mitigating neurodegeneration, and explore regenerative approaches to restore neural function are discussed. This includes both pharmacological treatments targeting mitochondria, proteostasis, inflammation, epigenetics, senescent cells, and DNA repair and nonpharmacological or biotherapeutic interventions such as lifestyle modifications, stem cell therapies, and systemic rejuvenation via young blood factors (Table 2.4). Interventions aimed at the upstream drivers of aging seek not only to mitigate the symptomatic manifestations of neurodegenerative diseases such as Alzheimer's and Parkinson's, but to reshape the aging biological landscape that predisposes to their development and advancement (Figure 2.5; Wyss-Coray 2016).

2.3.1 Therapeutic Enhancement of Proteostasis in Neurodegeneration

Aging is characterized by a decline in proteostasis, resulting in the accumulation of misfolded and aggregated proteins that are toxic to neurons (Lopez-Otín et al. 2013). Therapeutic strategies to enhance protein homeostasis involve upregulating protein degradation pathways and improving protein folding mechanisms. A central mechanism is autophagy, the cell's lysosomal degradation pathway for clearing damaged organelles and protein aggregates. Autophagy enhancers have shown considerable promise in preclinical studies. In particular, the mTOR pathway, a nutrient-sensing regulator that inhibits autophagy, has been targeted, mTOR inhibition via rapamycin or rapalog compounds reliably induces autophagy. In models of Alzheimer's and Huntington's disease, rapamycin-mediated autophagy upregulation facilitated the clearance of aggregated Aβ, tau, and huntingtin proteins, leading to improved synaptic and cognitive function (Caccamo et al. 2014). However, chronic rapamycin must be used cautiously, as excessive suppression of mTOR in aged brains can impair lysosomal function over time. Other

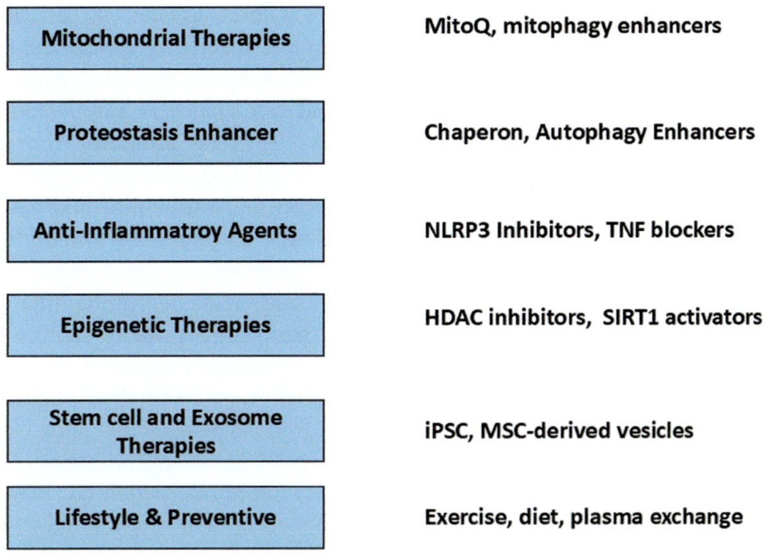

FIGURE 2.5 Therapeutic and Regenerative Strategies.

autophagy stimulators under investigation include caloric restriction mimetics, e.g., spermidine that induce a similar autophagic response. Preclinical data increasingly support the therapeutic utility of autophagy modulation in neurodegenerative diseases, both by clearing toxic proteins and regulating inflammation (Rubinsztein, Bento, and Deretic 2015). Indeed, autophagy can suppress the activation of inflammasomes and the release of pro-inflammatory cytokines, linking proteostatic enhancement to reduced neuroinflammation (Deretic and Levine 2018). Beyond autophagy, activating the ubiquitin-proteasome system is another approach to improve proteostasis. Small molecules that activate proteasome activity or increase the expression of chaperones can remove misfolded proteins. For example, experimental compounds that induce HSPs help refold or target aberrant proteins for degradation. In models of synucleinopathy and tauopathy, overexpression or pharmacological upregulation of HSP70 and other chaperones reduced protein aggregation and protected neurons (Klucken et al. 2004). Similarly, chemical chaperones such as tauroursodeoxycholic acid (TUDCA) and 4-phenylbutyrate can stabilize protein conformation and ease endoplasmic reticulum stress. These have shown neuroprotective effects in models of Parkinson's and Huntington's diseases by reducing misfolded protein stress. Many proteostasis-based therapies are still at the preclinical or early clinical stage, yet their rationale remains compelling. By enhancing the cell's capacity to regulate its proteome, these approaches aim to interrupt the pathological cascade of protein misfolding and aggregation that underlies neurodegenerative disorders.

2.3.2 Anti-Inflammatory and Immunomodulatory Interventions

Chronic inflammation in the aging brain is a critical contributor to neurodegenerative disease progression (Franceschi et al. 2018). Microglia and astrocytes in aged individuals often adopt pro-inflammatory phenotypes, secreting cytokines that exacerbate neuronal injury and amyloid/tau pathology (Heneka et al. 2015). The therapeutic strategies to counter neuroinflammation span from conventional anti-inflammatory agents to more precise immunomodulatory interventions designed to target specific pathways. NSAIDs were initially investigated given epidemiological observations that long-term NSAID use is associated with lower incidence of Alzheimer's dementia in some cohorts. However, clinical trials of NSAIDs in patients with established Alzheimer's disease did not show clear cognitive benefits, suggesting that timing and specificity are critical (Imbimbo et al. 2010). More specific anti-inflammatory strategies include inhibiting pivotal inflammatory signaling pathways. For example, the NLRP3 inflammasome in microglia has been implicated in Alzheimer's pathology by promoting IL-1β release, and mice lacking NLRP3 are protected from amyloid-driven neurodegeneration (Heneka et al. 2013). Small-molecule NLRP3 inhibitors such as MCC950 are being explored to suppress this pathway. Similarly, strategies that inhibit pro-inflammatory cytokines or their receptors are adapted from therapeutic paradigms in peripheral inflammatory diseases. Ongoing trials with TNF-α blockers and anti-IL-1 agents such as canakinumab are being conducted to determine whether reducing systemic inflammation can attenuate neuroinflammation and thereby slow cognitive decline (Naeem et al. 2024). Another therapeutic strategy centers on modulating glial activation states. Microglia exist along a dynamic spectrum ranging from pro-inflammatory (M1-like) to anti-inflammatory or reparative (M2-like) phenotypes, with their polarization critically influencing neuronal survival and tissue homeostasis. Drug candidates that shift microglia toward a neuroprotective state or reduce their chronic activation are under investigation. One example is the use of colony-stimulating factor-1 receptor (CSF1R) inhibitors, which can transiently deplete overactive microglia or alter their activation in animal models of neurodegeneration. They have been found to reduce neuroinflammation and synaptic loss (Spangenberg et al. 2016). Similarly, agonists of metabolic nuclear receptors like PPAR-γ (peroxisome proliferator-activated receptor gamma) have anti-inflammatory effects on glia. Pioglitazone, a PPAR-γ agonist used in the treatment of type 2 diabetes, has been shown to reduce biomarkers of neuroinflammation and may slow cognitive decline in mild Alzheimer's disease, although clinical outcomes have been inconsistent (Galimberti and Scarpini 2017). Lifestyle interventions also play a role in immunomodulation. Diets rich in anti-inflammatory components, e.g., the Mediterranean diet are associated with reduced dementia risk, presumably by lowering systemic inflammation (Lourida et al. 2019). Regular exercise is well documented to mitigate chronic inflammation in older adults. It has been shown to lower circulating inflammatory mediators and enhance neuroinflammatory profiles in the brain, effects that correlate with improved memory function (Erickson et al. 2011). Collectively, anti-inflammatory interventions in neurodegenerative disorders are designed to interrupt the vicious cycle linking chronic inflammation with progressive neuronal injury. Early modulation of inflammatory cascades, whether through pharmacological agents or lifestyle-based approaches, offers the potential to attenuate feed-forward neurotoxic processes that compromise the integrity of neural networks in the aging brain (Wyss-Coray and Rogers 2012).

2.3.3 Epigenetic Modulation Therapies

Epigenetic alterations accumulate with age can dysregulate gene expression in neurons, contributing to synaptic dysfunction and degeneration. Aberrant DNA methylation patterns, posttranslational histone modifications, and chromatin remodeling have all been observed in aging brains and in neurodegenerative disease models (Mahgoub and Monteggia 2014). These observations have prompted growing interest in epigenetic therapeutics to restore a more homeostatic gene expression profile. One major target class is HDACs. HDACs generally repress gene expression by removing acetyl groups from histones, and in aging or disease, overactivity of certain HDACs, e.g., HDAC2 has been linked to silencing of genes important for memory and neuronal plasticity. In mouse models of Alzheimer's disease, treatment with HDAC inhibitors (HDACi) resulted in increased histone acetylation, increased expression of learning-associated genes, and consequent improvements in cognitive performance (Gräff et al. 2012). For example, inhibiting HDAC2 was shown to relieve the epigenetic blockade of neurotrophic and synaptic genes, enhancing synaptic plasticity and memory in AD mice (Gräff et al. 2012). Several HDAC inhibitors are already clinically approved for cancer therapy, and modified derivatives are under investigation in preclinical models of neurodegeneration. However, their potential off-target effects highlight the need for careful specificity and targeted delivery (Landgrave-Gómez et al. 2015). Also, the sirtuins that are NAD^+-dependent deacetylases are associated with longevity and cellular stress resistance. Among these regulators, SIRT1 mediates neuroprotective transcriptional responses by enhancing the expression of antioxidant and mitochondrial genes while suppressing pro-inflammatory signaling pathways (Herskovits and Guarente 2014). Increasing SIRT1 activity either by NAD^+ supplementation or small-molecule activators like resveratrol has yielded neuroprotective effects. Resveratrol treatment in AD models reduced amyloid accumulation and microglial activation in part via SIRT1, ultimately preserving neurons (Gomes et al. 2018). Conversely, inhibiting certain detrimental epigenetic regulators is also being explored; for example, SIRT2 can promote α-synuclein toxicity in Parkinson's disease (Outeiro et al. 2007). DNA methylation modifiers represent an emerging therapeutic strategy, although achieving targeted precision remains challenging. Aging is marked by characteristic shifts in methylation landscapes, the molecular basis of the epigenetic clock, which can silence neuroprotective genes and activate deleterious or pro-inflammatory pathways. Although DNA methyltransferase inhibitors and other methylation-modifying agents are clinically employed in oncology, their translation to the central nervous system remains constrained by challenges of target specificity and effective delivery across the blood–brain barrier. Consequently, alternative strategies such as dietary supplementation with methyl donors or the use of histone methylation inhibitors are currently being explored in preclinical models (Sherwani and Khan 2015). Broadly, epigenetic therapies aim to restore a more adaptive neuronal epigenomic state, focusing on reactivating gene expression programs that support proteostasis, metabolic homeostasis, and synaptic plasticity while concurrently suppressing pro-degenerative pathways.

2.3.4 Stem Cell-Based Therapeutics

The restricted regenerative potential of the adult human brain has positioned stem cell–based approaches as promising yet still cautiously appraised for the treatment of neurodegenerative disorders. The aim of such therapies is to replace lost neurons, provide support to degenerating circuits, or modulate the disease environment via trophic factor release.

2.3.4.1 Neural Progenitor Cell–Based Therapeutic Strategies in Neurodegenerative Diseases

Neural progenitor cells are lineage-determined precursors capable of differentiating into neurons, astrocytes, and oligodendrocytes. Transplantation studies in animal models of neurodegeneration demonstrate that these cells can engraft, undergo differentiation, and in some cases, achieve partial functional integration. In rodent models of Parkinson's disease, dopaminergic neural progenitors derived from either fetal tissue or pluripotent stem cells have been shown to survive following striatal transplantation and mitigate motor deficits by restoring dopaminergic neurotransmission. (Barker 2014). Unlike the early fetal midbrain tissue transplants performed in Parkinson's patients, this strategy utilizes more defined and reproducible cell sources, including iPSC-derived dopaminergic neurons.

Early clinical trials involving the transplantation of pluripotent stem cell–derived dopaminergic neurons in Parkinson's patients are being investigated (Barker et al. 2017). In preclinical mouse models of Alzheimer's disease, NSCs have been transplanted into the hippocampus, although these grafted cells show limited integration, they secrete trophic and immunomodulatory factors that support host neuronal survival and enhance synaptic plasticity (Blurton-Jones et al. 2009). Consequently, the therapeutic effects of NSC transplantation may arise as much from paracrine signaling as from direct neuronal replacement.

2.3.4.2 Mesenchymal Stromal Cells and Exosome-Based Therapies in Neurodegenerative Disease

These multipotent stromal cells, derived from bone marrow or adipose tissue, do not readily differentiate into neurons but exert strong immunomodulatory and trophic effects. In animal models of ALS and multiple system atrophy, MSC infusions have been shown to reduce inflammation and slow disease progression, likely through the secretion of anti-inflammatory cytokines and neurotrophic factors (Uccelli, Laroni, and Freedman 2011). Some early-phase clinical trials of MSC infusion in Alzheimer's disease and ALS have indicated that the approach is safe and may confer modest functional benefits, although robust evidence for efficacy has yet to be established. An emerging strategy involves the use of stem cell–derived exosomes or extracellular vesicles (EVs) for cell-free therapy. Exosomes from mesenchymal stem cells contain miRNAs and proteins that modulate neuroinflammation and support regenerative pathways. In aged mouse models, administration of EVs from young MSCs enhanced hippocampal synaptic plasticity, reduced brain inflammation, and contributed to extended health span (Dorronsoro et al. 2021). Similarly, NSC-derived exosomes enriched with brain-specific microRNAs have shown promise in reducing the pathology in models of Alzheimer's and Parkinson's disease. EVs can cross the blood–brain barrier and circumvent several limitations associated with whole-cell transplantation, including risks of immune rejection and tumorigenesis.

2.3.4.3 Induced Pluripotent Stem Cells and Autologous Cell Therapy

The development of iPSC technology has enabled the reprogramming of a patient's somatic cells, such as skin fibroblasts, into pluripotent stem cells that can subsequently be differentiated into neural precursors. This approach provides a theoretically immune-compatible source of cells for transplantation. In 2018, researchers in Japan initiated the first transplantation of iPSC-derived dopaminergic progenitors into a Parkinson's disease patient. Autologous iPSC-based therapies for neurodegenerative disorders, however, remain in the experimental stage and face significant challenges, including high cost, lengthy production timelines, and the risk of genetic or epigenetic abnormalities arising during reprogramming. Nevertheless, this strategy represents a promising step towards personalized cell replacement therapy

2.3.4.4 In Situ Reprogramming and Neurogenesis

An alternative to exogenous cell transplantation is the direct reprogramming glial cells into neurons. Gene therapy approaches have demonstrated that astrocytes or NG2 glia can be converted into functional neurons in vivo. For instance, over expression of single transcription factors such as NeuroD1 or Ngn2 in astrocytes has been shown to induce their trans-differentiation into neurons in the mouse brain (Guo et al. 2014). In an Alzheimer's disease model, reprogramming astrocytes into neurons expanded the pool of functional neurons and improved memory performance, thereby enhancing endogenous neurogenesis which is otherwise characterized by neuronal loss. Although still distant from clinical translation, this line of research highlights the potential of regenerating neurons in situ, offering a strategy that could circumvent the need for exogenous cell grafts.

Despite encouraging progress, significant challenges constrain the clinical translation of stem cell–based therapies. Safety concerns include tumorigenic risk with pluripotent cells, inappropriate migration or differentiation, and immune rejection in nonautologous grafts (Lindvall and Kokaia 2010). A further obstacle is the requirement for transplanted cells to integrate into complex neural circuits to achieve functional benefit, which may necessitate adjunctive interventions. Ongoing clinical trials and technological refinements, such as gene editing to reduce immunogenicity in allogeneic cells, will address these issues. Overall, stem cell–based approaches, including transplantation, EV delivery, and in situ reprogramming embody a regenerative paradigm aimed at replenishing and repairing the aging brain.

2.3.5 Molecular and Systemic Regulation of Neurogenesis in the Aging Brain

In adult mammals, neurogenesis persists in restricted brain regions, most prominently the hippocampal dentate gyrus and the subventricular zone. Advancing age is accompanied by a marked decline in this process, reflected in the depletion and quiescence of NSC populations as well as in the progressive dysfunction of the neurogenic niche, including reduced vascularization and diminished trophic support (Katsimpardi et al. 2014).

Because hippocampal neurogenesis is closely associated with cognitive flexibility and memory, enhancing endogenous neurogenesis represents a promising strategy to mitigate age-related cognitive decline and neurodegenerative disease. Among the most potent physiological stimuli of adult neurogenesis is physical exercise, which activates a cascade of pro-neurogenic mechanisms, including upregulation of BDNF, improved cerebral perfusion, and the release of muscle- and liver-derived myokines and metabolites that act on the brain. For instance, aerobic exercise induces expression of the membrane protein FNDC5 in skeletal muscle, which is cleaved to produce the hormone irisin. The circulating irisin crosses into the brain and has been shown to increase hippocampal BDNF expression, thereby promoting neurogenesis and synaptic plasticity (Wrann et al. 2013).

In aged mice, exercise has been shown to restore hippocampal neurogenesis and improve spatial memory performance. Parallel evidence in humans supports these findings. In a landmark randomized controlled trial, Erickson et al. (2011) reported that 1 year of moderate aerobic exercise in older adults increased hippocampal volume by approximately 2%, effectively offsetting nearly 2 years of age-related atrophy, and this volumetric gain correlated with improved memory performance. Collectively, these findings establish structured physical activity as a potent nonpharmacological intervention capable of enhancing the neurogenic niche and mitigating age-related cognitive decline.

Cognitive enrichment and environmental stimulation are additional modulators of adult neurogenesis. In animal models, enriched environments have been shown to increase hippocampal neuron formation, enhance neuronal survival, and strengthen synaptic connectivity. Although direct measurement in humans is more challenging, epidemiological and clinical studies indicate that engagement in cognitively demanding activities and lifelong learning is associated with greater cognitive resilience during aging (Scarmeas et al. 2001), potentially through the preservation of neurogenesis and synaptic networks. These observations highlight the continued plasticity of the aging brain and emphasize the capacity of lifestyle factors to either support or compromise this neurogenic potential.

At the molecular level, several growth factors and hormones that regulate neurogenesis are under investigation for therapeutic application. Experimental infusion of BDNF in aged animals has been shown to stimulate neurogenesis and improve memory, although the challenge of delivering large proteins to the brain limits translational potential. Other trophic factors, including fibroblast growth factor-2 and vascular endothelial growth factor, contribute to the neurogenic niche by expanding progenitor populations and enhancing vascularization. While many of these factors are upregulated by exercise, direct administration strategies such as intranasal delivery of insulin-like growth factor or FGF-2 are currently being tested in trials for cognitive impairment.

Endocrine regulators like gonadotropin-releasing hormone (GnRH) decline with aging, and restoring GnRH in old mice have been shown to enhance olfactory neurogenesis and improve cognition (Zhang et al. 2022). A pilot clinical study in individuals with Down syndrome suggested that GnRH analog treatment may improve cognitive performance (Leboyer et al. 2022), thereby linking endocrine regulation to neurogenesis and cognition. Perhaps the most striking demonstration of neurogenic rejuvenation arises from heterochronic parabiosis experiments, in which the circulatory systems of young and old mice are joined, revealing systemic factors that can restore neurogenic capacity in the aged brain.

Heterochronic parabiosis experiments demonstrate that systemic milieu strongly influences brain plasticity. Aged mice exposed to young circulation show enhanced hippocampal neurogenesis, improved synaptic plasticity, and better cognitive performance (Villeda et al. 2014; Katsimpardi et al. 2014), whereas young mice joined to aged partners exhibit suppressed neurogenesis, implicating both pro-youthful and pro-aging blood factors. Specific mediators have been identified, growth differentiation factor restores cerebrovascular function and neurogenesis in aged mice (Katsimpardi et al. 2014), and tissue inhibitor of metalloproteinases 2, enriched in human umbilical cord plasma, improves cognition in old mice (Castellano et al. 2017). In contrast, age-associated factors such as β2-microglobulin and C-C motif chemokine 11 (CCL11) impair neurogenesis, with their neutralization partially reversing cognitive decline (Villeda et al. 2011). The AMBAR trial demonstrated that therapeutic plasma exchange slowed cognitive decline in Alzheimer's disease (Boada et al. 2020), and young plasma infusion in a small Alzheimer's cohort showed preliminary functional benefits (Sha et al. 2019). Collectively, this line of research underpins the concept of systemic rejuvenation therapy, whereby modulating circulating factors may enhance endogenous neurogenesis and brain repair (Castellano, Kirby, and Wyss-Coray 2015). Integrated with lifestyle interventions and protective strategies, such as exercise, cognitive stimulation, anti-inflammatory therapies, and clearance of toxic proteins has been found to restore the regenerative capacity, thereby extending cognitive healthspan in aging

2.3.6 Multidomain Preventive Strategies in Cognitive Decline and Neurodegeneration

It is increasingly evident that no single intervention counteracts the complex and multifactorial nature of brain aging. Consequently, the holistic strategies that simultaneously target multiple hallmarks of aging are being pursued (Jiang et al. 2025). Within this framework, lifestyle interventions remain central owing to their broad systemic effects and low-risk profiles. Caloric restriction and intermittent fasting, for instance, activate a range of cellular defense pathways, including suppression of insulin/IGF-1 signaling, enhancement of autophagy, improvement of mitochondrial function, and attenuation of inflammatory processes (Mattson, Longo, and Harvie 2017).

In animal models, lifelong caloric restriction extends lifespan and preserves cognitive function, while intermittent fasting paradigms enhance synaptic plasticity and improve resilience to neurotoxic stress (Mattson, Longo, and Harvie 2017). Although sustained caloric restriction is often impractical in humans, periodic fasting and fasting-mimicking diets are under investigation for feasibility and potential benefits in older adults. Beyond caloric intake, dietary composition also plays a critical role in brain aging. Diets rich in essential nutrients and low in processed sugars attenuate multiple aging mechanisms. The Mediterranean diet has been consistently associated with reduced risk of Alzheimer's disease and slower cognitive decline (Lourida et al. 2019). Its protective effects are attributed to bioactive components, including antioxidants that counter oxidative DNA damage, omega-3 fatty acids that mitigate neuroinflammation, and polyphenols such as resveratrol, which may activate sirtuin-mediated longevity pathways.

The MIND diet (Mediterranean–DASH Intervention for Neurodegenerative Delay), developed as a hybrid of the Mediterranean and DASH diets, was specifically designed to promote brain health and has shown efficacy in observational studies in lowering Alzheimer's disease risk (Morris et al. 2015). Importantly, combining lifestyle interventions appears to exert additive protective effects. The Lancet Commission on dementia prevention estimated that up to 40% of dementia cases could be prevented by addressing modifiable risk factors such as physical inactivity, hypertension, obesity, smoking, and diabetes (Livingston et al. 2020). These risk factors converge with core mechanisms of aging biology. For example, midlife hypertension and metabolic syndrome accelerate vascular aging, disrupt cerebral perfusion, and promote neuroinflammation. Accordingly, management of cardiovascular health through diet, physical activity, and pharmacological treatment when indicated is recognized as a key neuroprotective strategy in aging populations. The Finnish Geriatric Intervention Study to Prevent Cognitive Impairment and Disability (FINGER) trial provided proof-of-concept that multidomain interventions, including diet, exercise, cognitive training, and vascular risk monitoring improve cognitive performance in at-risk elderly individuals (Ngandu et al. 2015). Collectively, these findings underscore that integrated, multicomponent strategies are more effective than single interventions in sustaining brain function during aging.

Parallel to lifestyle strategies, pharmacological geroprotectors are being investigated to target molecular pathways of aging. Compounds such as senolytics, metformin, rapamycin analogues, and blood-based therapies are investigated in preclinical and early clinical stages. In mouse studies, a combination of acarbose and phenylbutyrate synergistically extended lifespan and improved health span more effectively than monotherapy, by simultaneously modulating distinct aging pathways (Jiang et al. 2022). Multitarget pharmacological strategies have also demonstrated superior efficacy in Alzheimer's disease models, where combinatorial treatments improved cognitive outcomes and reduced pathological hallmarks more effectively than single agents. Going forward, clinical translation may involve tailored regimens that integrate geroprotective drugs with lifestyle interventions. For example, an elderly patient at elevated dementia risk might receive senolytic therapy to eliminate senescent cells, combined with structured exercise, dietary counseling, metformin to enhance metabolic and proteostatic function, and adjunctive neurotrophic or anti-inflammatory treatments. Such personalized, multimodal interventions, guided by principles of geroscience, represent a promising avenue for optimizing brain healthspan in aging populations.

REFERENCES

Azam S, Haque ME, Balakrishnan R, Kim IS, Choi DK. The ageing brain: Molecular and cellular basis of neurodegeneration. *Front Cell Dev Biol.* 2021;9:683459.

Baker DJ, Petersen RC. Cellular senescence in brain aging and neurodegenerative diseases: Evidence and perspectives. *J Clin Invest.* 2018;128(4):1208–16. doi:10.1172/JCI95145

Baker DJ, Wijshake T, Jeganathan T, van Deursen JM. Clearance of p16Ink4a-positive senescent cells delays ageing-associated disorders. *Nature.* 2016;530(7589):184–9.

Barker RA. Developing stem cell therapies for Parkinson's disease: Waiting until the time is right. *Cell Stem Cell.* 2014;15(5):539–42.

Barker RA, Parmar M, Studer L, Takahashi J. Human trials of stem cell-derived dopamine neurons for Parkinson's disease: Dawn of a new era. *Cell Stem Cell.* 2017;21(5):569–73.

Bartels AL, Leenders KL. Cyclooxygenase and neuroinflammation in Parkinson's disease neurodegeneration. *Curr Neuropharmacol.* 2007;5(1):57–70.

Blurton-Jones M, et al. Neural stem cells improve cognition via BDNF in a transgenic model of Alzheimer disease. *Proc Natl Acad Sci U S A.* 2009;106(32):13594–9.

Boada M, et al. A randomized, controlled clinical trial of plasma exchange with albumin replacement for Alzheimer's disease: Primary results of the AMBAR study. *Alzheimer's & Dementia.* 2020;16(10):1412–25.

Bonab MM, Sahraian MA, Aghsaie A, Karvigh SA, Hosseinian SM, Nikbin B, Lotfi J, Khorramnia S, Motamed MR, Togha M, Harirchian MH, Moghadam NB, Alikhani K, Yadegari S, Jafarian S, Gheini MR. Autologous mesenchymal stem cell therapy in progressive multiple sclerosis: An open label study. *Curr Stem Cell Res Ther.* 2012;7(6):407–14.

Bussian TJ, Aziz JM, Lu X, et al. Clearance of senescent glial cells prevents tau-dependent pathology and cognitive decline. *Nature.* 2018;562(7728):578–82. doi:10.1038/s41586-018-0543-y

Caccamo A, et al. Molecular interplay between mammalian target of rapamycin (mTOR), amyloid-β, and tau: Effects on cognitive impairments. *J Biol Chem.* 2014;289(43):29400–11.

Campisi J. Aging, cellular senescence, and cancer. *Annu Rev Physiol.* 2013;75:685–705. doi:10.1146/annurev-physiol-030212-183653

Castellano JM, Kirby ED, Wyss-Coray T. Blood-borne revitalization of the aged brain. *JAMA Neurol.* 2015;72(10):1191–94.

Castellano JM, et al. Human umbilical cord plasma proteins revitalize hippocampal function in aged mice. *Nature.* 2017;544(7651):488–92.

Chinta SJ, Woods MT, Rane J, et al. Cellular senescence is induced by the environmental neurotoxin paraquat and contributes to neuropathology linked to Parkinson's disease. *Cell Rep.* 2018;22(4):930–40. doi:10.1016/j.celrep.2017.12.092

Choi SH, Bylykbashi SH, Chatila TM, et al. Combined adult neurogenesis and BDNF mimic exercise effects on cognition in an Alzheimer's mouse model. *Science.* 2008;361(6406). doi:10.1126/science.aan8821

Costantini E, D'Angelo C, Reale M, Di Napoli A. The role of immunosenescence in neurodegenerative diseases. *Mediators Inflamm.* 2018;2018:6039171. doi:10.1155/2018/6039171

Croteau DL, Fang EF, Simpson PJ, et al. DNA damage, DNA repair, aging, and neurodegeneration. *Cold Spring Harb Perspect Med.* 2015;5(10):a025130. doi:10.1101/cshperspect.a025130

Deretic V, Levine B. Autophagy balances inflammation in innate immunity. *Autophagy.* 2018;14(2):243–51. doi: 10.1080/15548627.2017.1402992

Dorronsoro A, Santiago FE, Grassi D, Zhang T, Lai RC, McGowan SJ, Angelini L, Lavasani M, Corbo L, Lu A, Brooks RW, Garcia-Contreras M, Stolz DB, Amelio A, Boregowda SV, Fallahi M, Reich A, Ricordi C, Phinney DG, Lim SK, Niedernhofer LJ, Robbins PD. Mesenchymal stem cell-derived extracellular vesicles reduce senescence and extend health span in mouse models of aging. *Aging Cell.* 2021;20(4):e13337.

Encinas JM, Michurina TV, Peunova N, et al. Division-coupled astrocytic differentiation and age-related depletion of neural stem cells in the adult hippocampus. *Cell Stem Cell.* 2011;8(5):566–79. doi:10.1016/j.stem.2011.03.012

Erickson KI, et al. Exercise training increases size of hippocampus and improves memory. *Proc Natl Acad Sci USA.* 2011;108(7):3017–22.

Exner N, Lutz P, Haass T, Schrader J. Mitochondrial dysfunction in Parkinson's disease: Molecular mechanisms and pathophysiological consequences. *EMBO J.* 2012;31(14):3038–62. doi:10.1038/emboj.2012.170

Fang EF, et al. NAD+ in brain aging and neurodegenerative disorders. *Cell Metab.* 2019;30(4):630–55.

Franceschi C, Bonafè M, Valensin S, et al. Inflamm-aging: An evolutionary perspective on immunosenescence. *Ann N.Y. Acad Sci.* 2007;1114:40–54. doi:10.1196/annals.1396.023

Franceschi C, Garagnani P, Parini P, Giuliani C, Santoro A. Inflammaging: A new immune–metabolic viewpoint for age-related diseases. *Nat Rev Endocrinol.* 2018;14(10): 576–90.

Frankland PW, Köhler S, Josselyn SA. Hippocampal neurogenesis and forgetting. *Trends Neurosci.* 2013;36(9):497–503. doi:10.1016/j.tins.2013.05.002

Galimberti D, Scarpini E. Pioglitazone for the treatment of Alzheimer's disease. *Expert Opin Investig Drugs.* 2017;26(1):97–101.

Gomes BAQ, et al. Neuroprotective mechanisms of resveratrol in Alzheimer's disease: Role of SIRT1. *Oxid Med Cell Longev.* 2018;2018:8152373.

Gonzales MM, Garbarino VR, Pollet E, Palavicini JP, Kellogg DL, Jr, Kraig E, Orr ME. Biological aging processes underlying cognitive decline and neurodegenerative disease. *J Clin Invest.* 2022;132(1):e158453.

Guo Z, Zhang L, Wu Z, Chen Y, Wang F, Chen G. In vivo direct reprogramming of reactive glial cells into functional neurons after brain injury and in an Alzheimer's disease model. *Cell Stem Cell* 2014;14(2):188–202.

Gräff J, Rei D, Guan JS, et al. An epigenetic blockade of cognitive functions in the neurodegenerating brain. *Nature.* 2012;483(7388):222–6. doi:10.1038/nature10849

Heneka MT, et al. NLRP3 is activated in Alzheimer's disease and contributes to pathology in APP/PS1 mice. *Nature.* 2013;493(7434):674–78.

Heneka MT, Carson MJ, El Khoury J, et al. Neuroinflammation in Alzheimer's disease. *Lancet Neurol.* 2015;14(4):388–405. doi:10.1016/S1474-4422(15)70016-5

Herskovits AZ, Guarente L. SIRT1 in neurodevelopment and brain senescence. *Neuron.* 2014;81(3):471–83. doi:10.1016/j.neuron.2014.01.028

Hipp MS, Park SH , Ulrich Hartl F. Proteostasis impairment in protein-misfolding and -aggregation diseases. *Trends Cell Biol.* 2014;24(9):506–14. doi:10.1016/j.tcb.2014.05.003

Höhn A, König M, Grune S, Grune T. Proteostasis failure in neurodegenerative diseases: Focus on oxidative stress. *Oxid Med Cell Longev.* 2020;2020:1403082. doi:10.1155/2020/1403082

Hong B, Ohtake Y, Itokazu T, Yamashita T. Glial senescence enhances α-synuclein pathology owing to its insufficient clearance caused by autophagy dysfunction. *Cell Death Disco.* 2024;10:50. doi:10.1038/s41420-024-01816-8

Horvath S, Raj K. DNA methylation-based biomarkers and the epigenetic clock theory of ageing. *Nat Rev Genet.* 2018;19(6):371–84.

Hou Y, Dan X, Babbar M, Wei Y, Hasselbalch SG, Croteau DL, Bohr VA. Ageing as a risk factor for neurodegenerative disease. *Nat Rev Neurol.* 2019;15(10):565–81. doi:10.1038/s41582-019-0244-7.

Imbimbo BP, et al. Solanezumab for the treatment of mild-to-moderate Alzheimer's disease. *Expert Rev Clin Immunol.* 2010;6(4):573–84.

Jiang Q, Liu J, Huang S, Wang X-Y, Chen X, Liu G-H, et al. Antiaging strategy for neurodegenerative diseases: From mechanisms to clinical advances. *Signal Transduct Target Ther*. 2025;10(1): Article 21 (published online). doi:10.1038/s41392-025-02145-7

Jiang Z, Wang J, Imai D, Snider T, Klug J, Mangalindan R, Morton J, Zhu L, Salmon AB, Wezeman J, Hu J, Menon V, Marka N, Neidernhofer L, Ladiges W. Short term treatment with a cocktail of rapamycin, acarbose and phenylbutyrate delays aging phenotypes in mice. *Sci Rep*. 2022;12(1):7300.

Kalamakis G, Brüne D, Ravichandran S, Bolz J, Fan W, Ziebell F, Stiehl T, Catalá-Martinez F, Kupke J, Zhao S, Llorens-Bobadilla E, Bauer K, Limpert S, Berger B, Christen U, Schmezer P, Mallm JP, Berninger B, Anders S, Del Sol A, Marciniak-Czochra A, Martin-Villalba A. Quiescence modulates stem cell maintenance and regenerative capacity in the aging brain. *Cell*. 2019 Mar 7;176(6):1407–19.e14. doi:10.1016/j.cell.2019.01.040

Kalmar B, et al. The heat shock response and Huntington's disease: Prospects for therapy. *Trends Pharmacol Sci*. 2014;35(3):171–77. doi:10.1016/j.tips.2014.01.004

Katsimpardi L, et al. Systemic activin signaling attenuates age-related neurogenic decline in the hippocampus. *Am J Pathol*. 2016;186(9):2387–96.

Katsimpardi L, et al. Vascular and neurogenic rejuvenation of the aging mouse brain by young systemic factors. *Science*. 2014;344(6184):630–34.

Kaushik S, Cuervo AM. Proteostasis and aging. *Nat Med*. 2015;21(12):1406–15.

Klucken J, Yim JS, McLean LP, Kopito, PM, Selkoe D. HSP70 reduces alpha-synuclein aggregation and toxicity. *J Neurosci*. 2004;24(33):7380–87.

Labbadia J, Morimoto RI. The biology of proteostasis in aging and disease. *Annu Rev Biochem*. 2015;84:435–64. doi:10.1146/annurev-biochem-060614-033955

Landgrave-Gómez J, Mercado-Gómez O, Guevara-Guzmán R. Epigenetic mechanisms in neurological and neurodegenerative diseases. *Front Cell Neurosci*. 2015;9:58.

Latorre A, Albani D, Vecchione A, et al. Epigenetic mechanisms of aging and aging-associated diseases. *Int J Mol Sci*. 2023;24(9):4200. doi:10.3390/ijms24094200

Lautrup S, Sinclair DA, Mattson MP, Fang EF. NAD+ in brain aging and neurodegenerative disorders. *Cell Metab*. 2019;30(4):630–55.

Leboyer M, Godin O, Llorca PM, Aubin V, Bellivier F, Belzeaux R, Courtet P, Costagliola D, Dubertret C, M'Bailara K, Haffen E, Henry C, Laouamri H, Passerieux C, Pelletier A, Polosan M, Roux P, Schwan R, Samalin L; FondaMental Advanced Center of Expertise for Bipolar Disorders (FACE-BD) collaborators; Olié E, Etain B. Key findings on bipolar disorders from the longitudinal FondaMental Advanced Center of Expertise-Bipolar Disorder (FACE-BD) cohort. *J Affect Disord*. 2022;307:149–56. doi:10.1016/j.jad.2022.03.053

Levine ME, et al. Epigenetic aging rates in the development of Alzheimer's disease. *J Gerontol A*. 2015;70(11):1304–8. doi:10.1093/gerona/glv062

Lin MT, Flint Beal M. Mitochondrial dysfunction and oxidative stress in neurodegenerative diseases. *Nature*. 2006;443(7113):787–95. doi:10.1038/nature05292

Lindvall O, Kokaia Z. Stem cell research in stroke: How far from the clinic? *Stroke*. 2010;42(8):2369–75.

Livingston G, et al. Dementia prevention, intervention, and care: 2020 report of the lancet commission. *Lancet*. 2020;396(10248):413–46.

Lodato MA, Rodin RE, Bohrson CL, Coulter ME, Barton AR, Kwon M, et al. Aging and neurodegeneration are associated with increased mutations in single human neurons. *Science*. 2018;359(6375):555–9. doi:10.1126/science.aao4426

López-Otín C, Blasco MA, Partridge L, Serrano M, Kroemer G. The hallmarks of aging. *Cell*. 2013;153(6):1194–217. doi:10.1016/j.cell.2013.05.039

Lourida I, et al. Association of lifestyle and genetic risk with incidence of dementia. *JAMA*. 2019;322(5):430–37.

Lu T, et al. Gene regulation and DNA damage in the ageing human brain. *Nature*. 2004;429(6994):883–91. doi:10.1038/nature02661

Ma T, et al. Suppressing eIF2α kinase and Atf4 activity enhances memory and synaptic plasticity in Alzheimer's disease model mice. *Nat Neurosci*. 2013;16(9):1299–305.

Madabhushi R, Pan L, Tsai LH. DNA damage and its links to neurodegeneration. *Neuron*. 2014 Jul 16;83(2):266–82.

Mahgoub M, Monteggia LM. Epigenetics and psychiatry. *Neurotherapeutics*. 2014;11(4):708–20.

Mattson MP, Longo VD, Harvie M. Impact of intermittent fasting on health and disease processes. *Ageing Res Rev.* 2017;39:46–58.

Maynard S, Fang EF, Scheibye-Knudsen M, Croteau DL, Bohr VA. DNA damage, DNA repair, aging, and neurodegeneration. *Cold Spring Harb Perspect Med.* 2015;5(10):a025130.

Mayne K, White JA, McMurran CE, Rivera FJ, de la Fuente AG. Aging and neurodegenerative disease: Is the adaptive immune system a friend or foe? *Front Aging Neurosci.* 2020;12:572090.

Meyers EA, Doetsch F. Distinct roles for BMP signaling in adult neural stem cell Fate and maintenance. *Stem Cells.* 2016;34(10):2391–400.

Morris MC, Tangney CC, Wang Y, Sacks FM, Barnes LL, Bennett DA, Aggarwal NT. MIND diet slows cognitive decline with aging. *Alzheimers Dement.* 2015;11(9):1015–22. doi:10.1016/j.jalz.2015.04.011

Mir FA, Amanullah A, Jain BP, Hyderi Z, Gautam A. Neuroepigenetics of ageing and neurodegeneration-associated dementia: An updated review. *Ageing Res Rev.* 2023;85:102067.

Naeem A, Prakash R, Kumari N, Khan MA, Khan AQ, Uddin S, Verma S, Robertson AAB, Boltze J, Raza SS. MCC950 reduces autophagy and improves cognitive function by inhibiting NLRP3-dependent neuroinflammation in a rat model of Alzheimer's disease. *Brain Behav Immun.* 2024;116:70–84.

Nagata T, Kobayashi T, Ishii S, Arai K.Association between DNA methylation of the BDNF promoter region and clinical presentation in Alzheimer's disease. *Dement Geriatr Cogn Disord Extra.* 2015;5(1):64–73.

Nayak D, Roth TL, McGavern DB. Microglia development and function. *Annu Rev Immunol.* 2014;32:367–402. doi:10.1146/annurev-immunol-032713-120240

Neumann B, et al. Metformin restores CNS remyelination capacity by rejuvenating aged stem cells. *Cell Stem Cell.* 2019;25(4):473–85.e8.

Ngandu T, et al. A 2 year multidomain intervention of diet, exercise, cognitive training, and vascular risk monitoring versus control to prevent cognitive decline in at-risk elderly people (FINGER): A randomised controlled trial. *Lancet.* 2015;385(9984):2255–63.

Nicaise AM, Wagstaff LJ, Willis CM, et al. Cellular senescence in progenitor cells contributes to diminished remyelination potential in progressive multiple sclerosis. *Proc Natl Acad Sci USA.* 2019;116(19):9030–9.

Nicaise AM, Willis CM, Crocker SJ, Pluchino S. Stem cells of the aging brain. *Front Aging Neurosci.* 2020;12:247.

Niccoli T, Partridge L. Ageing as a risk factor for disease. *Curr Biol.* 2012;22(17):R741–52.

Nixon RA. The role of autophagy in neurodegenerative disease. *Nat Med.* 2013;19(8):983–97.

Outeiro TF, et al. Sirtuin 2 inhibitors rescue α-synuclein-mediated toxicity in models of Parkinson's disease. *Science.* 2007;317(5837):516–19.

Plemel JR, et al. Remyelination after spinal cord injury: A target for repair. *Nat Rev Neurol.* 2020;16(12):769–84. doi:10.1038/s41582-020-00421-3

Reddy PH, Manczak M, Kandimalla R. Mitochondria-targeted small molecule SS31: A potential candidate for the treatment of Alzheimer's disease. *Hum Mol Genet.* 2017;26(8):1483–96.

Rubinsztein DC, Bento CF, Deretic V. Therapeutic targeting of autophagy in neurodegenerative and infectious diseases. *J Exp Med.* 2015;212(7):979–90.

Scarmeas N, et al. Influence of leisure activity on the incidence of Alzheimer's disease. *Neurology.* 2001;57(12):2236–42.

Sha SJ, et al. Safety, tolerability, and feasibility of young plasma infusion in the plasma for Alzheimer symptom amelioration study: A randomized clinical trial. *JAMA Neurol.* 2019;76(1):35–40.

Sherwani SI, Khan HA. Role of 5-hydroxymethylcytosine in neurodegeneration. *Gene.* 2015;570(1):17–24.

Spangenberg EE, et al. Eliminating microglia in Alzheimer's mice prevents neuronal loss without modulating amyloid-β pathology. *Brain.* 2016;139(Pt 4):1265–81.

Uccelli A, Laroni F, Freedman MS. Mesenchymal stem cells as treatment for MS – Progress to date. *Mult Scler J.* 2011;17(5):515–19.

Villeda SA, et al. The aging systemic milieu negatively regulates neurogenesis and cognitive function. *Nature.* 2011;477(7362):90–4.

Villeda SA, et al. Young blood reverses age-related impairments in cognitive function and synaptic plasticity in mice. *Nat Med.* 2014;20(6):659–63.

Walker KA, et al. Systemic inflammation during midlife and cognitive change over 20 years: The ARIC study. *Neurology.* 2019;92(11):e1256–67.

Wang Y, Xu E, Musich PR, Lin F. Mitochondrial dysfunction in neurodegenerative diseases and the potential countermeasure. *CNS Neurosci Ther.* 2019 Jul;25(7):816–24.

Waseem R, Shamsi A, Kazim SN, Islam A. An insight into mitochondrial dysfunction and its implications in neurological diseases. *Curr Drug Targets.* 2021;22(14):1548–64.

Wrann CD, et al. Exercise induces hippocampal BDNF through a PGC-1α/FNDC5 pathway. *Cell Metab.* 2013;18(5):649–59.

Wyss-Coray T. Ageing, neurodegeneration and brain rejuvenation. *Nature.* 2016;539(7628):180–86.

Wyss-Coray T, Rogers J. Inflammation in Alzheimer disease—A brief review of the basic science and clinical literature. *Cold Spring Harb Perspect Med.* 2012;2(1):a006346.

Yin X, Manczak M, Reddy PH. Mitochondria-targeted molecules MitoQ and SS31 reduce mutant huntingtin-induced mitochondrial toxicity and synaptic damage in Huntington's disease. *Hum Mol Genet.* 2016;25(9):1739–53.

Yousef H, et al. Aged blood impairs hippocampal neural progenitor activity and activates microglia via brain endothelial cell VCAM1. *Nat Med.* 2019;25(6):988–1000.

Zhang Z, et al. Reversing microglial activation prevents cognitive impairment in aged mice and Alzheimer's models. *Nat Neurosci.* 2022;25(9):1260–75.

3 Extracellular Vesicles for Neuroregeneration in Neurodegenerative Disease

3.1 INTRODUCTION

Extracellular vesicles (EVs), including exosomes, are 30–150 nm endosome-derived vesicles. These are membrane-bound particles released by cells and mediate intercellular communication in the nervous system. EVs carry proteins, lipids, mRNAs, and microRNAs from their cell of origin and travel between cells and cross the blood–brain barrier (BBB). In neurodegenerative diseases, EVs have a double-edged role. They can not only propagate pathological proteins, e.g., amyloid-beta, tau, α-synuclein between cells but also transmit neuroprotective signals and therapeutic cargo (Ananbeh et al. 2021). Increasing evidence suggests that the beneficial effects of stem cell therapies for neurodegeneration are largely mediated by their secreted EVs rather than cell replacement (Fayazi et al. 2021). Thus, stem cell-derived exosomes have emerged as a promising cell-free strategy to promote neuroregeneration and neuroprotection in neurodegenerative disorders (Fayazi et al. 2021; Quan et al. 2025).

This chapter provides review of EVs and exosomes in neuroregeneration, with a focus on stem cell-derived EV applications across major neurodegenerative diseases. It summarizes the biological mechanisms by which EVs influence neural cells, disease-specific findings from preclinical studies, experimental models that are used to evaluate EV therapies, and translational advances including early clinical trials. For each disease, both preclinical and clinical perspectives are integrated to underscore the translational progression of EV-based therapeutic strategies. Overall, the therapeutic potential of EVs as natural nanocarriers of neuro-regenerative signals is examined in the context of Alzheimer's disease, Parkinson's disease, amyotrophic lateral sclerosis (ALS), Huntington's disease, and multiple sclerosis.

3.2 BIOLOGICAL MECHANISMS OF EV-MEDIATED NEUROREGENERATION

Exosomes are formed as intraluminal vesicles within multivesicular bodies that fuse with the plasma membrane, releasing the vesicles externally. They are enriched in tetraspanins (CD9, CD63, CD81), heat shock proteins, adhesion molecules, and nucleic acids from the parent cell. Because of the presence of surface molecules like integrins and phosphatidylserine, EVs are internalized by recipient cells via endocytic pathways or direct membrane fusion, enabling functional delivery of bioactive proteins, lipids, and RNAs. In CNS, neurons, astrocytes, microglia, oligodendrocytes, and endothelial cells all secrete EVs, which circulate in cerebrospinal fluid and blood. Further, EVs can cross the BBB, making them attractive natural vehicles for delivery of therapeutic molecules to brain (Ananbeh et al. 2021). Importantly, EVs are

DOI: 10.1201/9781003582687-3

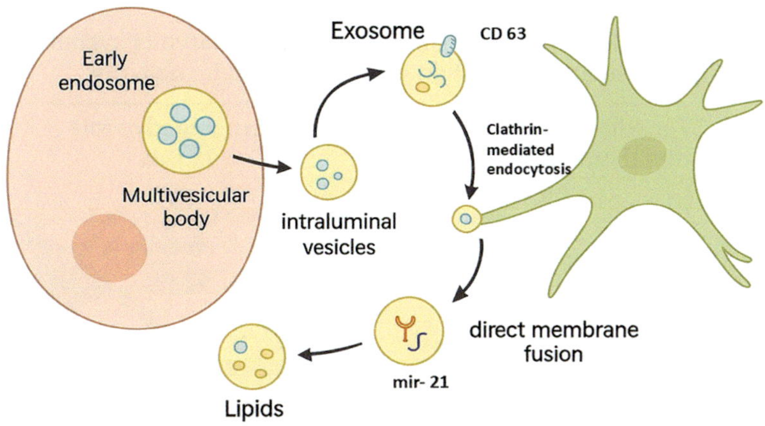

FIGURE 3.1 Biogenesis, Release, and Cellular Uptake of Extracellular Vesicles (EVs).

relatively stable in biological fluids due to surface proteins that protect them from complement and enzymatic destruction. These properties enable EVs to mediate long-distance signaling in the brain and periphery (Figure 3.1).

EVs contribute to CNS homeostasis and response to injury by transferring biomolecules between cells. They carry neurotrophic factors, growth factor receptors, enzymes, and noncoding RNAs that can modulate neuronal survival, inflammation, and plasticity (Fayazi et al. 2021). For instance, MSC-derived exosomes are enriched in miR-133b, which promotes neurite outgrowth and neuronal recovery and also contains miR-21 and miR-22 that modulate apoptotic and inflammatory pathways. (Heris et al. 2022; Wang et al. 2021). EVs derived from neural stem cells contain proteins like synapsin I that protect neurons from oxidative stress and promote synaptic integrity. Glia-derived exosomes carry antioxidant enzymes, e.g., catalase and anti-inflammatory cytokines, that contribute to neuroprotection in toxic environments. Notably, EV's molecular repertoire can be altered by preconditioning of the parent cells. For example, MSCs exposed to hypoxia or inflammatory stimuli secrete exosomes with heightened levels of neuroprotective miRNAs and immunomodulatory factors (Fayazi et al. 2021). Through paracrine mechanisms, EVs from stem cells have been shown to support neuronal survival, stimulate angiogenesis and neurogenesis, reduce oxidative stress, and modulate immune responses in the brain (Quan et al. 2025). These multifaceted actions underlie the regenerative capacity of EVs.

On other hand, in neurodegenerative disease states, EVs also carry and spread pathological molecules. Neurons undergoing degeneration release EVs that contains misfolded or toxic proteins such as oligomeric Aβ, hyperphosphorylated tau, mutant huntingtin, or TDP-43. These EVs are internalized by neighboring cells, promoting the propagation of protein aggregation and the transmission of pathological processes. (Ananbeh et al. 2021). For example, in Alzheimer's disease models, multivesicular bodies in neurons produce exosomes loaded with Aβ peptides, which accumulate extracellularly and promote plaque formation. Similarly, exosomal tau has been detected and implicated in trans-synaptic propagation of tau tangles (Ruan et al. 2021). In PD, α-synuclein released via exosomes contribute to Lewy body pathology spreading through neural networks. In ALS, EVs secreted by stressed neurons or glia carry misfolded SOD1 and TDP-43, which induce neurotoxic effects in recipient cells (Wang et al. 2021). Thus, EVs can be vehicles of disease propagation. At the same time, EVs from cells that are enriched with protective biomolecules may be exploited for therapeutic applications. The paradoxical nature of EVs is now well recognized. They may accelerate neurodegeneration

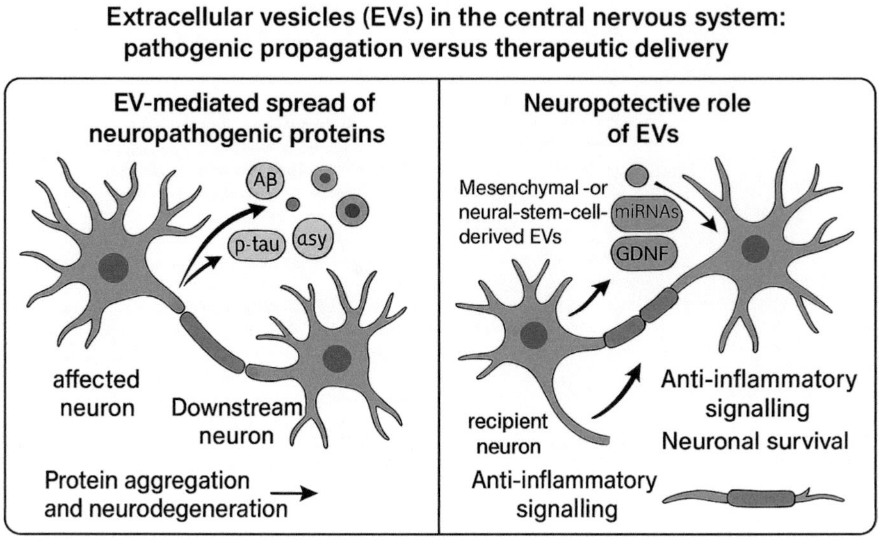

FIGURE 3.2 Extracellular Vesicles in the Central Nervous System: Pathogenic Propagation Versus Therapeutic Delivery.

via dissemination of misfolded proteins, but conversely, they can attenuate it by shuttling protective factors (Ananbeh et al. 2021). This has generated interest in therapeutic applications of EVs, either through employing vesicles from healthy or engineered cells to deliver beneficial molecules, or by reprogramming EV cargo to neutralize pathological species (Figure 3.2). The following sections examine how these concepts are applied in each major neurodegenerative disease, leveraging stem cell-derived exosomes for neuroregeneration. The key regenerative cargos identified in stem-cell EVs are summarized in Table 3.1.

TABLE 3.1
Principal Regenerative Cargo in Stem-Cell EVs

EV cargo class	Key examples	Principal CNS targets	Neuro-regenerative actions
Neurotrophic factors	BDNF, GDNF, NGF, CNTF	Neurons, oligodendrocytes	Promote neurite outgrowth, survival, synaptic plasticity
Anti-inflammatory cytokines	IL-10, TGF-β	Activated microglia, infiltrating T cells	Shift microglia to M2, expand Tregs, dampen inflammation
Antioxidant/repair enzymes	Catalase, SOD2, MMP-2	Neurons, extracellular α-syn fibrils	Neutralise ROS; degrade misfolded proteins
miRNAs (neuritogenic/anti-apoptotic)	miR-133b, miR-124, miR-21/22	Neurons, neural progenitors	Axonal sprouting, anti-apoptosis, neurogenesis
miRNAs (oligodendrogenic)	miR-219, miR-338	OPCs	Drive maturation and remyelination

3.3 EVS IN ALZHEIMER'S DISEASE

Alzheimer's disease is characterized by extracellular amyloid-β (Aβ) plaques and intracellular neurofibrillary tangles of tau protein in the brain, leading to synaptic loss and cognitive decline. EVs have been implicated in the pathogenesis of AD while also being investigated as potential therapeutic tools. Neurons generate Aβ peptide within multivesicular bodies and its release via exosomes externalizes Aβ, thereby facilitating plaque deposition. Exosomes isolated from the brains of AD patients and from cellular models carrying Aβ and phosphorylated tau induce aggregation of these proteins in recipient cells (Ananbeh et al. 2021). This prion-like spread via EVs is thought to contribute to disease progression. On the other hand, EVs also facilitate Aβ clearance. The neuron-derived exosomes have been shown to bind Aβ and promote its uptake by microglia for degradation. For instance, exosomes expressing surface prion protein sequester Aβ and neutralize its toxicity while also stimulating phagocytosis of Aβ by microglia. These findings demonstrate that EVs in Alzheimer's disease have a complex role, simultaneously spreading pathological agents while also facilitating their removal. A growing number of studies demonstrate that stem cell-derived exosomes attenuate AD pathology in experimental models.

Mesenchymal stem cell–derived exosomes have been extensively studied because they can cross the BBB and deliver enzymes and trophic factors that degrade Aβ and protect neurons. In one study, intracerebral injection of bone marrow MSC exosomes into transgenic AD model mice (APP/PS1 model) led to reduced amyloid plaque and fewer dystrophic neurites in the hippocampus and cortex. The neuroprotective of MSC-EVs is mediated by two complementary mechanisms, by directly binding to plaque with subsequent microglial uptake, and proteolytic degradation of Aβ through neprilysin delivery (Figure 3.3). Another study found that administering MSC-derived EVs increased hippocampal neurogenesis and improved cognitive function in AD mice, likely by delivering growth factors and microRNAs that activate pro-survival pathways. EVs from mesenchymal stem cells, especially when the cells are preconditioned with cytokines or hypoxia, contain anti-inflammatory cytokines such as IL-10 and TGF-β together with antioxidants like catalase. These cargos reduce neuroinflammation and oxidative stress in Alzheimer disease. For example, injecting MSC-EVs in an AD model leads to reduction of proinflammatory IL-1β and TNF-α levels while increasing anti-inflammatory IL-10, leading to the transition of microglia into a neuroprotective state. EV treatment also restores synaptic protein levels and mitochondrial function in AD mice, indicating overall neuroprotective effects (Wang et al. 2021). MSC-derived EVs from diverse origins, such as bone marrow, adipose tissue, Wharton's jelly, umbilical cord, and periodontal tissue, consistently mitigate AD pathology in preclinical studies. (Fayazi et al. 2021). These studies consistently highlight reductions in soluble Aβ oligomers and plaque load, together with synaptic preservation and memory enhancement (Fayazi et al. 2021). Notably, exosomes derived from neural stem cells (NSCs) have also shown therapeutic potential. In one study, stereotactic delivery of NSC-derived exosomes activated SIRT1 signaling, enhanced synaptic plasticity, and attenuated neuroinflammation in an AD mouse model, resulting in improved cognitive performance. Collectively, such findings underscore that EV-based interventions are consistently associated with cognitive benefits in Alzheimer's disease models, suggesting that their efficacy may stem from mechanisms of neuroregeneration and neuroprotection (Table 3.2).

Alzheimer's disease currently represents the most advanced clinical model for testing EV-based therapeutics. A phase I clinical trial evaluating intravenously delivered mesenchymal stem cell–derived exosomes in patients with Alzheimer's disease has demonstrated a favorable safety profile. Importantly, the study also yielded preliminary evidence of biological activity and potential therapeutic benefit, thereby providing a foundation for subsequent phase II trials aimed at establishing efficacy and optimizing dosing strategies.

In a recent Phase I/II open-label trial, patients with mild-to-moderate Alzheimer's disease received intranasal administration of allogeneic adipose-derived MSC exosomes twice weekly for 12 weeks

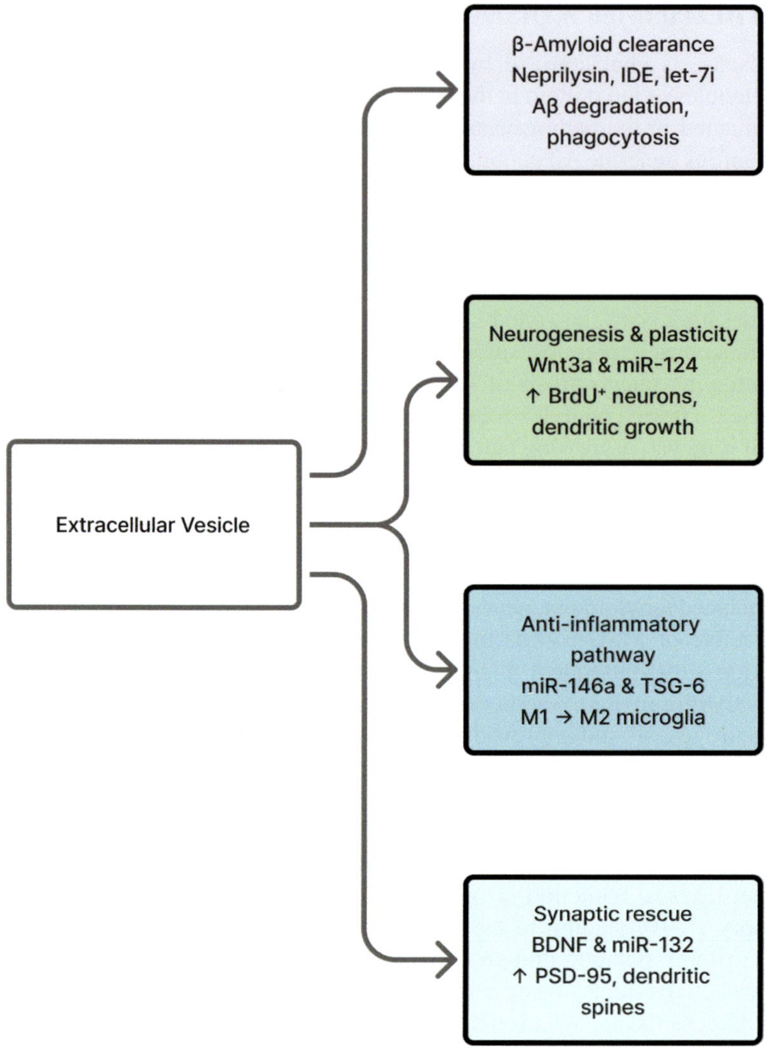

FIGURE 3.3 Extracellular Vesicles in Alzheimer's Disease.

TABLE 3.2
Preclinical and Early-Phase Trials of Exosomes in Alzheimer's Disease

Study/model	EV source and pre-conditioning	Delivery route/regimen	Core outcome(s)
APP/PS1 mouse	Bone-marrow MSC (normoxic)	Stereotactic hippocampal injection, single dose	↓ Aβ plaques (-38%), ↑ neprilysin, improved memory
5xFAD mouse	Hypoxia-preconditioned Wharton's-jelly MSC	IV, weekly ×4	Shifted microglia to M2, ↑ IL-10, rescued LTP
APP/PS1 mouse	NSC exosomes enriched in SIRT1	Intranasal, thrice weekly ×4 wk	↑ Synapsin I, ↓ Iba1, improved cognition
Phase I/II human	Adipose MSC exosome (allogenic)	Intranasal, biweekly ×12 wk	Safe; ADAS-Cog –2.3; slower hippocampal atrophy

(Xie et al. 2023). The intervention was safe and well tolerated, with no adverse events or immunological reactions reported. Notably, patients treated with a medium dose (~4 × 10^8 EV particles per dose) demonstrated cognitive improvement, ADAS-Cog scores improved by an average of 2.3 points from baseline at 12 weeks, with continued benefit reaching nearly a 4-point reduction by 36 weeks. Although no significant changes were detected in amyloid- or tau-PET imaging, MRI analysis suggested a slower rate of hippocampal atrophy compared with expected progression.

These findings provide the first clinical proof-of-concept that intranasal delivery of EVs can reach the brain and potentially ameliorate cognitive decline in AD. While preliminary, the safety profile and early signals of efficacy support further investigation in larger, controlled studies. Ongoing preclinical studies are actively developing exosome engineering approaches, including loading vesicles with siRNAs targeting tau or secretase enzymes, and functionalizing exosomal membranes with targeting ligands to enhance delivery to affected brain regions.

Exosome-based therapy for Alzheimer's disease has progressed rapidly from animal models to early clinical application, driven by robust preclinical evidence demonstrating reduced pathology and cognitive improvement (Fayazi et al. 2021; Xie et al. 2023). It exemplifies both the promise and complexity of EVs, which act not only as carriers of pathogenic proteins but also as transporters of therapeutic signals. Despite this duality, the net impact of exosome-based interventions in preclinical AD models has been strongly positive, positioning EVs as a potential disease-modifying strategy in a field where conventional small-molecule approaches have largely been unsuccessful.

3.4 EVS IN PARKINSON'S DISEASE

Parkinson's disease involves progressive loss of dopaminergic neurons in the substantia nigra and accumulation of intracellular α-synuclein aggregates. EVs play roles in PD pathogenesis and are being investigated as neurorestorative agents. It has been known that neuron- and glia-derived EVs transfers α-synuclein between cells. Laboratory studies have shown that pathogenic α-syn oligomers are released in exosomes and induce aggregation of α-syn in recipient neurons, suggesting a prion-like dissemination of Lewy pathology via EVs (Haney et al. 2015). This mechanism explains how Parkinson's disease pathology disseminates progressively across interconnected brain regions. EVs isolated from the cerebrospinal fluid and blood of PD patients frequently contain misfolded α-synuclein, and their abundance correlates with disease severity, reinforcing the view that EVs are active participants in PD pathogenesis. At the same time, EVs also transport regulatory RNAs and enzymes with potential neuroprotective functions. In light of this paradoxical function, current research is increasingly directed toward harnessing stem cell–derived EVs to support and protect dopaminergic neurons. Stem cell-derived exosomes have shown notable neuroprotective effects in PD models. MSCs and other stem cells secrete a plethora of neurotrophic factors e.g. GDNF, BDNF, and regenerative miRNAs, many of which are packaged into exosomes (Heris et al. 2022). Studies have compared direct stem cell grafts versus acellular secretome or exosomes in PD animal models. Remarkably, administering MSC-conditioned medium or purified exosomes can often recapitulate the benefits of cell transplantation. In a 6-hydroxydopamine (6-OHDA) rat model of PD, the transplantation of human MSCs improved motor function and dopaminergic neuron survival; however, infusion of the MSC secretome achieved similar or greater neuroprotective effects (Heris et al. 2022). Specifically, MSC-derived exosomes were found to cleave extracellular α-synuclein aggregates via exosomal metalloproteinases. One study identified matrix metalloproteinase-2 (MMP-2) within MSC exosomes as a key enzyme that degraded α-syn oligomers, thereby reducing the buildup of insoluble α-syn in a PD mouse model. This led to preservation of dopaminergic neurons and improved behavioral outcomes, e.g., reduced bradykinesia in treated animals (Heris et al. 2022).

In another approach, exosomes from dental pulp stem cells of human exfoliated deciduous teeth (SHED) have been tested for PD therapy. It has been shown that SHED-derived exosomes, but not larger microvesicles, protect cultured human dopaminergic neurons from apoptosis induced by

6-OHDA toxin. Intranasal administration of SHED-derived exosomes in 6-OHDA–lesioned rats resulted in selective homing to the lesioned striatum and substantia nigra, accompanied by significant motor recovery. Treated animals exhibited restored gait and improved motor coordination, which correlated with normalization of tyrosine hydroxylase expression along the nigrostriatal pathway. These findings suggest that EVs can partially regenerate dopaminergic circuitry, likely through the delivery of neurotrophic factors and antiapoptotic signals to vulnerable neurons.

In Parkinson disease models, the neuroprotective actions of EVs are mediated in part by their microRNA cargo (Figure 3.4). Mesenchymal stem cell derived exosomes are particularly rich in miR-133b, a neuritogenic microRNA that is markedly reduced in the midbrain of individuals with Parkinson disease and is known to enhance neurite extension and axonal remodeling.

Delivery of MSC-derived exosomes enriched with miR-133b to Parkinson's disease animal models has been shown to stimulate axonal regeneration and functional recovery (Heris et al. 2022). Similarly, exosomes engineered to carry miR-124 have been used to reprogram resident brain cells. In one study, MSC exosomes containing miR-124 and miR-145 promoted differentiation of endogenous neural progenitor cells in a PD model and upregulated glutamate transporter expression in astrocytes, potentially mitigating excitotoxicity. There is also evidence that MSC-derived exosomal miR-17-92 clusters can drive oligodendrocyte maturation and myelination, thereby improving neuronal support within PD lesions.

The cumulative effect of these paracrine signals is neurorestoration in animal models of PD. EV treatment has been associated with higher dopaminergic neuron counts in the substantia nigra and striatum, increased dopamine concentrations, and improvements in motor outcomes such as reduced tremors and bradykinesia (Heris et al. 2022; Fayazi et al. 2021). Notably, EV therapy has been tested through several routes, including intra-striatal injection, intravenous infusion, and intranasal delivery with each demonstrating some efficacy. Among these, intranasal administration is particularly promising for PD, as it provides a noninvasive route for brain targeting. The intranasally delivered MSC exosomes reached the brain and improve motor performance in PD mouse models (Perets et al. 2019).

Although no clinical trials of exosome-based therapies have been completed in Parkinson's disease, the preclinical evidence has established a solid foundation for translation. Current research is focused on optimizing EV sources, comparing MSCs with other stem cell types as well as refining dosing regimens and cargo modifications. For example, exosomes loaded with catalase have successfully reduced oxidative stress in a mouse PD model by protecting nigral neurons from oxidative injury (Haney et al. 2015). Importantly, in some animal studies, MSC-derived EVs have demonstrated superior outcomes compared with MSC transplantation (Heris et al. 2022), providing a compelling rationale for clinical trials. Unlike cell-based grafts, EV therapy could bypass the risks of graft-induced dyskinesias or immune rejection, positioning it as a safer and potentially more effective regenerative strategy.

In PD models, stem cell–derived exosomes exert multiple therapeutic actions, they provide neurotrophic support to degenerating dopaminergic neurons, facilitate clearance of toxic α-synuclein, modulate glial responses, and promote regenerative processes, collectively leading to measurable motor recovery (Heris et al. 2022; Fayazi et al. 2021). These findings highlight the capacity of EVs to establish a neuroregenerative milieu. The next step is clinical translation, as underway in Alzheimer's disease, though major challenges remain, particularly achieving efficient EV delivery to brain structures in PD and sustaining motor improvement. Despite these hurdles, EVs represent a promising disease-modifying therapeutic avenue, with the potential not merely to complement current symptomatic treatments such as L-DOPA but to protect and restore dopaminergic neurons.

3.5 EVS IN ALS

ALS is a progressive motor neuron disease characterized by degeneration of upper and lower motor neurons, leading to muscle weakness and paralysis. Pathologically, protein aggregates of TDP-43, SOD1, and FUS are often found in degenerating motor neurons. EVs have been implicated in ALS

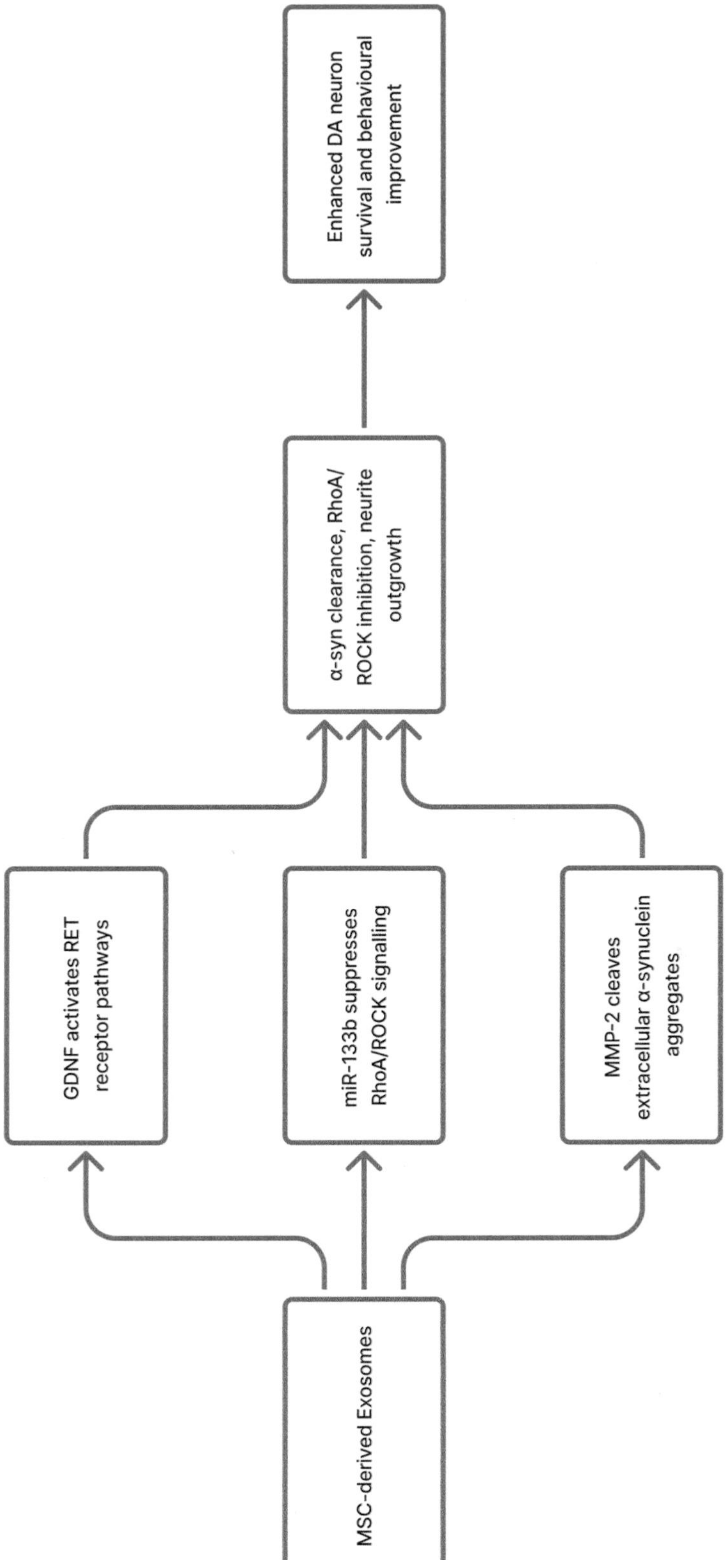

FIGURE 3.4 Neuroprotective Mechanisms of MSC Exosomes in Parkinson's Disease.

pathology and are being explored as therapeutic mediators, especially given the limited ability of transplanted cells to integrate into the motor system.

In ALS, EVs mediate the intercellular transfer of toxic proteins and RNA species. Astrocytes expressing mutant SOD1 release EVs that induce motor neuron death, implicating EVs in the transmission of toxic factors (Bonafede and Mariotti 2017). Likewise, misfolded TDP-43 has been identified in exosomes from the CSF of ALS patients, and these vesicles can propagate TDP-43 pathology to recipient cells in vitro (Iguchi et al. 2016). Such EV-mediated dissemination may underline the focal-to-generalized spread of ALS pathology. Conversely, stem cell–derived EVs represent a potential means of delivering protective factors to motor neurons, which are otherwise shielded from systemic therapies by the blood–spinal cord barrier.

Stem cell–based therapies for ALS using MSCs or neural progenitors have shown benefits in animal models and early trials, largely through trophic support rather than direct cell replacement. To capture these effects in an acellular format, researchers have investigated exosomes isolated from stem cells. MSC-derived exosomes have demonstrated robust neuroprotective activity on motor neurons both in vitro and in vivo. In a pivotal study, exosomes from human adipose-derived MSCs were applied to motor neuron–like cells expressing mutant SOD1 and stressed with hydrogen peroxide. Treatment significantly enhanced neuronal survival, reduced oxidative stress, and improved mitochondrial function (Bonafede et al. 2016), providing the first evidence that stem cell exosomes alone can directly protect ALS-model neurons. These exosomes carried multiple protective factors, including IGF-1, VEGF, antiapoptotic miRNAs, and antioxidant enzymes.

In vivo, MSC-derived EVs slowed disease progression and extended survival in a mouse model of ALS (Palanisamy et al. 2023). Treated animals exhibited delayed onset of motor weakness and preserved motor neuron counts in the spinal cord. Though improvements were modest, they were reproducible, confirming biological effect of EV therapy. Mechanistically, EVs appear to act by reshaping the hostile microenvironment in ALS. For example, intravenous infusion of human MSC exosomes reduced microglial and astrocytic activation and elevated anti-inflammatory cytokines such as IL-10 in the spinal cord (Wang et al. 2021). In parallel, levels of neurotrophic factors, including BDNF and NGF, were significantly increased, suggesting that exosomes either directly delivered these molecules or stimulated their endogenous production by resident glia. Consequently, motor neurons in exosome-treated ALS mice experienced less inflammatory stress and greater trophic support, correlating with improved motor performance.

Another promising strategy in ALS is the use of exosomes as vehicles for genetic therapies. Since many ALS cases involve toxic gain-of-function mutations such as SOD1 or C9ORF72 expansions, silencing these genes in motor neurons could be therapeutic. Exosomes can be engineered to carry small interfering RNAs (siRNAs) or antisense oligonucleotides, enabling targeted delivery. In one study, exosomes loaded with siRNA against SOD1 were injected into an ALS mouse model, achieving widespread neuronal uptake and a marked reduction of mutant SOD1 levels in the spinal cord (Basso et al. 2013). This intervention improved motor neuron survival and extended lifespan, demonstrating the feasibility of exosome-based gene therapy by effectively using EVs as nano-scale delivery vehicles to transfer RNA to motor neurons which are typically resistant to transfection in vivo.

Much of this evidence derives from the SOD1 mutant rodent model and from in vitro neuron–glia coculture systems employing patient-derived cells. In a commonly used paradigm, motor neurons or NSC-34 cells are exposed to toxic factors secreted by ALS patient astrocytes; under these conditions, MSC-derived exosomes have been shown to protect neuronal survival (Drago et al. 2017). Across these diverse experimental models, EVs consistently mitigate hallmarks of ALS pathology. They reduce neuronal apoptosis, oxidative damage, and neuroinflammation while enhancing neurite outgrowth and cell viability.

Although clinical translation is still in early stages, one small pilot trial evaluated administration of MSC secretome, though not purified exosomes. The intramuscular delivery of MSC secretions in ALS patients was reported to be safe and showed signs of functional stabilization (Petrou et al.

2016). Building on these preliminary findings, the future therapeutic strategies may involve intrathecal or intravenous administration of exosomes, enabling broad distribution across CNS to support vulnerable motor neurons.

In ALS, where conventional drug development has yielded limited benefit, EV–based therapy offers a novel disease-modifying strategy to slow neurodegeneration. Stem cell–derived exosomes provide trophic and anti-inflammatory factors that counteract the hostile ALS microenvironment and protect motor neurons (Wang et al. 2021). In addition, they represent a potential delivery platform for genetic therapeutics targeted directly at neurons. The principal challenge lies in ensuring adequate and sustained delivery of EVs to the large population of motor neurons throughout the course of disease. Nevertheless, preclinical data are sufficiently encouraging that exosome therapies are being actively pursued as candidate treatments to prolong survival and preserve motor function in ALS (Wang et al. 2021).

3.6 EVS IN HUNTINGTON'S DISEASE

Huntington's disease is characterized by progressive motor impairment, psychiatric disturbances, and cognitive decline. Its pathogenesis involves the intercellular propagation of mutant huntingtin (mHTT) and extensive neuroinflammatory responses, processes that may be modulated by EVs. EVs have been found to transport mHTT and might facilitate the propagation of disease. In cell culture models, neurons expressing mutant huntingtin release exosomes containing mHTT fragments, and neighboring cells uptakes these exosomes, acquiring aggregates of mHTT (Ananbeh et al. 2021). In mouse models of HD, transplanted wild-type cells eventually show inclusion bodies, suggesting that a spread of mHTT from diseased to healthy cells occurs, possibly via EVs. Further, one study observed that exosomes derived from fibroblasts of HD patients induce inclusion formation when added to normal neuronal cultures (Jeon et al. 2016). This prion-like dissemination contributes to the progressive neurodegeneration seen in HD. Conversely, EVs may also exert neuroprotective functions in Huntington's disease. Notably, EVs derived from cells experiencing mHTT-induced stress have been shown to contain elevated levels of heat shock proteins and antioxidant enzymes. This phenomenon may represent an adaptive cellular response aimed at mitigating proteotoxicity, either through the export of potentially harmful aggregates and debris or by disseminating protective molecules to neighboring cells.

Although research on EV-based therapies for Huntington's disease remains at an early stage, emerging findings are encouraging. A pivotal study demonstrated that astrocyte-derived exosomes can ameliorate Huntington's pathology in vivo. In an HD mouse model, striatal administration of exosomes isolated from healthy astrocytes significantly reduced the burden of mutant huntingtin aggregates within the region. Treated animals exhibited improved motor coordination and attenuated neuronal atrophy relative to controls (Bonafede et al. 2020). Mechanistically, these effects are attributed to the exosomal transfer of astrocyte-derived neuroprotective factors, including apolipoprotein D and BDNF, both of which support neuronal survival and facilitate aggregate clearance. Collectively, these findings suggest that harnessing glial exosomes may provide a therapeutic avenue for detoxifying the neuronal milieu in Huntington's disease.

Another innovative strategy exploits exosomes as vehicles for gene-silencing therapeutics targeting mutant huntingtin. In one proof-of-concept study, exosomes were engineered to encapsulate an antisense oligonucleotide directed against HTT mRNA. Systemic administration of these modified vesicles in Huntington's disease mouse model achieved robust knockdown of mutant HTT expression in the brain, which was accompanied by significant improvements in both motor function and neuropathological markers (Didiot et al. 2016). This work underscores the feasibility of exosome-mediated delivery of macromolecular therapeutics agents that would otherwise be unable to traverse the BBB. The application of EVs in this context is particularly compelling given that Huntington's disease arises from a single-gene defect, necessitating widespread delivery of gene-silencing molecules across diverse neural cell types.

Stem cell–derived EVs represent another promising therapeutic avenue in Huntington's disease, acting through both neurotrophic and anti-inflammatory mechanisms. In an in vitro HD system, administration of human mesenchymal stem cell-derived exosomes attenuated apoptotic neuronal death and reduced oxidative stress markers (Gowen et al. 2020). Notably, exosome-treated neurons exhibited upregulated expression of BDNF and other molecular indicators of neuronal integrity, suggesting the induction of intrinsic repair pathways. In parallel, MSC exosomes modulated microglial activity toward an anti-inflammatory phenotype, a shift that, in vivo, could mitigate the chronic neuroinflammation known to exacerbate disease progression. Collectively, these findings highlight the dual neuroprotective and immunomodulatory potential of MSC-derived exosomes in counteracting Huntington's pathology.

An emerging line of investigation explores the rejuvenating potential of exosomes derived from young blood or stem cells. Given that Huntington's disease is often conceptualized as a state of accelerated neuronal stress and premature cellular aging, such approaches are particularly compelling. Experimental studies have demonstrated that exosomes isolated from the circulation of young, healthy mice, when used in neuronal models of HD, ameliorated cellular phenotypes by reducing mutant huntingtin aggregates and enhancing mitochondrial function. These findings suggest that circulating EVs harbor systemic factors, such as specific microRNAs or proteins, that are capable of modifying HD pathology. Identification of the precise bioactive cargo could unveil novel therapeutic targets amenable to EV-based delivery. Going forward, the monogenic nature of HD raises the possibility of combinatorial EV therapeutics, wherein exosomes might be simultaneously engineered to carry gene-silencing agents, e.g., siRNAs targeting HTT together with neuroprotective molecules such as brain-derived neurotrophic factor. Such multifunctional vesicles could suppress toxic protein production while concurrently supporting neuronal survival.

Huntington's disease remains a formidable neurodegenerative disorder for which no disease-modifying therapies are currently available. Emerging research on EVs in HD, though still in its infancy, demonstrates their capacity to reduce mutant huntingtin aggregates and to provide trophic support to striatal neurons. Owing to their innate ability to cross the BBB and distribute widely within the neural parenchyma, exosomes are particularly well suited as therapeutic carriers for HD, a condition that necessitates broad central nervous system delivery of therapeutic agents. The prospect of employing stem cell–derived or engineered exosomes to both silence mutant HTT expression and concurrently deliver neuroprotective cargo raises the possibility of slowing or even halting the neurodegenerative cascade. Continued progress in EV engineering, coupled with rigorous preclinical validation, will be essential to defining the translational potential of this approach (Ananbeh et al. 2021).

3.7 EVS IN MULTIPLE SCLEROSIS

EVs are increasingly recognized as modulators of immune dysregulation and myelin repair, highlighting their relevance to pathogenesis and therapeutic strategies against multiple sclerosis. EVs participate in the intricate immune signaling that underlies multiple sclerosis, as it has been shown that immune cells release EVs that can either propagate or suppress inflammation. Within active MS lesions, microglia and macrophages secrete EVs enriched in pro-inflammatory cytokines and myelin antigens, thereby amplifying immune-mediated myelin injury (Manu, Hohjoh, and Yamamura 2021). Circulating EVs in MS patients also carry central nervous system–derived proteins and microRNAs reflective of disease activity; for example, exosomes containing myelin oligodendrocyte glycoprotein (MOG) have the capacity to trigger demyelinating immune responses. Conversely, EVs may exert regulatory influences, as elevated levels of IL-10–bearing vesicles have been observed during remission phases. These dual functions position EVs as orchestrators of immune imbalance in MS.

Therapeutically, considerable attention has focused on stem cell–derived EVs as modulators of immunity and facilitators of repair. Extensive investigations in experimental autoimmune encephalomyelitis (EAE), which is the prototypical rodent model of MS characterized by CNS demyelination, paralysis, and immune infiltration have demonstrated robust immunomodulatory and neuroregenerative effects of mesenchymal stem cell-derived exosomes. Rajan et al. (2016) reported that exosomes secreted by human periodontal ligament stem cells, delivered via conditioned medium to EAE mice, promoted substantial spinal cord remyelination, and attenuated clinical disease severity. Treated animals exhibited increased levels of anti-inflammatory cytokines (IL-10, TGF-β) and concomitant reductions in pro-inflammatory mediators (IFN-γ, TNF-α), indicating a shift from a Th1-driven inflammatory milieu toward a Th2/regulatory phenotype. Collectively, these findings suggest that stem cell–derived exosomes induce immune tolerance and partially reverse demyelination, underscoring their therapeutic promise in MS.

Exosomes derived from placenta-derived mesenchymal stem cells (PMSCs) have also been evaluated in EAE. High-dose administration of PMSC exosomes elicited therapeutic effects comparable to direct transplantation of the parent stem cells (Kråkenes et al. 2024). Proteomic profiling revealed that these vesicles are enriched in hepatocyte growth factor and vascular endothelial growth factor, both of which are implicated in tissue repair and the expansion of regulatory T cells (Tregs). Consistent with this cargo, PMSC-exosome treatment in EAE enhanced Treg frequency and functional activity, thereby suppressing autoreactive immune responses. Histopathological analyses further demonstrated improved remyelination in exosome-treated mice, suggesting that oligodendrocyte precursor cells were maturing and restoring axonal myelin. These reparative effects are likely mediated, at least in part, by exosomal HGF, which is known to promote oligodendrocyte survival and differentiation.

Exosomes derived from adipose mesenchymal stem cells have similarly demonstrated therapeutic efficacy in EAE. Intravenous administration of human adipose MSC-derived EVs at the onset of paralysis accelerated recovery and significantly reduced disease severity compared with untreated controls. Mechanistically, EV treatment attenuated immune cell infiltration into the central nervous system, with fewer T cells and macrophages present in demyelinating lesions and suppressed their activation states. Correspondingly, pro-inflammatory cytokines such as IL-17, IFN-γ, and TNF-α were diminished in both the CNS and peripheral lymphoid tissues, while anti-inflammatory IL-10 was elevated. These changes fostered an immunological environment permissive of endogenous repair which is reflected in increased oligodendrocyte marker expression and clear histological evidence of remyelination in the spinal cord.

Across diverse MSC-EV studies, a recurring theme is the induction of regulatory immune phenotypes. EVs appear to promote the expansion of regulatory T cells and type II macrophages, thereby restoring immune tolerance (Fayazi et al. 2021). This tolerogenic effect is mediated in part by the upregulation of immune checkpoint and immunoregulatory molecules such as PD-L1 and galectin-1 on recipient immune cells. In one experiment, MSC-derived microvesicles induced apoptosis in autoreactive splenic mononuclear cells from EAE mice while enhancing IL-10 and TGF-β secretion from surviving populations. Notably, the vesicles displayed PD-L1 and carried TGF-β, directly reinforcing Treg induction and the establishment of immune tolerance.

Beyond immunomodulation, EVs also create a reparative environment in the CNS. MSC-derived exosomes have been shown to polarize microglia toward the M2 phenotype, characterized by arginase-1 expression and reduced iNOS activity, thereby promoting axonal repair and remyelination while minimizing bystander injury (Riazifar et al. 2019). In addition, EV cargo includes miRNAs such as miR-219 and miR-338, which are key regulators of oligodendrocyte differentiation. MSC exosomes enriched in these miRNAs accelerated the maturation of oligodendrocyte precursor cells in vitro and are predicted to enhance remyelination in vivo. Collectively, these findings demonstrate that MSC-derived EVs not only suppress pathogenic immune responses but also actively promotes remyelination, underscoring their dual potential as immunomodulators and neuroregenerative agents in MS.

Preclinical investigations of mesenchymal stem cell-derived EVs in multiple sclerosis models have produced consistent outcomes and have already undergone formal meta-analytical evaluation. A 2022 meta-analysis encompassing 12 independent studies of MSC-EVs in rodent EAE demonstrated a significant improvement in clinical scores and neurological function, with a pooled large effect size (Xun et al. 2022). Although heterogeneity in dosing regimens and timing of administration was noted, the aggregate evidence confirmed that MSC-EV therapy consistently attenuates disease severity and enhances functional recovery across models. Importantly, the analysis also identified critical knowledge gaps, including the need to optimize dosing strategies besides establishing standardized manufacturing protocols.

To date, no clinical trials have tested exosome therapy in patients with MS, but translational efforts are anticipated given the strength of preclinical data. The therapeutic rationale centers on leveraging EVs to achieve durable remission by balancing immune responses while simultaneously promoting remyelination of demyelinated lesions. However, there are still key challenges, such as scalable production, dosing standardization, and rigorous safety validation that remain to be resolved before clinical application. Nonetheless, the dual-action profile of EV therapy, combining immunomodulation with neuroregeneration, offers a compelling strategy to address both the inflammatory drivers and degenerative sequelae of MS. If successfully translated, this approach may fill a critical therapeutic gap, particularly in progressive forms of MS where current immunotherapies fail to prevent neurodegeneration.

3.8 THERAPEUTIC APPLICATIONS AND STRATEGIES OF EVS IN NEURO-REGENERATION

EVs, particularly exosomes, are increasingly recognized as versatile therapeutic platforms for neurodegenerative disorders. Their nanoscale dimensions, intrinsic biocompatibility, and capacity to traverse biological barriers render them well suited as natural delivery vehicles for diverse therapeutic cargos. Several strategic applications are now being pursued:

A) **Cell-Free Alternative to Stem Cell Therapy**: As highlighted in preceding sections, exosomes recapitulate many of the beneficial effects of stem cell transplantation while circumventing risks associated with administering live cells, such as immune rejection, embolism, or tumorigenicity (Fayazi et al. 2021). Exosomes encapsulate the stem cell secretome, including growth factors, cytokines, and regulatory RNAs, thus conferring neurotrophic and immunomodulatory support without direct cellular engraftment. For instance, rather than transplanting mesenchymal stem cells into injured neural tissue, infusion of MSC-derived exosomes can achieve similar outcomes by promoting neuronal survival and modulating inflammatory pathways (Heris et al. 2022). Furthermore, exosomes are inherently less immunogenic than their parent cells, as they lack MHC class II molecules and express complement regulatory proteins such as CD55 and CD59, enabling them to evade immune surveillance. Also, clinical investigations, including those in Alzheimer's disease, have thus far reported no significant adverse immune responses to allogeneic MSC-derived exosomes (Xie et al. 2023).

B) **Drug Delivery Vehicles:** Exosomes are being actively developed as biomimetic nanocarriers capable of delivering therapeutic molecules, including small drugs, siRNAs, microRNAs, and peptides across the BBB with high efficiency. This strategy leverages the body's natural intercellular communication system to achieve targeted drug delivery. In preclinical models of Parkinson's disease, for example, exosomes loaded with the antioxidant enzyme catalase, administered intravenously, successfully traversed the BBB, and released their cargo into the brain, attenuating neuroinflammation and oxidative stress within the nigrostriatal pathway and

TABLE 3.3
Engineering Strategies for Therapeutic EVs

Modification Strategy	Methodology	Typical modification	Therapeutic objective	Disease model examples
Passive loading via producer-cell overexpression	Transfect parent MSCs	siRNA-BACE1, miR-124	Knock-down amyloid; neurogenesis	AD mouse, PD rat
Ex vivo electroporation	Electroporate isolated EVs	Antisense HTT, CRISPR-Cas9	Silence mutant genes	HD mouse
Surface ligand display (genetic fusion)	Fuse RVG peptide to Lamp2b	RVG-EVs target neurons	Neuron-specific delivery	Tau-transgenic AD mouse
Chemical surface click-chemistry	Conjugate PEG-RGD post-isolation	αvβ3-targeted EVs	Home to ischemic endothelium	Stroke mouse
Enzyme encapsulation	Load catalase with saponin	Catalase-EVs	Reduce oxidative burden	PD mouse (MPTP)

providing significant neuroprotection (Haney et al. 2015). Similarly, in Alzheimer's disease, exosomes engineered to carry siRNA against β-secretase (BACE1) effectively reduced amyloid production in mouse models (Alvarez-Erviti et al. 2011).

Beyond cargo encapsulation, exosomes exhibit an inherent tropism for injured or inflamed tissues, enabling preferential accumulation at sites of neuropathological damage (Perets et al. 2019). Bioengineering approaches further enhance this targeting capacity, for instance, exosomes functionalized with the rabies virus glycoprotein (RVG) peptide display markedly improved uptake by neurons, thereby amplifying therapeutic efficacy in models of Alzheimer's and Parkinson's disease. Such advances in EV engineering broaden the therapeutic spectrum, opening avenues to deliver otherwise impermeable macromolecules, including antisense oligonucleotides for ALS or CRISPR–Cas9 constructs for gene editing directly to the central nervous system. Table 3.3 summarizes the principal exosome-engineering strategies that are currently under investigation.

C) **Enhancing Endogenous Repair:** EV-based therapies not only deliver exogenous trophic factors but also stimulate intrinsic repair mechanisms within the central nervous system. Several preclinical studies illustrate this dual role, exosomes have been shown to promote neurogenesis in Alzheimer's disease, facilitate axonal sprouting following stroke (Xin et al. 2013), and enhance remyelination in multiple sclerosis by activating oligodendrocyte precursor cells. The regenerative potential of EVs can be tailored by selecting specific donor cell types or by genetic engineering to modify their molecular cargo. For example, neural stem cell–derived exosomes inherently contain pro-neurogenic factors that drive precursor proliferation and neuronal differentiation, making them attractive candidates for conditions such as Huntington's disease or chronic neurodegeneration where neuronal loss is profound.

Moreover, combinatorial approaches integrate EVs with biomaterial scaffolds to provide both sustained release and structural support for tissue regeneration. In spinal cord injury models, a fibrin gel embedded with MSC-derived exosomes enhanced axonal regrowth and functional recovery by delivering vesicular cargo in a controlled manner while simultaneously offering a substrate for repair (He et al. 2022). Such bioengineered strategies could be extended to demyelinating lesions in MS or traumatic CNS injuries, providing a synergistic platform that couples immunomodulation, neuroprotection, and structural regeneration.

D) **Biomarkers and Diagnostics:** EVs present in patient biofluids are investigated as biomarkers for neurodegenerative diseases. Neuron- and glia-derived exosomes are isolated from blood or cerebrospinal fluid (CSF) and are analyzed for disease-associated proteins, such as phosphorylated tau, α-synuclein, or mutant huntingtin, as well as disease-relevant RNAs (Fayazi et al. 2021). These vesicles provide a minimally invasive window into central nervous system pathology. For example, plasma-derived neuronal exosomes enriched in tau and Aβ have demonstrated high sensitivity and specificity in distinguishing Alzheimer's disease patients at early stages. In Parkinson's disease, exosomal microRNAs are under active study as predictors of disease progression (Heris et al. 2022).

Although distinct from therapeutic applications, EV-based diagnostics plays a critical complementary role. By enabling patient stratification, longitudinal tracking of disease activity, and real-time monitoring of treatment response, EV biomarkers optimize clinical decision-making. Looking forward, theranostic strategies may integrate these functions, with engineered exosomes simultaneously delivering therapeutic cargo and carrying imaging tracers or molecular reporters to track biodistribution and efficacy.

3.9 COMBINING EV THERAPY WITH EXISTING TREATMENTS

EVs also hold potential as adjunctive agents to enhance the efficacy of existing pharmacotherapies. In multiple sclerosis, for example, exosome-based interventions designed to induce immune tolerance could be paired with remyelinating drugs to achieve synergistic benefits. In Alzheimer's disease, exosomes carrying anti-inflammatory cargo complement amyloid-targeting antibodies by mitigating treatment-associated neuroinflammation. Owing to their multimodal mechanisms of action, EVs serve as a therapeutic backbone across diverse strategies, enabling dose reduction of conventional drugs while minimizing adverse effects.

3.10 TRANSLATIONAL CHALLENGES AND FUTURE DIRECTIONS

While the therapeutic promise of EVs and exosomes in neuroregeneration is compelling, translation from bench to bedside requires overcoming significant challenges. Unlike conventional small-molecule drugs, exosomes are complex biologics, and their clinical development requires rigorous standardization. One major hurdle is the absence of consensus on dosing strategies. Current studies quantify EV doses using disparate metrics-protein concentration, particle counts, or vesicle number, making cross-study comparisons difficult (Quan et al. 2025). In murine models, effective doses have ranged widely, from 10^9 to 10^{11} particles per injection, yet the appropriate scaling to human applications remains uncertain. For example, an ongoing Alzheimer's disease trial has employed intranasal delivery of ~4×10^8 particles twice weekly (Xie et al. 2023), but whether this regimen is optimal is unknown. Establishing dose–response relationships through systematic preclinical titration and early-phase clinical trials will be essential.

Another critical challenge lies in the characterization and standardization of EV preparations. EV populations are heterogeneous, comprising exosomes, microvesicles, and other vesicle subtypes with potentially distinct biological activities. The ultracentrifugation and size-exclusion chromatography are most widely used EV isolation method further variability in yield, purity, and vesicle composition. To ensure reproducibility and potency, validated quality control parameters must be applied, including particle size distribution, surface marker profiling, and purity assessments. Regulatory agencies are likely to require such criteria, analogous to lot-release standards for biologic therapeutics. Without harmonized protocols for manufacturing and characterization, clinical translation of EV-based therapies will remain constrained.

Also, scaling up EV production for clinical use remains a formidable challenge. Most EVs are harvested from conditioned media of cultured cells, yet the long-term treatment of chronic neurodegenerative diseases will likely require repeated dosing over extended periods, necessitating the generation of vast quantities of vesicles. For example, in an Alzheimer's disease trial, patients received 24 intranasal doses over 12 weeks (Xie et al. 2023). The extrapolation to larger trials or routine clinical application implies the need to manufacture tens of trillions of exosomes under good manufacturing practice (GMP) conditions. However, fulfilling this demand requires optimization of cell culture platforms, such as bioreactors incorporating three-dimensional MSC cultures, and the development of genetically engineered producer cell lines with enhanced exosome output. Immortalized MSC lines or alternative stable producer cells are also being investigated as scalable EV biofactories.

Further, downstream processing represents an additional bottleneck. The purification methods must efficiently scale to large volumes while ensuring high purity and reproducibility. Presently, the ultrafiltration, tangential flow filtration, and advanced size-exclusion chromatography are being adapted for clinical-grade EV isolation, while commercial kits are under development for large-scale applications. However, rigorous quality controls are essential, as coisolation of contaminating proteins, nucleic acids, or viral particles could compromise both safety and efficacy. Unlike small-molecule drugs or even monoclonal antibodies, which are chemically or structurally well defined, EV therapeutics are characterized by complex biological activity arising from heterogeneous cargo. This makes standardization and regulatory approval particularly challenging, underscoring the need for continued innovation in bioprocess engineering and analytical characterization to achieve consistent, scalable EV manufacturing.

Although exosomes possess an inherent ability to traverse the BBB, the efficiency of targeting specific neural cell types or regions remains a critical limitation. Following peripheral administration, such as intravenous delivery, a substantial fraction of EVs is sequestered by the mononuclear phagocyte system in the liver and spleen, thereby reducing CNS bioavailability. Several strategies have been proposed to overcome this challenge. Intranasal administration, which exploits direct access to the CNS via olfactory pathways, has been implemented in the Alzheimer's disease clinical trial (Xie et al. 2023). Intra-arterial injections have shown enhanced brain accumulation in stroke models, while surface engineering of exosomes with neuron-specific peptides or antibodies significantly increase neuronal uptake relative to other cell types (Choi et al. 2022). Focused ultrasound combined with microbubbles offers another innovative approach, enabling transient and localized BBB opening to facilitate targeted EV entry, for example, to the motor cortex in ALS or the hippocampus in Alzheimer's disease.

Pharmacokinetics present an additional hurdle, as EVs are rapidly cleared from circulation, with half-live measured in minutes to hours. Consequently, repeated dosing regimens are likely required to maintain therapeutic concentrations within the CNS. Indeed, the Alzheimer's trial suggested that sustained dosing was necessary to preserve cognitive benefits (Xie et al. 2023). Defining optimal administration routes, dosing frequencies, and delivery-enhancing strategies will be essential for the successful translation of EV therapies, necessitating detailed pharmacokinetic and pharmacodynamic studies.

3.11 FUTURE DIRECTIONS

Despite existing challenges, the future of EVs in neuroregeneration is highly promising. Ongoing research is increasingly centered on engineered exosomes with tailored cargo, either through genetic modification of parent cells or by postisolation loading of therapeutic molecules, to maximize disease-specific efficacy. Such strategies include modifying EVs with targeting ligands to enhance cell-type or region-specific delivery, and developing multifunctional vesicles capable of codelivering complementary therapeutics, for instance, an siRNA to silence pathogenic genes together with a neuroprotective protein to support neuronal survival. Also, future therapeutic strategies may also involve synergistic combinations of EVs with complementary interventions such as neurorehabilitation, electrical stimulation to enhance EV distribution, or concurrent pharmacological agents. Given the programmable nature of extracellular vesicle cargo, vesicles are engineered to align with an individual's molecular pathology. For example, selective incorporation of microRNAs that are demonstrably deficient in a patient permits targeted correction of dysregulated gene networks underlying disease progression. In parallel, identifying alternative EV sources, including induced pluripotent stem cell–derived neurons or immune cell–derived vesicles, such as tolerogenic dendritic cell exosomes holds promise. The latter could be particularly valuable in multiple sclerosis, where vesicles carrying disease-relevant antigens may induce targeted immune tolerance.

3.12 CONCLUSION

This chapter examines EVs, particularly those derived from stem cells exosomes as emerging therapeutic tools for neurodegenerative diseases. EVs possess the unique ability to traverse the BBB and influence critical processes such as neuroinflammation, neuronal survival, synaptic plasticity, and myelination. Preclinical studies have demonstrated that EVs confer neuroprotective and regenerative benefits in models of Alzheimer's disease, Parkinson's disease, ALS, Huntington's disease, and multiple sclerosis, largely via delivery of neurotrophic factors, anti-inflammatory cytokines, and disease-modifying RNAs. Also, EVs can be engineered to transport therapeutic molecules, including small drugs, siRNAs, and gene-silencing constructs, positioning them as versatile delivery platforms for targeted interventions within the central nervous system. Early clinical evidence, most notably in Alzheimer's disease, supports their safety and suggests cognitive benefit, reinforcing their translational promise. At the same time, key challenges must be addressed for clinical implementation, including optimization of large-scale manufacturing, dosing regimens, delivery efficiency, and regulatory standardization. Taken together, the evidence reviewed underscores the potential of EVs as a cell-free, multimodal therapeutic strategy capable of both modulating disease mechanisms and promoting neural repair. With continued refinement of engineering approaches and resolution of translational hurdles, EV-based therapies hold considerable promise for advancing neuroregeneration and disease modification in currently intractable neurological disorders.

REFERENCES

Ananbeh H, Vodicka P, Kupcova Skalnikova H. Emerging roles of exosomes in Huntington's disease. *Int J Mol Sci*. 2021;22(8):4085. doi:10.3390/ijms22084085

Alvarez-Erviti L, Seow Y, Yin H, Betts C, Lakhal S, Wood MJ. Delivery of siRNA to the mouse brain by systemic injection of targeted exosomes. *Nat Biotechnol*. 2011;29(4):341–5. doi:10.1038/nbt.1807

Basso M, Pozzi S, Tortarolo M, Fiordaliso F, Bisighini C, Pasetto L, Spaltro G, Lidonnici D, Gensano F, Battaglia E, Bendotti C, Bonetto V. Mutant copper-zinc superoxide dismutase (SOD1) induces protein secretion pathway alterations and exosome release in astrocytes: Implications for disease spreading and motor neuron pathology in amyotrophic lateral sclerosis. *J Biol Chem*. 2013 May 31;288(22):15699–711. doi: 10.1074/jbc.M112.425066

Bonafede R, Mariotti R. ALS pathogenesis and therapeutic approaches: The role of mesenchymal stem cells and extracellular vesicles. *Front Cell Neurosci*. 2017;11:80. doi:10.3389/fncel.2017.00080

Bonafede R, Scambi I, Peroni D, Potrich V, Boschi F, Benati D, Bonetti B, Mariotti R. Exosome derived from murine adipose-derived stromal cells: Neuroprotective effect on in vitro model of amyotrophic lateral sclerosis. *Exp Cell Res*. 2016;340(1):150–8. doi:10.1016/j.yexcr.2015.12.009

Bonafede R, Turano E, Scambi I, Busato A, Bontempi P, Virla F, Schiaffino L, Marzola P, Bonetti B, Mariotti R. ASC-exosomes ameliorate the disease progression in SOD1(G93A) murine model underlining their potential therapeutic use in human ALS. *Int J Mol Sci*. 2020;21(10):3651. doi:10.3390/ijms21103651

Choi H, Choi K, Kim D-H, Oh B-K, Yim H, Jo S, Choi C. Strategies for targeted delivery of exosomes to the brain: Advantages and challenges. *Pharmaceutics*. 2022;14(3):672. doi:10.3390/pharmaceutics14030672

Didiot MC, Hall LM, Coles AH, Haraszti RA, Godinho BM, Chase K, Sapp E, Ly S, Alterman JF, Hassler MR, Echeverria D, Raj L, Morrissey DV, DiFiglia M, Aronin N, Khvorova A. Exosome-mediated delivery of hydrophobically modified siRNA for huntingtin mRNA silencing. *Mol Ther*. 2016;24(10):1836–47. doi:10.1038/mt.2016.126

Drago F, Lombardi M, Prada I, Gabrielli M, Joshi P, Cojoc D, Franck J, Fournier I, Vizioli J, Verderio C. ATP modifies the proteome of extracellular vesicles released by microglia and influences their action on astrocytes. *Front Pharmacol.* 2017;8:910. doi:10.3389/fphar.2017.00910

Fayazi N, Sheykhhasan M, Soleimani Asl S, Najafi R. Stem cell-derived exosomes: A new strategy of neurodegenerative disease treatment. *Mol Neurobiol.* 2021;58(7):3494–514. doi:10.1007/s12035-021-02324-x

Gowen A, Shahjin F, Chand S, Odegaard KE, Yelamanchili SV. Mesenchymal stem cell-derived extracellular vesicles: Challenges in clinical applications. *Front Cell Dev Biol.* 2020;8:149. doi:10.3389/fcell.2020.00149

Haney MJ, Klyachko NL, Zhao Y, Gupta R, Plotnikova EG, He Z, Patel T, Piroyan A, Sokolsky M, Kabanov AV, Batrakova EV. Exosomes as drug delivery vehicles for Parkinson's disease therapy. *J Control Release.* 2015;207:18–30. doi:10.1016/j.jconrel.2015.03.033. Erratum in: J Control Release. 2021 Nov 10;339:232–234. doi:10.1016/j.jconrel.2021.09.027

He X, Yang L, Dong K, Zhang F, Liu Y, Ma B, Chen Y, Hai J, Zhu R, Cheng L. Biocompatible exosome-modified fibrin gel accelerates the recovery of spinal cord injury by VGF-mediated oligodendrogenesis. *J Nanobiotechnology.* 2022;20(1):360. doi:10.1186/s12951-022-01541-3

Iguchi Y, Eid L, Parent M, Soucy G, Bareil C, Riku Y, Kawai K, Takagi S, Yoshida M, Katsuno M, Sobue G, Julien JP. Exosome secretion is a key pathway for clearance of pathological TDP-43. *Brain.* 2016;139(Pt 12):3187–201. doi:10.1093/brain/aww237

Jeon I, Cicchetti F, Cisbani G, et al. Human-to-mouse prion-like propagation of mutant huntingtin protein. *Acta Neuropathol.* 2016;132:577–92. doi:10.1007/s00401-016-1582-9

Kråkenes T, Sandvik CE, Ytterdal M, Gavasso S, Evjenth EC, Bø L, Kvistad CE. The therapeutic potential of exosomes from mesenchymal stem cells in multiple sclerosis. *Int J Mol Sci.* 2024;25(19):10292. doi:10.3390/ijms251910292

Manu MS, Hohjoh H, Yamamura T. Extracellular vesicles as pro- and anti-inflammatory mediators, biomarkers and potential therapeutic agents in multiple sclerosis. *Aging Dis.* 2021;12(6):1451–61. doi:10.14336/AD.2021.0513

Heris RM, Shirvaliloo M, Abbaspour-Aghdam S, Hazrati A, Shariati A, Youshanlouei HR, Niaragh FJ, Valizadeh H, Ahmadi M. The potential use of mesenchymal stem cells and their exosomes in Parkinson's disease treatment. *Stem Cell Res Ther.* 2022;13:371. doi:10.1186/s13287-022-03050-4

Palanisamy CP, Pei J, Alugoju P, Anthikapalli NVA, Jayaraman S, Veeraraghavan VP, Gopathy S, Roy JR, Janaki CS, Thalamati D, Mironescu M, Luo Q, Miao Y, Chai Y, Long Q. New strategies of neurodegenerative disease treatment with extracellular vesicles (EVs) derived from mesenchymal stem cells (MSCs). *Theranostics.* 2023;13(12) 4138–65. doi:10.7150/thno.83066

Petrou P, Gothelf Y, Argov Z, et al. Safety and clinical effects of mesenchymal stem cells secreting neurotrophic factor transplantation in patients with amyotrophic lateral sclerosis: Results of phase 1/2 and 2a clinical trials. *JAMA Neurol.* 2016;73(3):337–44. doi:10.1001/jamaneurol.2015.4321

Perets N, Betzer O, Shapira R, Brenstein S, Angel A, Sadan T, Ashery U, Popovtzer R, Offen D. Golden exosomes selectively target brain pathologies in neurodegenerative and neurodevelopmental disorders. *Nano Lett.* 2019;19(6):3422–31. doi:10.1021/acs.nanolett.8b04148

Quan J, Liu Q, et al. Mesenchymal stem cell exosome therapy: Current research status in the treatment of neurodegenerative diseases and the possibility of reversing normal brain aging. *Stem Cell Res Ther.* 2025;16:76. doi:10.1186/s13287-025-04160-5

Rajan T, Giacoppo S, Diomede F, et al. The secretome of periodontal ligament stem cells from MS patients protects against EAE. *Sci Rep.* 2016;6:38743.

Riazifar M, Mohammadi MR, Pone EJ, Yeri A, Lässer C, Segaliny AI, McIntyre LL, Shelke GV, Hutchins E, Hamamoto A, Calle EN, Crescitelli R, Liao W, Pham V, Yin Y, Jayaraman J, Lakey JRT, Walsh CM, Van Keuren-Jensen K, Lotvall J, Zhao W. Stem cell-derived exosomes as nanotherapeutics for autoimmune and neurodegenerative disorders. *ACS Nano.* 2019;13(6):6670–88. doi:10.1021/acsnano.9b01004

Ruan Z, Pathak D, Venkatesan Kalavai S, Yoshii-Kitahara A, Muraoka S, Bhatt N, Takamatsu-Yukawa K, Hu J, Wang Y, Hersh S, Ericsson M, Gorantla S, Gendelman HE, Kayed R, Ikezu S, Luebke JI, Ikezu T. Alzheimer's disease brain-derived extracellular vesicles spread tau pathology in interneurons. *Brain.* 2021 Feb 12;144(1):288–309. doi:10.1093/brain/awaa376. Erratum in: *Brain.* 2021 May 7;144(4):e42. doi:10.1093/brain/awaa452

Wang H, Liu Y, Li J, et al. Tail-vein injection of MSC-derived small extracellular vesicles facilitates the restoration of hippocampal neuronal morphology and function in APP/PS1 mice. *Cell Death Discov.* 2021;**7**:230. doi:10.1038/s41420-021-00620-y

Wang K, Li Y, Ren C, Wang Y, He W, Jiang Y. Extracellular vesicles as an innovative treatment strategy for amyotrophic lateral sclerosis. *Front Cell Dev Biol.* 2021;9:754630. doi:10.3389/fcell.2021.754630

Xie X, Song Q, Dai C, et al. 2023. Clinical safety and efficacy of allogenic human adipose mesenchymal stromal cells-derived exosomes in patients with mild to moderate Alzheimer's disease: A phase I/II clinical trial. *Gen Psychiatry.* 36(5):e101143. doi:10.1136/gpsych-2023-101143

Xin H, Li Y, Cui Y, Yang JJ, Zhang ZG, Chopp M. Systemic administration of exosomes released from mesenchymal stromal cells promote functional recovery and neurovascular plasticity after stroke in rats. *J Cereb Blood Flow Metab.* 2013 Nov;33(11):1711–15. doi:10.1038/jcbfm.2013.152

Xun C, Deng H, Zhao J, Ge L, Hu Z. Mesenchymal stromal cell extracellular vesicles for multiple sclerosis in preclinical rodent models: A meta-analysis. *Front Immunol.* 2022;13:972247. doi:10.3389/fimmu.2022.972247

4 Role of Mesenchymal Stem Cells in Neurodegenerative Disorders

4.1 INTRODUCTION

Neurodegenerative disorders such as Alzheimer's disease (AD), Parkinson's disease (PD), amyotrophic lateral sclerosis (ALS), and multiple sclerosis (MS) are characterized by progressive loss of neurons or glial cells, leading to irreversible functional decline. These conditions pose enormous societal burdens and lack curative treatments, which has led to exploring regenerative strategies including stem cell therapies. Mesenchymal stem cells (MSCs), also known as mesenchymal stromal cells, have emerged as a promising strategy for neuro-regeneration due to their ease of isolation, multipotency, immunomodulatory properties, and secretion of neurotrophic factors (Caplan and Correa 2011). MSCs can be derived from adult tissues, including bone marrow, adipose tissue, umbilical cord, etc. and cultivated in culture for therapeutic use. Unlike neural stem cells or embryonic stem cells, MSCs do not inherently form neurons at high efficiency, but they can influence the neural microenvironment in beneficial ways (Uccelli, Laroni, and Freedman 2020). Preclinical studies over the past two decades have provided encouraging evidence that MSC transplantation can improve neuropathology and improve function in models of AD, PD, ALS, MS, and other neurodegenerative conditions (Shin et al. 2014; McCoy et al. 2008). These findings have prompted early-phase clinical trials to test MSC therapy in human patients. In this chapter, we review the current understanding of how MSCs act in neurodegenerative disorders and discuss the applications and outcomes of MSC-based therapies in major neurodegenerative diseases. The chapter also examines the limitations and challenges of MSC therapy and highlights future directions, including novel strategies to enhance the effectiveness of MSCs or their derivatives. The chapter reviews the role of MSCs in neurodegenerative disease modulation and their potential as next-generation treatments for these neurodegenerative disorders (Soares et al. 2022).

4.2 MECHANISMS OF MSC ACTION IN NEURODEGENERATIVE DISEASES

MSCs are thought to exert therapeutic effects in neurodegenerative disorders primarily through paracrine signaling and immunomodulation rather than direct replacement of lost neurons. Upon transplantation or delivery to a diseased nervous system, MSCs migrate to sites of injury or inflammation and secrete a broad array of growth factors, cytokines, and extracellular vesicles that influence the local environment (Caplan and Correa 2011; Phinney and Pittenger 2017; Figure 4.1). The key mechanisms of action of MSCs are summarized in Table 4.1.

DOI: 10.1201/9781003582687-4

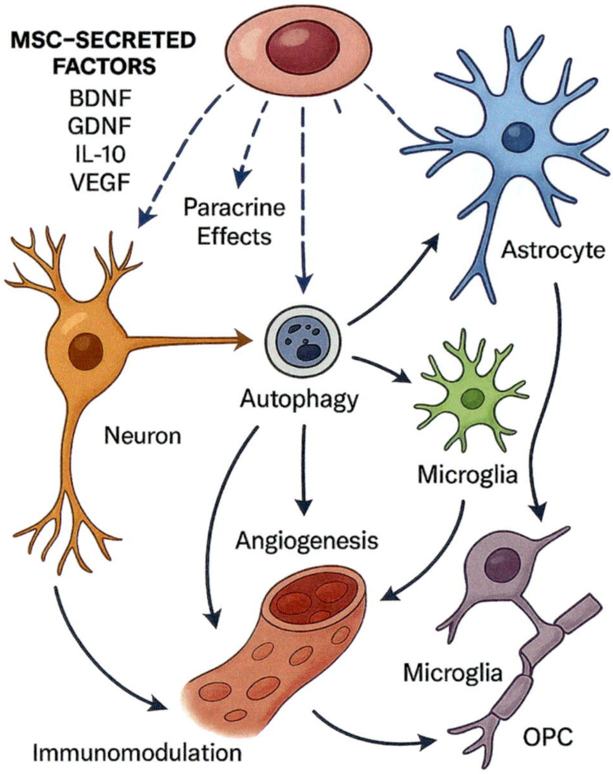

FIGURE 4.1 Multimodal Mechanisms of MSCs in Neurodegenerative Disease Repair.

TABLE 4.1
Mechanisms of MSC Action in Neurodegeneration

Mechanism	Description
Neurotrophic factor secretion	Release of BDNF, GDNF, NGF, VEGF to support neuronal survival and regeneration
Immunomodulation	Shift from pro-inflammatory to anti-inflammatory microglial and astrocyte phenotypes
Aggregate clearance	Enhancement of autophagy and phagocytosis to remove protein aggregates
Antioxidant effects	Secretion of enzymes and molecules reducing oxidative stress
Angiogenesis	Promotion of vascular repair via VEGF and other angiogenic factors
Stimulation of endogenous progenitors	Activation of neural stem cells and OPCs to promote repair

4.2.1 Secretion of Neurotrophic and Neuroprotective Factors

MSCs releases growth factors such as brain-derived neurotrophic factor (BDNF), glial cell line–derived neurotrophic factor (GDNF), nerve growth factor (NGF), and vascular endothelial growth factor (VEGF) that support the survival and regeneration of neurons (Figure 4.1). These factors promote neurogenesis, enhance synaptic connectivity, and protect neurons from apoptotic cell

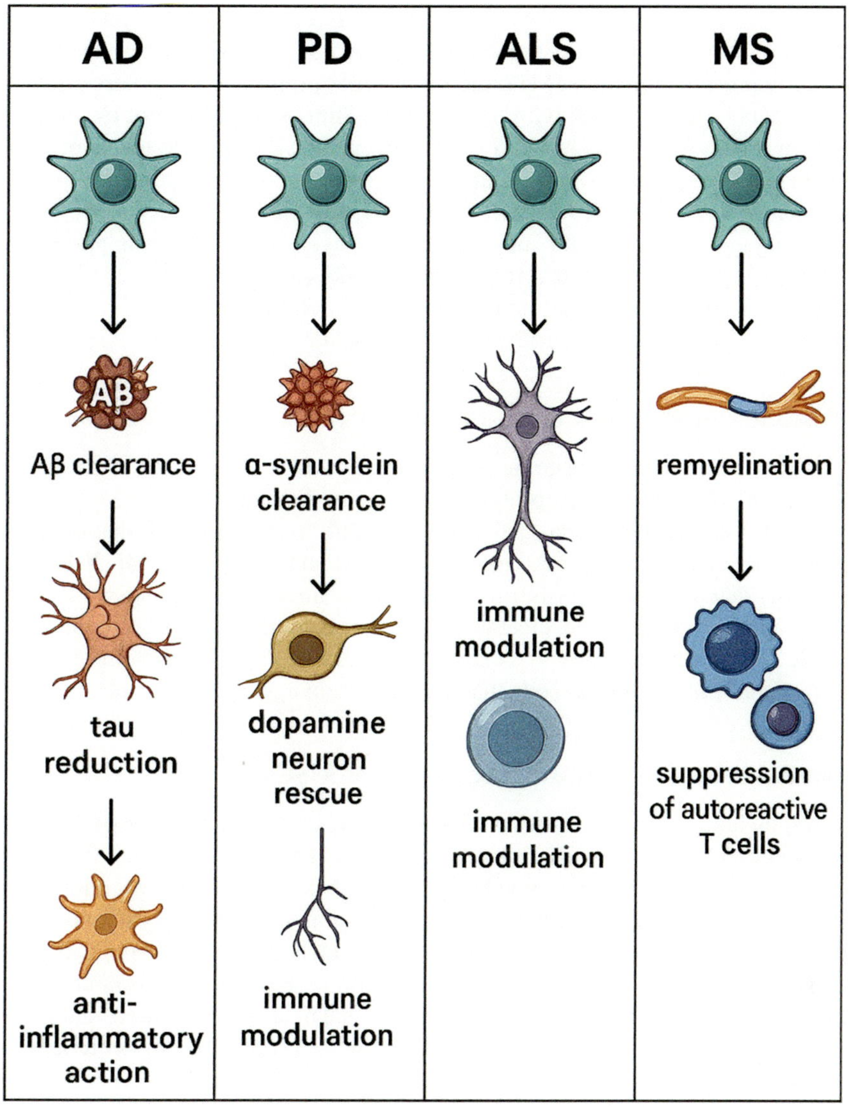

FIGURE 4.2 MSC Therapeutic Pathways Across Major Neurodegenerative Diseases (AD, PD, ALS, MS).

death. For example, MSC-derived BDNF and GDNF have been shown to enhance the survival of cortical and dopaminergic neurons in vitro and in vivo. Genetically engineered MSCs overexpressing GDNF or other trophic factors amplify these neuroprotective effects, as demonstrated in models of PD and ALS (Figure 4.2; Soares et al. 2022). The trophic support provided by MSCs can counteract neuronal atrophy and promote repair of injured circuits in neurodegenerative diseases.

4.2.2 Immunomodulation and Anti-Inflammatory Effects of MSCs

MSCs have a well-documented capacity to modulate immune responses, which is crucial given the role of neuroinflammation in most neurodegenerative disorders. MSCs interact with microglia, astrocytes, and infiltrating immune cells to shift the balance from a pro-inflammatory state to an anti-inflammatory or tissue-reparative state (Figure 4.1; Uccelli, Laroni, and Freedman 2020). They secrete anti-inflammatory cytokines, e.g. IL-10, TGF-β and express surface molecules that induce immune cell anergy or apoptosis. In the central nervous system (CNS), transplanted MSCs can inhibit classically pro-inflammatory (M1) microglia while promoting the polarization of microglia toward an M2 phenotype associated with tissue repair (Shin et al. 2014). For example, administration of MSCs in rodent models of AD and PD has been shown to reduce levels of tumor necrosis factor-alpha (TNF-α) and other pro-inflammatory mediators while concurrently upregulating anti-inflammatory interleukins such as IL-10 and IL-4. (McCoy et al. 2008). MSCs also enhance populations of regulatory T cells and suppress autoreactive T cells, which is particularly relevant in autoimmune neurodegenerative disorders such as MS (Karussis et al. 2010). Through these mechanisms, MSCs attenuate chronic neuroinflammation and promote a more permissive microenvironment conducive to neuronal survival and functional recovery.

4.2.3 Promotion of Protein Clearance and Neurotoxic Aggregate Removal

MSCs can facilitate the clearance of toxic protein aggregates in the context of proteinopathies like amyloid plaques in AD or misfolded proteins in ALS. Studies indicate that MSC-secreted factors stimulate phagocytosis and lysosomal activity in resident microglia/macrophages, which helps to remove extracellular deposits such as Aβ plaques in AD (Figure 4.1; Shin et al. 2014). For example, Shin et al. (2014) demonstrated that intravenous MSCs enhanced autophagy in the brains of AD model mice, leading to accelerated clearance of beta-amyloid and improved neuronal survival. Similarly, MSC treatment has been associated with reduced alpha-synuclein aggregation in PD models, likely via activation of glial cells that engulf protein aggregates (Yasuhara et al. 2017).

4.2.4 Antiapoptotic and Antioxidative Effects of MSCs

MSC secretions include antioxidants and antiapoptotic molecules that counteract the oxidative stress and programmed cell death pathways that are active in degenerating neurons (Figure 4.1). MSCs can reduce the production of reactive oxygen species and elevate endogenous antioxidant enzymes in injured neural tissue. They also secrete factors like stanniocalcin-1 and osteopontin that directly inhibit apoptosis in neurons and oligodendrocytes. In models of ALS and spinal cord injury, MSC administration was shown to downregulate proapoptotic caspases in motor neurons and decrease markers of oxidative damage, thereby preserving functional neurons (Boucherie and Hermans 2015). By protecting neurons from further damage, MSCs slow the progression of neurodegeneration.

4.2.5 Role of MSCs in Angiogenesis

Neurodegenerative disorders are frequently associated with cerebrovascular dysfunction and diminished cerebral perfusion. MSCs contribute to vascular repair and regeneration by secreting a repertoire of pro-angiogenic mediators, including VEGF, hepatocyte growth factor (HGF), and angiopoietin-1. These factors not only stimulate neovascularization but also enhance the stabilization and maturation of existing vascular networks, thereby improving cerebral microcirculation and supporting neuronal survival (Bronckaers et al. 2014). Enhanced angiogenesis augments cerebral perfusion in hypoxic or injured regions of the brain, thereby ensuring a more effective delivery of oxygen and essential nutrients while simultaneously promoting the clearance of metabolic byproducts and neurotoxic waste. This restoration of vascular function creates a supportive microenvironment that favors neuronal survival and functional recovery. In MS lesions and traumatic CNS injuries, MSCs have been shown to facilitate repair of the blood–brain barrier and enhance vascular density, effects that are closely associated with improved structural integrity and functional recovery of the affected tissue.

4.2.6 Role of MSCs in Stimulating Endogenous Stem and Progenitor Cells

MSC-derived trophic factors have the capacity to recruit and activate endogenous neural stem and progenitor cells, promoting their proliferation and differentiation into functional neural lineages. Preclinical evidence indicates that MSC transplantation enhances neurogenesis within the hippocampus and subventricular zone, particularly in experimental models of AD and ischemic stroke. (Blurton-Jones et al. 2009). Similarly, in demyelinating conditions, MSCs secrete signaling molecules that facilitate the maturation of oligodendrocyte progenitor cells (OPCs) and promote axonal remyelination (Jaramillo et al. 2013). Thus, MSC-based therapy, in part, exerts its effects by reactivating intrinsic regenerative mechanisms within the CNS.

4.2.7 Trophic Support for Glial Cells and Axons

In addition to supporting neuronal health, MSCs contribute to the maintenance of white matter integrity by providing trophic and immunomodulatory support to oligodendrocytes and astrocytes. Studies in MS models have shown that MSC-secreted factors like BDNFs foster oligodendrocyte survival and stimulate remyelination of demyelinated axons (Figure 4.2; Razavi et al. 2017). MSCs also produce extracellular matrix components and enzymes that remodel scar tissue. For example, they secrete matrix metalloproteinases that repair glial scars, possibly aiding axonal regeneration in spinal injury or chronic MS lesions (Qu and Zhang 2017).

4.2.8 Immunomodulatory and Paracrine Actions of MSCs

Notably, MSCs themselves are largely immuno-evasive. They express low levels of MHC class II and costimulatory molecules, allowing allogeneic MSCs to be used without immediate rejection (Aggarwal and Pittenger 2005). This property enables MSCs to persist transiently following transplantation and exert paracrine bystander effects that influence the host tissue environment. However, their persistence in the CNS is typically transient. Studies have shown that MSCs often do not permanently engraft or differentiate into neurons in significant numbers (Caplan and Correa 2011). Rather than exerting long-term engraftment effects, the principal influence of MSCs manifests within the initial weeks following transplantation through their paracrine secretory activity. Caplan's medicinal signaling cell paradigm underscores this notion, proposing that MSCs act as site-specific, biologically responsive drug repositories that release regulatory signals to activate and coordinate endogenous repair processes. (Caplan and Correa 2011).

4.3 MSC THERAPY IN NEURODEGENERATIVE DISORDERS

MSC therapy has emerged as a promising regenerative and immunomodulatory approach for neurodegenerative disorders. They support neural repair through anti-inflammatory effects, trophic factor secretion, modulation of oxidative stress, and protection of damaged neuronal networks. In neurodegenerative diseases, where progressive neuronal loss and chronic inflammation drive clinical decline, MSC-based interventions offer a potential strategy to slow disease progression and improve functional outcomes. The following sections discuss the role and applications of MSC therapy in specific neurodegenerative disorders.

4.3.1 MSC Therapy in AD

AD is a chronic neurodegenerative dementia marked by extracellular amyloid-β (Aβ) plaque deposition, intraneuronal tau protein tangles, synaptic loss, and widespread neuronal death in the cortex and hippocampus (Table 4.2). Neuroinflammation mediated by microglia and astrocytes is increasingly recognized as a contributor to AD progression. Given these complex pathologies, MSCs

TABLE 4.2
MSC Applications across Major Neurodegenerative Disorders

Disease	Pathological Target	MSC Therapeutic Mechanism	Delivery Methods
Alzheimer's disease	Aβ plaques, tau pathology	Aβ clearance, anti-inflammatory, neurotrophic support	Intracerebral, intravenous
Parkinson's disease	Dopaminergic neuron loss, α-synuclein	Neuroprotection, α-synuclein clearance, anti-inflammation	Intrastriatal, intranasal, intravenous
ALS	Motor neuron degeneration	Trophic factor delivery, immune modulation	Intrathecal, intramuscular, intravenous
Multiple sclerosis	Demyelination, autoimmunity	Immunoregulation, remyelination support	Intrathecal, intravenous

have been investigated as a multi-target approach for AD, aiming to provide neurotrophic support, enhance Aβ clearance, and modulate inflammation (Kim et al. 2020; Figure 4.2). Numerous animal studies have demonstrated beneficial effects of MSC administration in AD models. In transgenic mouse models of AD, both intracerebral and systemic delivery of MSCs have resulted in reduced Aβ plaque formation and improved cognitive function. For example, human bone marrow–derived MSCs injected into APP/PS1 transgenic AD mice migrated to the brain and significantly decreased Aβ levels, which was accompanied by improvements in memory tests. Mechanistically, MSCs appear to facilitate Aβ clearance by activating the brain's innate phagocytic cells. Microglial activation and autophagy are central to this process, as transplanted MSCs have been shown to recruit microglia to amyloid plaques and promote their polarization toward a phenotype capable of efficiently engulfing and degrading amyloid-beta (Aβ). (Shin et al. 2014). Shin et al. (2014) reported that MSC treatment enhanced autophagosome formation and lysosomal activity in microglia, accelerating the removal of Aβ deposits and leading to synaptic preservation and increased neuronal survival in AD mice. Similarly, MSC-secreted exosomes carrying enzymatic and miRNA have been found to promote proteostatic mechanisms that remove misfolded proteins (Candelario et al. 2013). In addition to reducing amyloid plaque burden, MSCs have also been shown to attenuate tau pathology in certain studies; for example, intravenous administration of MSCs led to decrease in the levels of hyperphosphorylated tau and conferred neuroprotection in tau transgenic models (Kim et al. 2015).

The cognitive benefits observed in MSC-treated AD models are significant. As treated animals show improved performance in maze tests, object recognition, and avoidance tasks relative to untreated AD controls (Blurton-Jones et al. 2009). These functional improvements correlate with neuroprotective changes such as increased synaptic density and reduced neuronal apoptosis in the hippocampus. For instance, MSCs have been shown to upregulate synaptic support proteins and stimulate neurogenesis within the dentate gyrus of AD mouse models, effects that may underlie observed improvements in learning and memory performance (Blurton-Jones et al. 2009). Crucially, the immunomodulatory effects of MSCs contribute to the attenuation of neuroinflammation in AD. Multiple studies have reported reduced levels of pro-inflammatory cytokines such as TNF-α and IL-1β, alongside elevated expression of the anti-inflammatory cytokine IL-10 in the brains of MSC-treated AD models (Neves et al. 2021). By shifting microglia and astrocytes to a neuroprotective phenotype, MSCs mitigate the chronic inflammatory damage that accelerates progression of AD.

Different sources of MSCs have been tested in AD models with broadly similar outcomes. Adipose-derived MSCs, for example, have shown the ability to migrate to sites of Aβ deposition and secrete neprilysin, an Aβ-degrading enzyme, thus lowering levels of plaque (Wei et al. 2018). Umbilical cord MSCs, which are fetal in origin and highly secretory, have demonstrated potent reduction of amyloid burden and reversal of cognitive deficits in AD mice (Boutajangout et al. 2017). These observations indicate that the therapeutic efficacy of MSCs in AD is robust and not

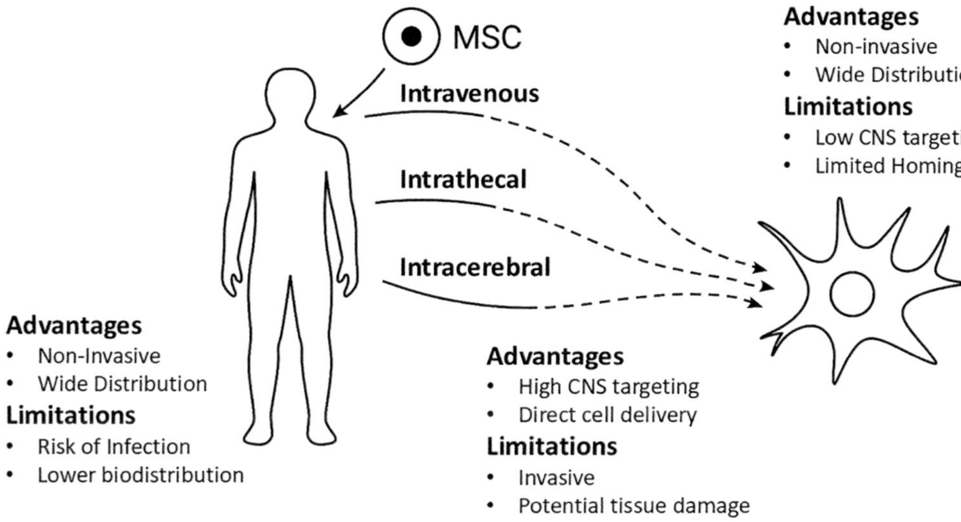

FIGURE 4.4 Routes of MSC Administration and CNS Targeting Efficiency.

confined to a specific tissue source. Nonetheless, the route of administration remains a critical determinant of effectiveness, i.e., direct intracerebral or intraventricular delivery ensures targeted migration of MSCs into the brain parenchyma, whereas intravenous administration depends primarily on the capacity of MSC-derived secretome and extracellular vesicles, such as exosomes, to traverse the blood–brain barrier and exert neuroprotective effects (Figure 4.4). Some studies have achieved success with peripheral MSC delivery, likely because the secretome of MSCs enters circulation and influence the brain even if few cells migrate across the blood–brain barrier (Alvites et al. 2022).

Although early-phase clinical investigations of MSC-based therapies in AD are limited, yet their number is steadily increasing as accumulating safety data provide support for broader clinical application. An initial open-label trial in 2014 in South Korea provided intracerebral (intraventricular) injections of autologous adipose-derived MSCs to AD patients via an Ommaya reservoir. This Phase I trial indicated that the procedure was technically feasible and safe as out of nine patients with moderate AD who received three MSC doses into the ventricular system, none experienced serious adverse events, and some showed stabilization of cognitive decline over a 1-year follow-up (Kim et al. 2015). Encouraged by these results, a Phase I/II trial by Kim et al. (2021) used allogeneic human umbilical cord blood–derived MSCs delivered intraventricularly in patients with mild-to-moderate AD. They reported that the multiple injections were well tolerated and relatively safe, with transient mild headaches being the main side effect (Kim et al. 2021). Although not designed to assess efficacy conclusively, the study noted trends of reduced neuroinflammation via PET imaging and slight cognitive improvements in some patients. The intraventricular route is thought to enable MSCs to directly interact with brain parenchyma and cerebrospinal fluid, thereby potentially enhancing their therapeutic efficacy in AD (Kim et al. 2021).

Intravenous MSC administration has also been tested in AD. A notable recent trial is the Phase IIa CLEAR MIND study evaluating *Lomecel-B* (allogeneic bone marrow MSCs) in mild AD. In this randomized placebo-controlled trial (36 treated vs. 12 placebo), patients received MSC infusions monthly for 4 months. The results, published in 2025, indicated that the MSC therapy met safety endpoints and showed signs of efficacy on cognitive measures and brain atrophy rates (Rash et al. 2025). Treated patients had a slower decline on a composite Alzheimer's score combining cognition and function, and MRI analysis suggested reduced loss of brain volume compared to placebo (Hare et al. 2025). No cases of amyloid-related imaging abnormalities (ARIA) were seen, in contrast to amyloid antibody therapies, highlighting a safety advantage (Rash et al. 2025). Despite its exploratory design, the trial demonstrated promising results that have garnered significant attention and led to the FDA's Fast Track designation of the MSC-based therapeutic candidate. Another trial by a biopharmaceutical company (Longeveron) used MSCs in AD also reported improved quality-of-life outcomes and maintained cognitive function in treated individuals over 9 months (Brody et al. 2023).

It should be noted that, to date, no MSC therapy for AD has received regulatory approval and all human studies remain in experimental phases. The sample sizes have been small, and some have lacked placebo controls, making it hard to distinguish treatment effects from disease variability or placebo effects. Nonetheless, the consistent findings demonstrated safety and preliminary indications of cognitive stabilization are encouraging. Ongoing studying are focused on improving protocols, such as comparing intrathecal versus intravenous delivery of MSCs in dementia or using repeated dosing schedules. For instance, an EU consortium is conducting a trial of adipose-derived MSCs delivered intravenously in early AD (NCT04040348), and a separate trial is examining placenta-derived MSC-like cells in AD (NCT02899091). These studies will provide more data on efficacy endpoints like memory testing, daily function, and biomarker changes (Cummings et al. 2022).

The translational research so far suggests that MSCs in AD patients act through similar mechanisms as observed in animals. Cerebrospinal fluid analyses from AD patients treated with MSCs have shown increased levels of anti-inflammatory cytokines and trophic factors (Kim et al. 2015). Some patients exhibited a reduction in phosphorylated tau and inflammatory markers in CSF after MSC therapy. Neuroimaging has hinted at metabolic improvements, the FDG-PET scans in a small MSC-treated AD cohort revealed elevated glucose uptake in certain brain regions, possibly reflecting restored neuronal activity (Vaquero et al. 2017). These findings support the concept that MSCs confer neuroprotective and pro-regenerative benefits in the AD brain by mitigating inflammatory damage, enhancing neuronal viability, and potentially facilitating the clearance of neurotoxic protein aggregates.

Thus, MSC-based therapy for AD has progressed from promising animal models to early human trials that demonstrate safety and potential cognitive benefits. By secreting a broad range of neurotrophic and immunomodulatory factors (Figure 4.3), MSCs target multiple facets of AD pathology, including plaque clearance, inflammation, and neural network dysfunction. While larger controlled

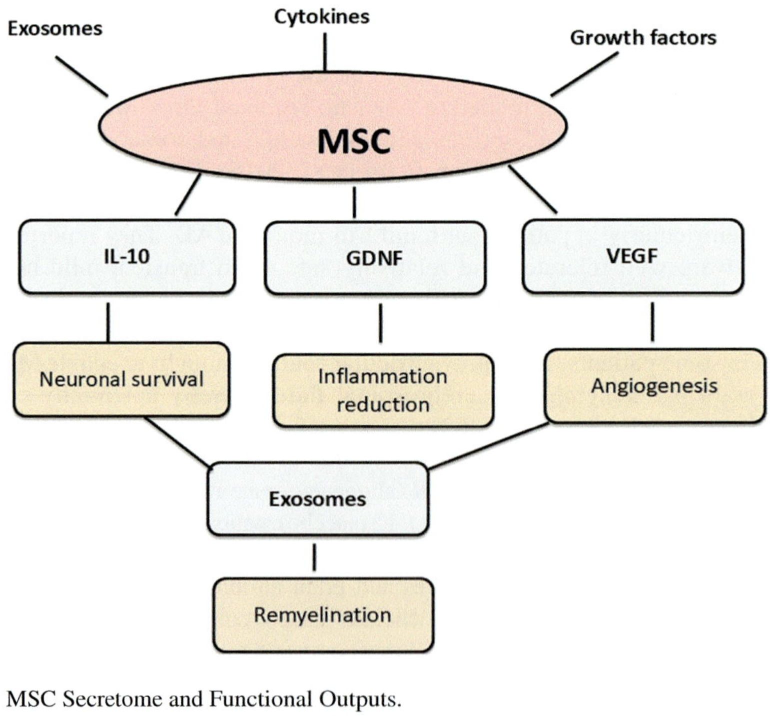

FIGURE 4.3 MSC Secretome and Functional Outputs.

trials are needed to confirm efficacy, MSCs offer a novel biological approach that contrasts with conventional single-target drugs. Ongoing and future studies will be critical in determining whether MSC-based interventions can translate into clinically meaningful attenuation of neurodegeneration or improvement in cognitive and functional outcomes in AD. If so, MSCs could become part of combination strategies to modify the course of AD, especially at early stages of the disease (Kim et al. 2020).

4.3.2 MSC Therapy in PD

PD is a movement disorder caused primarily by degeneration of dopaminergic neurons in the substantia nigra pars compacta, leading to dopamine deficiency in the striatum. Pathologically, PD features intracellular aggregates of α-synuclein (Lewy bodies) and extensive microglial activation in affected brain regions (Table 4.2). The hallmarks of PD are resting tremors, bradykinesia, rigidity, and gait disturbance. The current treatments are mainly symptomatic e.g. L-DOPA without halting disease progression. MSC therapy is being explored in PD with the aim of neuroprotection for dopaminergic neurons, reduction of α-synuclein toxicity, and possibly restoration of striatal dopaminergic function (Figure 4.2; Mendes et al. 2019).

The transplantation of MSC in animal models of PD has shown neuroprotective and anti-inflammatory effects that translate into improved motor function. A commonly used preclinical model of PD involves induction of nigrostriatal pathway lesions in rodents using 6-hydroxydopamine (6-OHDA), which selectively replicates dopaminergic neuronal degeneration. In this model, the transplantation of MSCs derived from bone marrow or adipose tissue confers neuroprotective effects by attenuating dopaminergic neuron loss. These cellular interventions are further associated with improved functional outcomes, as noted by a reduction in amphetamine-induced rotational behavior, indicative of preserved striatal dopaminergic activity (McCoy et al. 2008). In the same study, it was shown that the MSCs delivered to the striatum led to attenuation of microglial activation around the site of injury and upregulated GDNF expression, correlating with about 50% higher survival of tyrosine hydroxylase–positive neurons in the substantia nigra compared to controls. Similarly, Berg et al. (2015) found that adipose-derived MSCs transplanted in a PD rat model secreted neurotrophic factors that safeguarded nigral neurons and improved motor deficits. Notably, MSCs have been implicated not only in neuroprotection but also in regenerative processes. Experimental studies in PD models have demonstrated that MSC promotes neurite outgrowth and facilitates regeneration of residual dopaminergic neurons, thereby contributing to circuit remodeling and repair. These findings suggest that MSCs create a growth-permissive neural microenvironment conducive to structural plasticity and functional recovery (d'Angelo, Cimini, and Castelli 2020).

Beyond their neuroprotective role, MSCs in PD models have also been shown to contribute to neurochemical restoration. Preclinical studies report that MSC-treated animals exhibit elevated striatal dopamine levels and increased concentrations of their metabolites compared with untreated counterparts (Venkataramana et al. 2010). These effects are thought to arise either from MSC-induced preservation and functional support of surviving host dopaminergic neurons and to a lesser extent from the differentiation of transplanted MSCs into dopamine-producing phenotypes. Authentic neural differentiation of MSCs within in vivo environment remains a rare phenomenon, although experimental evidence indicates that under defined conditions MSCs may acquire neuron-like phenotypes. More consistently, their therapeutic efficacy appears to derive from intercellular communication processes, including the transfer of mitochondria or extracellular vesicles to metabolically compromised neurons, thereby augmenting bioenergetic capacity and resilience (Tseng et al. 2021). Notably, in vitro studies have documented mitochondrial transfer from MSCs to injured dopaminergic neurons, resulting in enhanced neuronal survival and restoration of functional activity (Hayakawa et al. 2016). In models of synucleinopathy, MSC therapy has shown potential to mitigate the toxic accumulation of α-synuclein. For example, in transgenic mice overexpressing human α-synuclein, intravenous MSC administration led to reduced oligomeric α-synuclein in the brain and preservation of neurons in the substantia nigra (Wang et al. 2020). The proposed mechanism is that MSCs release exosomes containing enzymes like neprilysin or IDE and regulatory microRNAs that enhance the clearance of α-synuclein aggregates by neurons and glia. Additionally, MSCs' anti-inflammatory actions counteract the self-perpetuating cycle of microglial activation triggered by α-synuclein. MSC-treated PD models consistently exhibit reduced levels of pro-inflammatory mediators and a phenotypic shift in microglia toward a neuroprotective state, as neuroinflammation is thought to increase dopaminergic neuronal loss in PD. In certain studies, these benefits approach those achieved with direct GDNF therapy, highlighting the robust trophic support conferred by MSCs. Notably, safety assessments in these models have revealed no evidence of tumorigenesis or ectopic tissue formation within the brain, underscoring the favorable safety profile of MSC-based interventions.

The application of MSCs in human PD is still in early experimental stages, but a handful of clinical studies have been conducted. A pioneering pilot study by Venkataramana et al. (2010) transplanted autologous bone marrow MSCs into the sublateral ventricular zone of seven PD patients. The procedure was found to be safe, and over a 1-year follow-up the patients exhibited modest improvements in their Unified Parkinson's Disease Rating Scale (UPDRS) motor scores and activities. Although open label, this study suggested possible symptomatic benefit, with some patients reducing their required dopaminergic medications (Venkataramana et al. 2010). Another early study in Italy infused autologous MSCs intra-arterially via the carotid artery in PD patients. Giordano et al. (2014) reported that intra-arterial delivery of MSCs in five patients was feasible and safe, and preliminary outcomes showed stable or slightly improved motor function over 6 months postinfusion (Giordano et al. 2014). Positron emission tomography (PET) imaging further suggested an elevation in striatal dopaminergic activity among patients receiving MSC) treatment; however, the limited sample size precluded definitive statistical interpretation. Subsequently, Canesi et al. (2016) conducted an open-label study in which eight patients with advanced PD received intravenous infusions of MSCs. The treatment was well tolerated, and the modest improvements in nonmotor symptoms such as sleep quality and gastrointestinal function was reported, along with stabilization of motor manifestations in several patients over a follow-up period of several months (Canesi et al. 2016). Although lacking control groups, these early-phase clinical trials established the safety of autologous MSC administration, with no evidence of immune rejection or serious adverse events. These findings provide an important foundation and rationale for advancing to larger, rigorously controlled clinical investigations.

More recently, researchers have conducted controlled trials. Boika et al. (2020) in Belarus conducted a controlled pilot study combining intravenous and intranasal delivery of autologous bone marrow MSCs in PD. They reported a statistically significant reduction in both motor and nonmotor symptom severity in the MSC-treated cohort relative to baseline values, whereas the control group exhibited progressive clinical deterioration over the same period. (Boika et al. 2020). Patients treated with MSCs exhibited sustained improvements in UPDRS scores and quality-of-life indices over a 12-month follow-up period.

One challenge in PD trials is determining the optimal route of MSC delivery. The blood–brain barrier limits CNS availability after intravenous infusion, leading some studies to explore intrathecal MSC administration, like approaches used in ALS and MS. Intrathecal administration of MSCs introduces them directly into the cerebrospinal fluid, allowing circulation within the CNS and potentially improving delivery to brain regions. An example is an ongoing Phase I trial in Iran delivering intrathecal Wharton's jelly–derived MSCs to PD patients (NCT04876326). On the other hand, intra-arterial infusion can target cells to the brain but carries a risk of embolism or vascular irritation. Thus far, no serious complications like stroke have been reported from intra-arterial MSC infusions in PD.

Although human data remain limited, some studies have reported functional improvements following MSC therapy in patients with PD. Treated individuals have demonstrated enhanced motor performance, reflected by reductions in UPDRS Part III scores, decreased rigidity and tremor, and increased gait velocity (Venkataramana et al. 2010; Boika et al. 2020). Furthermore, nonmotor manifestations that represent a significant contributor to disease burden also appear to exhibit favorable responses to MSC-based interventions. For instance, Canesi et al. (2016) reported improved sleep quality and reduced fatigue following MSC therapy. Although no clinical investigation has yet provided conclusive evidence for an increase in dopaminergic neuron numbers, an outcome that would require verification through PET imaging or post-mortem histological analysis. However, an autopsy report of a PD patient who had received MSC transplantation revealed preservation of host dopaminergic neurons without any signs of abnormal proliferation or tumorigenesis (Jang et al. 2020). These findings lend support to the long-term safety profile of intracerebral MSC administration.

Although current imaging and biomarker data are limited, a case series demonstrated increased cerebrospinal fluid levels of dopamine and its metabolites in patients with PD following MSC therapy, implying a potential enhancement of dopaminergic neuronal activity. In addition, MSC administration has been correlated with decreased concentrations of pro-inflammatory cytokines in both plasma and CSF (Petrou et al. 2022), thereby reinforcing the hypothesis that MSCs exert neuroprotective effects, at least in part, through modulation of neuroinflammatory pathways.

While the early findings are encouraging, it remains important to emphasize that no large-scale randomized controlled trial of MSC therapy in PD has yet been completed. Reported clinical improvements have generally been modest, and because PD is a slowly progressive neurodegenerative disorder, extended longitudinal follow-up will be required to determine whether MSC treatment truly modifies disease trajectory rather than providing only transient symptomatic benefits. Most studies to date have employed autologous MSCs derived from the patient's own bone marrow, which necessitates an invasive procedure and introduces variability in cell yield and quality influenced by donor age and health status. Future investigations may increasingly focus on standardized allogeneic MSC products to ensure reproducibility and scalability. Another promising direction involves enhancing MSCs through genetic modification, for example, engineering them to secrete dopamine or GDNF. In a recent preclinical study, MSCs engineered to express key dopamine-synthetic enzymes produced dopamine in situ following transplantation into PD models and led to measurable functional recovery (Li et al. 2022). If such approaches demonstrate safety and efficacy in rigorous trials, engineered MSCs could simultaneously protect surviving dopaminergic neurons and provide a biologically active source of dopamine replacement.

Thus, MSC therapy for PD has shown encouraging neuroprotective effects in animal models and acceptable safety in early human trials. Patients in early MSC trials have generally reported mild clinical improvements with no major adverse effects, supporting continued investigation. Mechanistically, MSCs act on two critical pathological axes in PD, they secrete neurotrophic factors that may slow progressive dopaminergic neuron loss and they modulate neuroinflammation and α-synuclein aggregation, both of these are key contributors to PD pathophysiology (Mendes et al. 2019).

While MSCs alone are unlikely to reconstruct the intricate nigrostriatal circuitry lost during the course of PD, they hold promise as disease-modifying adjuncts capable of prolonging responsiveness to dopaminergic pharmacotherapy and preserving functional capacity and quality of life. Ongoing clinical investigations and next-generation MSC-based strategies such as gene-modified MSCs engineered to secrete neurotrophic or anti-inflammatory factors, or combinatorial regimens integrating MSCs with molecular or gene therapies are anticipated to elucidate their precise therapeutic role and optimize their integration.

4.3.3 MSC Therapy in ALS

MSC-based therapy has been extensively investigated in ALS as a potential strategy to provide sustained neurotrophic support to degenerating motor neurons, mitigate neuroinflammatory processes within the spinal cord, and ultimately slow the trajectory of disease progression (Bonafede and Mariotti 2017). Among neurodegenerative disorders, ALS has advanced the furthest in clinical evaluation of MSCs, with several phase II and phase III trials conducted to assess safety, feasibility, and therapeutic benefit, making it a key reference model for therapeutic application of MSCs in neurodegeneration.

In ALS preclinical models, particularly transgenic rodents carrying mutant SOD1 alleles, administration of MSCs has produced encouraging results. Both intraspinal and intravenous transplantation of MSCs into mSOD1 mice significantly slowed the rate of motor decline and prolonged survival relative to untreated controls (Boucherie et al. 2009; Vercelli et al. 2008). In a landmark study, Vercelli et al. (2008) delivered human MSCs directly into the spinal cord of SOD1-G93A mice and observed a marked extension of lifespan, preservation of spinal motor neuron counts, and improved motor performance as measured by rotarod testing and grip strength, even in later disease stages. Histopathological analysis revealed attenuated microglial activation and reduced levels of pro-inflammatory cytokines within the spinal cord, strongly implicating immunomodulation as a key therapeutic mechanism of MSCs in ALS.

Another key mechanism by which MSCs exert benefit in ALS models is through paracrine secretion of neurotrophic factors, including GDNF, insulin-like growth factor-1, and VEGF, which collectively support motor neuron survival (Figure 4.3). Because these molecules have very short half-lives when delivered as recombinant proteins, MSCs act as localized, sustained-release vehicles within the spinal cord microenvironment. In a pivotal study, MSCs genetically engineered to overexpress GDNF were transplanted into SOD1 rats, resulting in significantly prolonged survival and greater preservation of motor neurons compared to animals receiving unmodified MSCs (Nagata et al. 2013). Even naïve MSCs secrete measurable levels of GDNF and IGF-1, and their presence in ALS models has been associated with a shift toward a pro-survival transcriptional profile in residual motor neurons (Sun et al. 2013). Additionally, transplanted MSCs exhibit plasticity toward glial differentiation, with a subset acquiring an astrocyte-like phenotype that may buffer extracellular glutamate and provide metabolic support (Boucherie and Hermans 2015). Although MSC-derived astroglial cells are incapable of regenerating lost motor neurons, they can partially reconstitute key astrocytic support functions such as glutamate clearance, trophic factor secretion, and maintenance of redox homeostasis that are disrupted in the pathophysiology of ALS.

A critical feature of ALS pathology is the breakdown of the blood–spinal cord barrier, which permits infiltration of peripheral immune cells into vulnerable motor neuron regions. MSCs can counteract this process by promoting barrier repair and reducing entry of aberrant immune cells. In ALS models, intravenously administered MSCs have been shown to be located to spinal cord tissue and significantly reduce infiltration of T lymphocytes and macrophages, accompanied by lower concentrations of pro-inflammatory cytokines such as TNF-α and IL-1β (Zhao et al. 2007). Furthermore, MSC therapy skews systemic immunity toward a neuroprotective phenotype, increasing the prevalence of anti-inflammatory M2 macrophages and regulatory T cells, thereby creating a microenvironment less hostile to surviving motor neurons (Donega et al. 2014).

The clinical translation of MSC therapy from preclinical ALS models to human subjects commenced in the late 2000s with small, first-in-human safety trials. Since then, this field has advanced to encompass larger, multicenter investigations, culminating in a recent phase III clinical trial aimed at rigorously evaluating therapeutic efficacy and long-term safety in ALS patients. To date, dozens of patients have received MSC transplants both autologous and allogeneic using bone marrow, adipose, or umbilical sources, delivered via intrathecal, intramuscular, or intravenous routes. Across multiple clinical investigations, MSC administration has demonstrated a consistently favorable safety profile. The treatment is generally well tolerated, with adverse events limited to transient, procedure-related effects such as localized discomfort at the injection site, mild headache, or low-grade fever, none of which have necessitated treatment discontinuation. (Staff et al. 2016). This is a crucial consideration given the fragility of ALS patients. A landmark early study by

Mazzini and colleagues in 2009 transplanted autologous bone marrow–derived MSCs directly into the spinal cord of ALS patients, finding no surgery-related disease acceleration and even observing a slower decline in muscle strength in a subset of treated patients (Mazzini et al. 2009). These results established both the feasibility and biological plausibility of MSC therapy in ALS, thus paving the way for subsequent controlled trials.

The intrathecal route has become the preferred and least invasive method for MSC administration in ALS. In a phase I trial, Staff et al. (2016) delivered autologous adipose-derived MSCs intrathecally to 10 ALS patients and demonstrated safety at doses up to 100 million cells, with no serious adverse events. Importantly, longitudinal follow-up revealed a modest attenuation in the rate of ALS Functional Rating Scale–Revised (ALSFRS-R) decline during the 6 months after infusion compared with each patient's pretreatment trajectory, suggesting a potential biological effect (Staff et al. 2016). These promising results laid the groundwork for several phase II studies designed to more rigorously evaluate efficacy, optimal dosing, and durability of clinical benefit.

A major advance in ALS therapeutics has been the development of neurotrophic factor–enhanced MSCs. BrainStorm Cell Therapeutics created NurOwn, an autologous MSC product derived from patient's bone marrow, expanded ex vivo, and transiently induced to secrete elevated levels of neurotrophic factors (MSC-NTF cells). In a multicenter, randomized Phase II trial, NurOwn demonstrated a favorable biological signal, although the primary endpoint was not achieved, a prespecified subgroup of patients with less advanced ALS exhibited a significantly higher proportion achieving slowed disease progression compared with placebo (Berry et al. 2019). Biomarker analyses confirmed target engagement, i.e., treated patients showed increased cerebrospinal fluid concentrations of VEGF, HGF, and IGF-1 along with decreased inflammatory mediators, indicating robust in vivo bioactivity of the administered cells. Clinically, some patients even demonstrated transient stabilization or improvement in ALSFRS-R scores, a remarkable finding given the typically relentless functional decline characteristic of ALS. These outcomes have provided the impetus for continued phase III clinical development of NurOwn and strengthen the rationale for employing trophic factor–augmented MSCs as a viable disease-modifying intervention in ALS, capable of delivering multifaceted neuroprotection through combined trophic support and immunomodulatory mechanisms.

A pivotal milestone in MSC-based therapy for ALS was the Phase III randomized controlled trial of NurOwn (MSC-NTF), which enrolled over 180 patients (Cudkowicz et al. 2022). Although the study did not achieve its primary endpoint in the overall cohort, exploratory analyses indicated potential clinical benefits in subgroups, particularly among patients with early-stage disease or favorable biomarker profiles. The safety findings were reassuring, with MSC-NTF well tolerated and adverse events largely limited to transient headache and back pain related to lumbar puncture procedures, with no significant differences compared with placebo. Also, the biomarker analyses confirmed biological activity, cerebrospinal fluid from treated patients showed elevated neurotrophic factors and reduced neuroinflammatory mediators, consistent with target engagement. While the responder proportion favored MSC-NTF (34%) over placebo (27%), this difference did not reach statistical significance (OR = 1.33, $p = 0.45$). Post hoc analyses suggested that therapeutic timing is critical, as patients with milder ALS at baseline were more likely to achieve clinically meaningful slowing of progression. These nuanced findings have driven calls for refined trial designs focusing on early-stage ALS populations, where MSC-NTF therapy may have the highest potential impact.

Beyond NurOwn, several studies have investigated both autologous and allogeneic MSC approaches in ALS. In a controlled clinical trial, Syková et al. (2017) administered repeated intrathecal doses of bone marrow-derived MSCs and observed a slower decline in both forced vital capacity and ALSFRS-R scores over six months compared with untreated controls, suggesting a potential disease-modifying effect. Barczewska et al. (2020) evaluated multiple intrathecal infusions of Wharton's jelly–derived MSCs, reporting a striking twofold increase in median survival time relative to historical controls, with good tolerability of up to three serial injections. Similarly, Siwek et al. (2020) administered autologous intrathecal MSCs to patients with rapidly progressive ALS and observed a transient slowing of functional decline following treatment; however, the limited cohort size prevented definitive interpretation of efficacy. Collectively, these studies reaffirm the safety and feasibility of repeated intrathecal MSC administration and provide supportive evidence for a modest attenuation of disease progression in ALS, thereby justifying further evaluation in larger, rigorously controlled clinical trials.

Although definitive efficacy has yet to be established, multiple clinical studies report encouraging biological and functional signals. Many patients experience a period of disease stabilization following MSC therapy. For example, Petrou et al. (2016) observed that 87% of treated individuals demonstrated a slower ALSFRS-R decline in the six months post-treatment compared with their own pretreatment trajectory, indicating potential modulation of disease progression. In some cases, transient improvements in muscle strength or respiratory capacity have been documented, changes rarely observed spontaneously in ALS, further suggesting a therapeutic effect. Remarkably, one patient from an early study maintained stable muscle strength for more than 4 years after MSC transplantation (Mazzini et al. 2018), a highly atypical outcome that underscores the possibility of strong responder phenotypes. While such cases remain uncommon, they provide proof-of-concept that MSC-based interventions, particularly when enriched for neurotrophic factor secretion, may alter disease course in at least a subset of ALS patients and justify continued refinement of these strategies in larger targeted trials.

Biomarker investigations in ALS MSC trials have provided some of the strongest mechanistic support for a therapeutic effect. Post-treatment cerebrospinal fluid analyses frequently demonstrate increased concentrations of neurotrophic factors such as VEGF and HGF, consistent with the intended paracrine action of MSCs as well as decreased levels of pro-inflammatory mediators including MCP-1 and soluble CD40 ligand (Berry et al. 2019). In parallel, neurofilament light chain (NfL), a robust marker of axonal degeneration, has been observed to rise more slowly in some MSC-treated cohorts relative to untreated natural history, suggesting a possible attenuation of neurodegeneration (Goutman et al. 2022). Imaging-based biomarkers have also yielded supportive signals; MRI and MR spectroscopy studies have reported patterns consistent with preservation of motor neuron integrity in a subset of treated patients (Staff et al. 2016). Together, these findings strengthen the biological plausibility that MSC therapy exerts a protective cellular effect in ALS.

Regulatory developments have also emphasized the therapeutic potential of MSC therapy. In 2014, South Korea granted conditional approval with orphan-drug designation to NeuroNata-R (lenzumestrocel), an autologous bone marrow–derived MSC preparation, for the treatment of ALS (Soares et al. 2022). This approval was based on national clinical trial outcomes and compassionate-use data indicating a favorable safety profile and preliminary functional improvement, establishing South Korea as the first country to authorize an MSC-based therapy for ALS. In contrast, in most other regions, including the United States and the European Union, the MSC therapy remains under investigation and is accessible only through regulated clinical trials or expanded access programs until efficacy is conclusively demonstrated in large-scale randomized studies.

4.3.4 MSC Therapy in MS

Current disease-modifying therapies for MS primarily act by modulating or suppressing aberrant immune activity to reduce the incidence of new relapses; however, they lack the capacity to repair pre-existing tissue damage or reverse established neurological disability. MSCs have therefore gained attention as a promising investigational strategy in MS due to their dual ability to regulate pathogenic immune mechanisms while promoting endogenous repair processes, including remyelination and neuroregeneration (Figure 4.2; Uccelli, Laroni, and Freedman 2020). In contrast to classical neurodegenerative disorders, MS is characterized by pronounced immune dysregulation, which renders MSCs particularly attractive as a therapeutic strategy. Their capacity to reset immune homeostasis offers the potential to attenuate aberrant immune responses, while simultaneously providing trophic support that protect or even replace oligodendrocytes, the myelin-producing cells that are essential for restoring functional myelination.

In preclinical models of MS, particularly experimental autoimmune encephalomyelitis (EAE) in rodents, MSC therapy has demonstrated robust anti-inflammatory and neuroprotective effects. Both intravenous and intrathecal administration of MSCs in EAE mice markedly attenuate disease severity, characterized by reduced paralysis scores and fewer relapses (Gerdoni et al. 2007). The therapeutic benefit is largely attributed to MSC-mediated immunomodulation, wherein they suppress the infiltration of encephalitogenic T cells into the CNS and enhance the expansion of regulatory T cells, along with the secretion of anti-inflammatory cytokines (Gerdoni et al. 2007; Zhang et al. 2020). Mechanistically, MSCs inhibit pathogenic Th17 and Th1 subsets, the key drivers of MS pathogenesis, while promoting their differentiation toward a less inflammatory phenotype (Müller et al. 2021). These immunological effects translate into smaller demyelinating lesions and reduce axonal damage, underscoring potential of MSC as both immunoregulatory and neuroprotective agents.

In addition to their immunomodulatory effects, MSCs have been shown to facilitate remyelination in experimental models of demyelination. In toxin-induced demyelination paradigms, MSC transplantation enhances myelin repair by secreting trophic factors that drive endogenous oligodendrocyte precursor cell differentiation and maturation, ultimately resulting in the formation of new myelin sheaths (Lublin et al. 2014). The key mediators include HGF and insulin-like growth factor, both of which have been demonstrated to accelerate remyelination in vivo (Bai et al. 2012). Notably, Laso-García et al. (2018) reported that the intravenous administration of human adipose-derived MSCs in a chronic demyelination model significantly increased the number of mature oligodendrocytes and remyelinated axons, translating into measurable improvements in motor function. The reparative effect of MSCs appears to involve modulation of the lesion microenvironment. MSCs secrete proteases that degrade inhibitory extracellular matrix molecules, thereby facilitating OPC migration and differentiation, while simultaneously depositing matrix components that promote tissue repair. Although some studies suggest that a small subset of MSCs can differentiate into oligodendrocyte-like cells, with human myelin proteins detected in transplanted demyelinated mouse spinal cord, the predominant mechanism of action is considered paracrine, i.e., stimulating the host's endogenous repair pathways rather than serving as direct myelin-replacing cells.

Beyond their effects on adaptive immunity, MSCs exert significant influence on innate immune cell populations in MS. Within the CNS, MSCs modulate microglial activation by shifting the balance away from the pro-inflammatory M1 phenotype associated with oligodendrocyte injury toward the reparative M2 phenotype, which facilitates myelin debris clearance and supports tissue regeneration (Li et al. 2019). MSCs also attenuate the activation of B cells and peripheral antigen-presenting cells, thereby reducing autoantibody production and dampening antigen presentation that perpetuates CNS inflammation (Glenn and Whartenby 2014). Collectively, these broad-spectrum immunomodulatory effects highlight MSCs' therapeutic potential to correct the complex immune dysregulation underlying MS pathogenesis, offering a strategy that targets both inflammatory injury and impaired repair mechanisms.

MSC-based therapy for MS has now progressed from early studies to multicenter Phase II clinical trials. The first small-scale investigations conducted in the late 2000s established the safety of autologous MSC infusion, with favorable tolerability profiles and preliminary signals of efficacy, including reductions in gadolinium-enhancing lesions on MRI scans (Mohyeddin Bonab et al. 2007). Building upon these encouraging findings, the past decade has seen the initiation of several randomized, controlled trials aimed at rigorously evaluating both the immunological effects and the clinical benefits of MSC therapy in MS. These trials have assessed diverse endpoints ranging from relapse rate reduction and disability progression to MRI lesion activity and biomarkers of neuroprotection that laid the groundwork for defining the optimal cell source, route of administration, and dosing regimen for therapeutic application.

For example, MESEMS Trial (Uccelli et al. 2021) which was Phase II randomized, placebo-controlled trial across Europe and Canada evaluated autologous bone marrow–derived MSCs in 144 patients with active MS. Patients received an IV infusion of MSCs in one period and placebo in another period. The primary outcome was MRI detection of new inflammatory lesions. The results showed that MSC treatment was safe and well tolerated, meeting the primary safety endpoint (Uccelli et al. 2021). However, the primary efficacy endpoint was not met as MSCs did not significantly reduce the number of new gadolinium-enhancing lesions over 24 weeks compared to placebo (Uccelli et al. 2021). In other words, as an anti-inflammatory disease-modifying therapy (DMTs), MSC infusion did not outperform placebo in this trial for short-term MRI outcomes. This outcome aligns with preclinical evidence that IV MSCs might have limited acute impact on the most aggressive inflammation. Importantly though, secondary analyses in MESEMS suggested positive trends, the patients on MSCs had no unexpected relapses and some functional measures like visual acuity in those with optic neuritis improved in the MSC group (Uccelli et al. 2021). Additionally, mechanistic studies showed increased levels of immunomodulatory molecules in blood and CSF after MSC infusion, indicating biological activity. Thus, while MESEMS did not lead to suppression of lesion at 6 months, it confirmed safety and hinted that MSCs might exert subtler benefits not captured by MRI in that timeframe.

A key finding emerging from MSC trials in MS is that the route of administration significantly influences clinical outcomes. In a randomized trial, Cohen et al. (2018) directly compared intrathecal (IT) and intravenous (IV) delivery of autologous MSCs (Figure 4.4). The IT group demonstrated superior outcomes across several endpoints, including a greater proportion of patients achieving no evidence of disease activity (NEDA) at 12 months compared with those receiving IV infusion. Similar trends were observed by Sahraian et al. (2019), who reported that 58.6% of patients treated via the IT route remained NEDA at 1 year, compared with only 40.6% in the IV cohort. Moreover, IT administration was associated with clinically meaningful improvements in functional parameters, such as timed walk performance and upper-limb dexterity, which were less pronounced following IV infusion.

Although these studies are limited by relatively small sample sizes, they collectively suggest that direct CNS delivery enhances therapeutic efficacy, likely by increasing MSC bioavailability at sites of neuroinflammation and enabling more direct interaction with meningeal immune populations and resident glial cells. This mechanistic rationale has led to a growing preference for intrathecal administration in contemporary clinical trials investigating MSC-based therapies for MS (Petrou et al. 2022).

Some clinical investigations have explored the use of MSC–derived neural progenitors (MSC-NPs) to enhance therapeutic efficacy in MS. In a Phase I clinical trial, Harris et al. (2018) harvested autologous MSCs from patients and directed their differentiation toward a neural progenitor–like phenotype ex vivo before intrathecal administration. This approach leveraged the dual potential of MSCs, i.e., retaining their robust immunomodulatory properties while enhancing their capacity to support neuroregeneration and remyelination. Early-phase findings indicated a favorable safety profile and suggested biological activity, with a subset of participants with progressive MS demonstrating a deceleration of clinical disease progression. These observations underscore the therapeutic flexibility of MSCs, which can be directed toward specific functional roles, potentially allowing for tailored interventions that are aimed at modulating neuroinflammation while simultaneously promoting neural repair.

In relapsing–remitting MS, the primary goal is to reduce relapse frequency and MRI lesion burden. Although highly effective DMTs are available, MSCs have a role in patients refractory to conventional therapy or in progressive MS, where inflammation is less prominent. Compassionate-use studies in progressive MS have reported stabilization of disability following MSC administration.

For instance, Lublin et al. (2014) described a secondary progressive MS patient who showed no EDSS worsening for two years after repeated MSC infusions, along with improvements in fatigue and ambulation. Such observations suggest MSCs may offer neuroprotection and repair even when overt inflammation is minimal. MRI findings from early trials support this hypothesis. Connick et al. (2012) reported increased magnetization transfer ratios within lesions and reduced brain atrophy rates in MSC-treated patients, indicating possible remyelination and tissue preservation.

Nearly all clinical investigations of MSC therapy in MS have demonstrated immunological changes indicative of an immune rebalancing that is characterized by shifts toward regulatory phenotypes and suppression of pro-inflammatory immune activity. Karussis et al. (2010) demonstrated that MSC-treated patients exhibited increased circulating regulatory T cells and reduced frequencies of pro-inflammatory myeloid dendritic cells expressing CD86 and HLA-DR. Similarly, Person et al. observed elevated cerebrospinal fluid levels of IL-10 and TGF-β following intrathecal MSC infusion, indicating a shift toward an anti-inflammatory milieu. Sahraian et al. (2019) further reported decreased CSF oligoclonal band intensity in some patients, suggesting reduced B-cell activity within the CNS. Together, these immunological findings support the mechanism by which MSCs attenuate autoimmune activity in MS. Ongoing Phase II/III studies are now concentrating on progressive MS, where neurorepair is a primary therapeutic goal. The SPRINT-MS trial has evaluated repeated intrathecal autologous MSCs in primary progressive MS, while the French METEMS trial has tested placenta-derived MSCs. Combination strategies are also being explored, including the coadministration of MSCs with low-dose interleukin-2 (IL-2) to enhance the expansion and function of regulatory T cells (Wang et al. 2022).

Overall, MSC therapy in MS has consistently demonstrated a favorable safety profile and robust immunomodulatory effects. Although the largest trial (Uccelli et al. 2021) did not show a statistically significant reduction in new inflammatory lesions, other trials suggest potential benefits in slowing disability progression and promoting tissue repair, particularly with direct delivery in CNS. Given the heterogeneity of MS subtypes and disease activity, MSCs may be best suited as adjunctive therapy, either to enhance repair in progressive MS or to treat aggressive, treatment-refractory disease. Their multimodal actions, including immunosuppression, neuroprotection, and promotion of remyelination, remain highly attractive. Future studies are aimed at refining therapeutic protocols, optimizing routes and schedules of administration, and exploring engineered MSCs or combination regimens. If efficacy is confirmed, MSC therapy may advance MS treatment toward the long-sought goal of regenerative intervention, restoring function lost to prior neurodegeneration.

4.4 LIMITATIONS AND CHALLENGES

Although MSCs hold considerable therapeutic promise, several key challenges limit their clinical translation. The therapeutic efficacy of MSCs remains inconsistent, with marked variability in patient responses. Some individuals experience meaningful clinical benefits, whereas others show little or no improvement. This variability likely reflects the underlying heterogeneity of neurodegenerative diseases and the influence of disease stage at treatment initiation. In ALS, the most pronounced slowing of progression has been observed in patients at earlier stages (Cudkowicz et al. 2022). Similarly, in MS, some patients remain lesion-free for extended periods after MSC infusion, while others continue to develop new lesions (Uccelli et al. 2021). These findings indicate that MSC therapy is likely to be most effective when administered early in the disease course, before irreversible neuroaxonal damage has occurred (Langer-Gould et al. 2023). Future progress in MSC therapy will rely on optimizing selection of patients and defining robust predictive biomarkers such as immune signatures or disease stage markers to enable stratified treatment approaches and maximize therapeutic efficacy (Berry et al. 2019).

MSC therapy is likely to achieve maximal benefit when administered early in the disease course, prior to the onset of extensive and irreversible neurodegeneration. In advanced stages, the limited availability of viable neural tissue may constrain the capacity of MSCs to exert therapeutic effects (Langer-Gould et al. 2023). Future efforts should focus on identifying reliable predictive biomarkers such as immune signatures or disease stage indicators to enable patient stratification and the development of personalized MSC-based treatment strategies (Berry et al. 2019). Although MSCs can slow neurodegeneration, they rarely halt or reverse it, particularly in advanced disease stages. Even in studies showing a slowing or functional decline, patients typically continued to exhibit gradual disease progression over time. This limited efficacy demonstrates that MSCs, while promising, cannot be regarded as curative therapy (Langer-Gould et al. 2023). In progressive MS, where the primary objective is to halt disability accumulation, current evidence suggests that MSCs are more likely to decelerate progression rather than fully arrest it (Li et al. 2022). Achieving true regeneration such as neuronal replacement or complete remyelination remains an ambitious goal that MSC monotherapy alone is unlikely to achieve. As a result, ongoing research is increasingly directed toward combination strategies, including coupling MSCs with bioengineered scaffolds or adjunctive pharmacological agents to amplify repair processes and enhance functional recovery.

The optimal MSC dose and dosage frequency remain undefined. Most clinical trials have employed a single administration, yet observed benefits are often transient, suggesting that repeated dosing may be necessary to sustain therapeutic effects. This approach would conceptualize MSC therapy as a recurrent treatment rather than a one-time intervention. Some studies have explored this strategy, for instance, an ALS trial administering three monthly intrathecal injections reported superior outcomes compared with a single dose (Barczewska et al. 2020). Nevertheless, practical and safety considerations impose potential limits on repeated dosage. Repeated intrathecal administration of MSCs entails cumulative procedural risks, such as infection and arachnoiditis, although serious adverse events have thus far been uncommon. Moreover, the limited in vivo persistence of transplanted MSCs provides a biological rationale for employing multiple dosing regimens in the context of chronic neurodegenerative disorders.

Determining the optimal balance between therapeutic efficacy and the procedural burden or risk of frequent MSC dosage remains a major challenge. Long-term studies are needed to assess whether repeated exposure, particularly to allogeneic MSCs, could eventually elicit immune responses. Although MSCs are generally considered immune-privileged, repeated administration of donor-derived cells carries a theoretical risk of antibody formation or sensitization. Indeed, in an early stroke trial, a subset of patients developed donor-specific antibodies following redosing, albeit without detectable clinical consequences (Lukomska et al. 2019).

As living cell–based therapeutic products, MSCs pose distinct manufacturing and regulatory challenges, encompassing issues of standardization, scalability, quality control, and compliance with stringent cell therapy guidelines. Their phenotypic and functional characteristics can vary depending on culture conditions, passage number, donor source, and expansion methods, complicating efforts to ensure batch-to-batch consistency in quality, potency, and identity. In autologous approaches, interpatient variability further influences cell behavior, whereas in allogeneic therapies, establishing a well-characterized, standardized cell bank is essential. Regulatory agencies classify MSCs as biologics and require stringent quality control, including sterility testing, viability assessment, and immunophenotypic marker verification.

A key gap in the field is the absence of standardized potency assays, which are robust, validated measures of therapeutic activity. Current surrogates, such as the ability to suppress T-cell proliferation in vitro, may not fully capture in vivo efficacy (Salmikangas et al. 2023). This variability was highlighted in the NurOwn trials, where subtle differences in manufacturing processes between batches and between phase II and III studies may have contributed to variability in clinical outcomes. To address this, manufacturers are moving toward centralized, GMP-compliant production using young-donor MSCs expanded into controlled bioreactor systems to improve reproducibility.

Additional challenges include the logistics of cryopreservation, storage, and transport, as freezing and thawing can impact MSC viability and function (Moll et al. 2020). Furthermore, the cost of

MSC production remains relatively high, raising concerns about accessibility and scalability for widespread clinical application. Overcoming these manufacturing, quality control, and economic hurdles will be essential to make MSC therapy broadly feasible.

Although short-term safety data for MSC therapy are reassuring, long-term monitoring remains essential. A theoretical risk stems from the mesenchymal differentiation capacity of MSCs, raising concern for ectopic tissue formation. In an early stroke trial, a small calcification was observed at a prior intracerebral MSC injection site, likely resulting from osteogenic differentiation (Kalladka et al. 2016). Importantly, such events have not been reported in neurodegeneration trials, where MSCs are typically delivered via intrathecal or intravenous routes rather than direct parenchymal injection, thereby reducing this risk.

Another important consideration is the potential interaction between MSCs and malignant processes. Although not inherently tumorigenic, MSCs may influence the tumor microenvironment by enhancing angiogenesis and modulating immune responses in a manner that could facilitate neoplastic progression. Preclinical and clinical data suggest that MSC administration could theoretically accelerate tumor progression in patients with active malignancies (Babajani et al. 2020). Although this risk is likely low in neurodegenerative disease populations, prudent clinical practice involves excluding individuals with known active cancers and implementing long-term surveillance for neoplastic complications. Reassuringly, follow-up studies of MSC-treated patients extending up to 5 years have not demonstrated an increased incidence of malignancy.

Although MSCs exhibit low immunogenicity, they are not entirely immune-evasive. With allogeneic MSCs, repeated dosing may induce immune rejection or diminish efficacy as host immunity develops against donor antigens. Evidence suggests that allogeneic MSCs may persist for shorter durations than autologous counterparts, despite exerting transient therapeutic effects. The choice between autologous and allogeneic MSCs remains debated. Autologous MSCs evade immune recognition but require individualized processing, adding procedural complexity and cost, whereas allogeneic products offer off-the-shelf availability but risk immune neutralization upon repeated administration. Approaches such as transient peri-infusion immunosuppression or gene editing to reduce immune antigen expression are being explored to mitigate these challenges (Pavlovic et al. 2020)

MSC therapy resides at the intersection of cellular therapeutics and regenerative medicine, making regulatory oversight inherently complex. Comprehensive informed consent is imperative, given the uncertainty surrounding long-term safety and efficacy profiles. High manufacturing costs and issues of timely access, particularly critical for patients with rapidly progressive diseases such as ALS, remain significant barriers to widespread implementation. Additionally, the proliferation of unregulated commercial stem cell clinics poses risks to patient safety and threatens the integrity of legitimate clinical research. Harmonized regulatory frameworks and longitudinal patient registries are crucial for systematically collecting safety and efficacy data, thereby enabling evidence-based assessment and informed clinical integration of MSC therapy (Sipp et al. 2018).

4.5 FUTURE DIRECTIONS

MSC-based therapy for neurodegenerative disorders is progressing rapidly, with diverse strategies being explored to enhance efficacy, durability, and clinical applicability. An emerging focus is the development of cell-free approaches that harness the bioactive secretome of MSCs, particularly extracellular vesicles such as exosomes. These nanoscale vesicles contain proteins, mRNAs, and microRNAs that replicate many of the therapeutic effects attributed to MSCs. Importantly, exosomes possess the ability to traverse the blood–brain barrier and offer advantages in standardization and long-term storage relative to viable cell preparations. Preclinical investigations have demonstrated that MSC-derived exosomes confer neuroprotective benefits comparable to those of MSC therapy in models of AD and PD (Losurdo et al. 2020). For instance, MSC exosomes enhanced memory performance and reduced amyloid pathology in AD mice, while in PD models, they preserved dopaminergic neurons and mitigated neuroinflammation. As nonviable particles, exosomes circumvent the risks of immune rejection and tumorigenicity associated with cell-based therapies (Zhu et al. 2025). Early-phase clinical trials are underway to assess MSC-derived exosome therapy in AD and autism. According to Zhu et al. (2025), MSC exosomes preserve the antioxidative, anti-inflammatory, and neurogenic functions of their parent cells while reducing the likelihood of adverse effects such as vascular embolism or aberrant differentiation. In the near future, off-the-shelf exosome formulations may be developed, potentially tailored to promote remyelination in MS or enhance the clearance of protein aggregates in AD. Nonetheless, achieving scalable manufacturing, purification, and standardized potency assays remains a major challenge, though these barriers are the focus of active translational efforts.

Genetic modification of MSCs represents a promising approach to enhance their therapeutic potential. Through targeted engineering to overexpress disease-relevant molecules, MSCs can be tailored to amplify neuroprotective and reparative functions. In ALS, MSCs modified to secrete higher levels of neurotrophic factors such as GDNF and IGF-1 have shown markedly improved neuroprotective efficacy in preclinical models (Nagata et al. 2013; Petrou et al. 2016). Comparable strategies have been proposed for neurodegenerative proteinopathies, including engineering MSCs to express neprilysin, an amyloid-degrading enzyme, in AD, or molecular chaperones that facilitate the clearance of misfolded SOD1 in ALS.

In PD, a novel concept involves generating DOPA-MSCs by introducing genes encoding the dopamine biosynthetic enzymes tyrosine hydroxylase (TH), aromatic L-amino acid decarboxylase (AADC), and GTP cyclohydrolase 1 (GCH1), thereby enabling MSCs to act as in situ dopamine-producing biological pumps. In a 2021 study, such engineered MSCs restored striatal dopamine levels and improved motor function in a PD rat model. These cells could complement, or potentially reduce, the need for exogenous dopamine-replacement therapies. Gene modification can also improve MSC homing and tissue targeting. For example, MSCs engineered to express chemokine receptors show enhanced migration to inflamed CNS sites (Yu et al. 2012). With the rapid evolution of CRISPR and other gene-editing technologies, the generation of customized next-generation MSCs tailored to specific neurodegenerative conditions is increasingly feasible. However, genetic modification introduces additional regulatory complexity, and rigorous safety testing is required to ensure that engineered MSCs remain nontumorigenic and clinically safe.

Further, deriving MSC-like cells from induced pluripotent stem cells represents an important future direction. iPSCs offer an effectively unlimited source of stem cells, overcoming the donor limitations of tissue-derived MSCs. Mesenchymal progenitors generated from iPSCs exhibit similar surface markers and functional properties to conventional MSCs (Xu et al. 2019) and can be manufactured in large, standardized batches for biobanking. This approach enables greater product consistency and scalability.

iPSC technology enables precise genetic manipulation, allowing patient-derived iPSCs to be corrected for pathogenic mutations, such as those implicated in ALS, followed by differentiation into genetically corrected autologous MSCs. Additionally, the development of universal donor lines has become feasible through targeted knockout of key immune recognition genes. This approach has produced hypoimmunogenic iPSCs capable of generating MSCs that evade host immune detection (Deuse et al. 2019), offering the prospect of rejection-free allogeneic therapy.

Several biotechnology companies are developing iPSC-derived MSC platforms for acute conditions such as stroke and acute respiratory distress syndrome (ARDS), with expansion into chronic neurodegenerative diseases representing a logical progression. The primary safety concern pertains to the potential presence of residual undifferentiated iPSCs capable of forming teratomas. Nonetheless, contemporary directed differentiation protocols are optimized to selectively generate mesenchymal lineage cells, thereby substantially reducing this risk.

Future strategies may integrate MSC therapy with complementary treatments to enhance efficacy. For example, combining MSC infusions with immunomodulatory drugs is under investigation in MS to augment therapeutic effects. In PD and ALS, MSCs may be paired with neurotrophic factor delivery or gene therapies such as AAV-mediated trophic factor expression to target multiple disease pathways. Another approach involves incorporating MSCs into biomaterial scaffolds or hydrogels to improve cell retention and sustained factor release. In preclinical spinal cord injury models, hydrogel-embedded MSCs promoted superior repair compared with MSCs alone (Mukhamedshina et al. 2019). Similar strategies could be applied to MS lesions to facilitate localized remyelination. Additionally, MSCs are being explored as cellular vectors engineered to deliver anti-inflammatory cytokines or therapeutic payloads, highlighting their versatility as a combinatorial platform.

A major limitation of current MSC-based therapy lies in the low proportion of transplanted cells that successfully reach target tissues and persist long enough to elicit meaningful therapeutic effects. Current research efforts are directed toward enhancing MSC survival and homing efficiency through various preconditioning strategies. Hypoxic preconditioning prior to administration has been shown to augment the secretion of prosurvival factors and chemokines, thereby improving therapeutic outcomes in preclinical models of stroke and potentially other CNS disorders (Beegle et al. 2015). Pharmacological priming with agents such as SDF-1α or CXCR4 agonists further enhances MSC homing by upregulating chemotactic receptor expression, promoting targeted migration to sites of neural injury (Wang et al. 2011). Additionally, encapsulation techniques that provide immune isolation of MSCs may prolong cell viability and enable sustained release of trophic factors over extended periods ranging from weeks to months (Lejis et al. 2017).

With increasing insight into patient-specific determinants of therapeutic response, MSC-based interventions are expected to evolve toward greater personalization. This approach may involve tailoring the MSC secretome to an individual's disease profile for instance, priming MSCs to counteract dominant inflammatory cytokines via engineered receptor expression or controlled release of antagonists. Biomarker-guided patient stratification could further enhance treatment precision, whereby highly immunomodulatory MSC formulations are directed toward individuals exhibiting active inflammation, while MSCs enriched in neurotrophic activity are prioritized for patients with primarily degenerative pathology. This emerging paradigm of condition-specific, engineered designer MSCs, often termed MSC 2.0, represents a significant and promising direction for future regenerative therapeutics (Galipeau and Sensebe 2018).

Future advancement in MSC therapy will rely on optimizing clinical trial designs and endpoints to more comprehensively capture its multidimensional therapeutic effects. Traditional outcome measures may inadequately reflect the breadth of benefit, underscoring the value of composite endpoints such as those recently employed in AD trials or patient-centered metrics including quality of life and caregiver burden. For slowly progressive neurodegenerative disorders, adaptive trial designs and response-adaptive protocols offer the potential to identify responders early and individualize treatment continuation.

Regulatory frameworks are concurrently evolving to support the development of advanced cell therapies. The U.S. FDA's granting of Regenerative Medicine Advanced Therapy (RMAT) designation to Lomecel-B for AD exemplifies an emerging pathway for accelerated clinical translation. As clinical and biomarker evidence grows, structured treatment algorithms are likely to emerge, for instance, incorporating MSC therapy at ALS diagnosis in combination with riluzole, or deploying MSCs in MS following relapse despite standard DMTs.

4.6 CONCLUSION

MSCs have emerged as one of the most versatile and promising therapeutic platforms for neurodegenerative disorders. Their unique combination of immunomodulatory, neurotrophic, and regenerative properties enables simultaneous targeting of multiple pathological pathways, a particularly valuable feature given the complex, multifactorial nature of conditions such as AD, PD, ALS, and MS. Over the past two decades, extensive preclinical research has established that MSCs can confer neuronal protection, promote tissue repair, and attenuate neuroinflammation across diverse experimental models. Early-phase clinical trials have successfully translated these findings to human studies, demonstrating the feasibility and safety of MSC therapy, along with encouraging indications of efficacy, including slowed functional decline and symptomatic improvement in select patient cohorts.

Despite these advances, the transition of MSC therapy from experimental intervention to established clinical modality remains underway. Current evidence suggests that MSCs are not curative but represent a novel, disease-modifying adjunct capable of complementing conventional pharmacological treatments. In ALS and progressive MS, where therapeutic options are severely limited even modest delays in disability progression achieved through MSC therapy may translate into meaningful clinical benefit. Similarly, in AD and PD, MSCs could form integral components of multimodal treatment paradigms, augmenting targeted pharmacotherapies aimed at amyloid clearance, synaptic repair, or dopaminergic restoration. Several critical challenges persist, including the optimization of delivery routes and dosage, improvement of product consistency and potency, and the demonstration of unequivocal efficacy through large-scale, randomized controlled trials. Ongoing research is addressing these gaps through innovative strategies such as repeated intrathecal dosing, genetic enhancement of MSCs, and biomarker-based patient stratification to improve therapeutic precision and predictability. Regulatory frameworks are likewise adapting to accommodate cell-based therapeutics, with certain MSC-derived products already approaching conditional approval in specific regions.

REFERENCES

Aggarwal S, Pittenger MF. 2005. Human mesenchymal stem cells modulate allogeneic immune cell responses. *Blood* 105(4):1815–22. doi:10.1182/blood-2004-04-1559

Alvites R, Branquinho M, Sousa AC, Lopes B, Sousa P, Maurício AC. Mesenchymal stem/stromal cells and their paracrine activity-immunomodulation mechanisms and how to influence the therapeutic potential. *Pharmaceutics*. 2022;14(2):381. doi:10.3390/pharmaceutics14020381

Babajani A, Soltani P, Jamshidi E, Farjoo MH, Niknejad H. Recent advances on drug-loaded mesenchymal stem cells with anti-neoplastic agents for targeted treatment of cancer. *Front Bioeng Biotechnol*. 2020;8:748. doi:10.3389/fbioe.2020.00748

Bai L, Lennon DP, Caplan AI, et al. Hepatocyte growth factor mediates mesenchymal stem cell–induced recovery in multiple sclerosis models. *Nat Neurosci*. 2012;15(6):862–70. doi:10.1038/nn.3109

Barczewska M, Maksymowicz S, Zdolińska-Malinowska I, Siwek T, Grudniak M. Umbilical cord mesenchymal stem cells in amyotrophic lateral sclerosis: An original study. *Stem Cell Rev Rep*. 2020;16(5):922–32. doi:10.1007/s12015-020-10016-7

Beegle J, et al. Hypoxic preconditioning of mesenchymal stromal cells induces metabolic changes, enhances survival, and promotes cell retention in vivo. *Stem Cells*. 2015;33(6):1818–28. https://doi.org/10.1002/stem.1976

Berg J, Roch M, Altschüler J, Winter C, Schwerk A, Kurtz A, Steiner B. Human adipose-derived mesenchymal stem cells improve motor functions and are neuroprotective in the 6-hydroxydopamine-rat model for Parkinson's disease when cultured in monolayer cultures but suppress hippocampal neurogenesis and hippocampal memory function when cultured in spheroids. *Stem Cell Rev Rep*. 2015;11(1):133–49. doi:10.1007/s12015-014-9551-y

Berry JD, Cudkowicz ME, Windebank AJ, et al. NurOwn, phase 2, randomized, clinical trial in patients with ALS: Safety, clinical, and biomarker results. *Neurology*. 93(24):e2294–305. doi:10.1212/WNL.0000000000008620

Blurton-Jones M, Kitazawa M, Martinez-Coria H, Castello NA, Müller FJ, Loring JF, Yamasaki TR, Poon WW, Green KN, LaFerla FM. Neural stem cells improve cognition via BDNF in a transgenic model of Alzheimer disease. *Proc Natl Acad Sci U S A*. 2009;106(32):13594–9. doi:10.1073/pnas.0901402106

Boika A, Lyashenko T, Vasiliev A, Shukevich A. Intravenous and intranasal delivery of autologous mesenchymal stem cells in Parkinson's disease: A pilot study. *J Pers Med*. 2020;10(4):93. doi:10.3390/jpm10040093

Bonafede R, Mariotti R. ALS pathogenesis and therapeutic approaches: The role of mesenchymal stem cells and extracellular vesicles. *Front Cell Neurosci.* 2017 Mar 21;11:80. doi:10.3389/fncel.2017.00080

Boucherie C, Hermans E. Adult stem cell therapies for neurological disorders: Beneficial effect of tissue plasminogen activator overexpression in mesenchymal stromal cells. *Brain Res Bull.* 2015;109:90–100. doi:10.1016/j.brainresbull.2014.10.007

Boucherie C, Schäfer P, Lavand'homme C, Maloteaux A, Hermans B. Chimerization of SOD1G93A transgenic mice with bone marrow–derived mesenchymal stem cells delays disease onset and prolongs survival. *J Neurosci Res.* 2009;87(9):2033–43. https://doi.org/10.1002/jnr.22021

Boutajangout A, Noorwali A, Atta H, Wisniewski T. Human umbilical cord stem cell xenografts improve cognitive decline and reduce the amyloid burden in a mouse model of Alzheimer's disease. *Curr Alzheimer Res.* 2017;14(1):104–111. doi:10.2174/1567205013666161004151416

Brody M, Agronin M, Herskowitz BJ, Bookheimer SY, Small GW, Hitchinson B, Ramdas K, Wishard T, McInerney KF, Vellas B, Sierra F, Jiang Z, Mcclain-Moss L, Perez C, Fuquay A, Rodriguez S, Hare JM, Oliva AA Jr, Baumel B. Results and insights from a phase I clinical trial of Lomecel-B for Alzheimer's disease. *Alzheimers Dement.* 2023;19(1):261–73. doi:10.1002/alz.12651

Bronckaers A, Hilkens P, Martens W, Gervois P, Ratajczak J, Struys T, Lambrichts I. Mesenchymal stem/stromal cells as a pharmacological and therapeutic approach to accelerate angiogenesis. *Pharmacol Ther.* 2014;143(2):181–96. doi:10.1016/j.pharmthera.2014.02.013

Candelario KM, Shuttleworth CW, Cunningham LA. Neural stem/progenitor cells display a low requirement for oxidative metabolism independent of hypoxia inducible factor-1alpha expression. *J Neurochem.* 2013;125:420–9. doi:10.1111/jnc.12204

Canesi M, Giordano R, Lazzari T, et al. Intravenous infusion of autologous mesenchymal stem cells in Parkinson's disease: A randomized, cross-over pilot study. *CNS Neurosci Ther.* 2016;22(10):870–8. doi:10.1111/cns.12568

Caplan AI, Correa D. The MSC: An injury drugstore. *Cell Stem Cell.* 2011;9(1):11–15. doi:10.1016/j.stem.2011.06.008

Cohen JA, Imrey PB, Planchon SM, Bermel RA, Fisher E, Fox RJ, Bar-Or A, Sharp SL, Skaramagas TT, Jagodnik P, Karafa M, Morrison S, Reese Koc J, Gerson SL, Lazarus HM. Pilot trial of intravenous autologous culture-expanded mesenchymal stem cell transplantation in multiple sclerosis. *Mult Scler.* 2018;24(4):501–11. doi:10.1177/1352458517703802

Connick P, Kolappan M, Crawley C, Webber DJ, Patani R, Michell AW, Du MQ, Luan SL, Altmann DR, Thompson AJ, Compston A, Scott MA, Miller DH, Chandran S. Autologous mesenchymal stem cells for the treatment of secondary progressive multiple sclerosis: an open-label phase 2a proof-of-concept study. *Lancet Neurol.* 2012 Feb;11(2):150–6. doi: 10.1016/S1474-4422(11)70305-2

Cudkowicz ME, Lindborg SR, Goyal NA, et al. A randomized placebo-controlled phase 3 study of mesenchymal stem cells induced to secrete high levels of neurotrophic factors in amyotrophic lateral sclerosis. *Muscle Nerve.* 2022;65(3):291–302. doi:10.1002/mus.27472.

Cummings J, Lee G, Nahed P, Kambar MEZN, Zhong K, Fonseca J, Taghva K. Alzheimer's disease drug development pipeline. *Alzheimers dement (N Y).* 2022;8(1):e12295. doi: 10.1002/trc2.12295

d'Angelo M, Cimini A, Castelli V. Insights into the effects of mesenchymal stem cell-derived secretome in Parkinson's disease. *Int J Mol Sci.* 2020;21(15):5241.

Deuse T, Hu X, Gravina A, Wang D, Tediashvili G, De C, Thayer WO, Wahl A, Garcia JV, Reichenspurner H, Davis MM, Lanier LL, Schrepfer S. Hypoimmunogenic derivatives of induced pluripotent stem cells evade immune rejection in fully immunocompetent allogeneic recipients. *Nat Biotechnol.* 2019;37(3):252–8. doi: 10.1038/s41587-019-0016-3. Epub 2019 Feb 18. Erratum in: Nat Biotechnol. 2022 Nov;40(11):1690. doi:10.1038/s41587-022-01426-8

Donega V, Nijboer CH, van Tilborg G, Dijkhuizen RM, Kavelaars A, Heijnen CJ. Intranasally administered mesenchymal stem cells promote a regenerative niche for repair of neonatal ischemic brain injury. *Exp Neurol.* 2014;261:53–64. doi:10.1016/j.expneurol.2014.06.009

Galipeau J, Sensébé L. Mesenchymal stromal cells: Clinical challenges and therapeutic opportunities. *Cell Stem Cell.* 2018;22 (6): 824–33. https://doi.org/10.1016/j.stem.2018.05.004

Gerdoni E, Gallo B, Casazza V, et al. Mesenchymal stem cells effectively modulate inflammatory autoimmune response in experimental autoimmune encephalomyelitis. *Ann Neurol.* 2007;61(3):219–27. doi:10.1002/ana.21076.

Giordano C, Iommarini L , Giordano L, Maresca A, Pisano A, Valentino ML, Caporali L, Liguori R, Deceglie S, Roberti M, Fanelli F, Fracasso F, Ross-Cisneros FN, D'Adamo P, Hudson G, Pyle A, Yu-Wai-Man P, Chinnery PF, Zeviani M, Salomao SR, Berezovsky A, Belfort R Jr, Ventura DF, Moraes M, Moraes Filho M, Barboni P, Sadun F, De Negri A, Sadun AA, Tancredi A, Mancini M, d'Amati G, Loguercio Polosa P, Cantatore P, Carelli V. Efficient mitochondrial biogenesis drives incomplete penetrance in Leber's hereditary optic neuropathy. *Brain*. 2014;137(Pt 2):335–53. doi:10.1093/brain/awt343

Glenn JD, Whartenby KA. Mesenchymal stem cells: Emerging mechanisms of immunomodulation and therapy. *World J Stem Cells*. 2014;6(5):526–39. doi:10.4252/wjsc.v6.i5.526

Goutman SA, Hardiman O, Al-Chalabi A, Chió A, Savelieff MG, Kiernan MC, Feldman EL. Recent advances in the diagnosis and prognosis of amyotrophic lateral sclerosis. *Lancet Neurol*. 2022;21(5):480–93. doi:10.1016/S1474-4422(21)00465-8

Hare JM, Baumel BJ, Oliva A. et al. Allogeneic mesenchymal stem cell therapy with laromestrocel in mild Alzheimer's disease: A randomized controlled phase 2a trial. *Nat Med*. 2025;31:1257–66. doi:10.1038/s41591-025-03559-0

Harris VK, Stark J, Vyshkina T, Blackshear L, Joo G, Stefanova V, Sara G, Sadiq SA. Phase I trial of intrathecal mesenchymal stem cell-derived neural progenitors in progressive multiple sclerosis. *EBioMedicine*. 2018;29:23–30. doi:10.1016/j.ebiom.2018.02.002

Hayakawa K, Esposito E, Wang X. et al. Transfer of mitochondria from astrocytes to neurons after stroke. *Nature*. 2016;535:551–5. doi:10.1038/nature18928.

Jang SE, Qiu L, Chan LL, Tan EK, Zeng L. Current status of stem cell-derived therapies for Parkinson's disease: From cell assessment and imaging modalities to clinical trials. *Front Neurosci*. 2020;14:558532. doi:10.3389/fnins.2020.558532

Jaramillo-Merchán J, Jones J, Ivorra JL, Pastor D, Viso-León MC, Armengól JA, Moltó MD, Geijo-Barrientos E, Martínez S. Mesenchymal stromal-cell transplants induce oligodendrocyte progenitor migration and remyelination in a chronic demyelination model. *Cell Death Dis*. 2013;4(8):e779. doi:10.1038/cddis.2013.304

Kalladka D, Sinden J, Pollock K, Haig C, McLean J, Smith W, McConnachie A, Santosh C, Bath PM, Dunn L, Muir KW. Human neural stem cells in patients with chronic ischaemic stroke (PISCES): A phase 1, first-in-man study. *Lancet*. 2016;388(10046):787–96. doi:10.1016/S0140-6736(16)30513-X

Karussis D, Karageorgiou I, Vaknin-Dembinsky T, et al. Safety and immunological effects of mesenchymal stem cell transplantation in patients with multiple sclerosis and amyotrophic lateral sclerosis. *Arch Neurol*. 2010;67(10):1187–94. doi:10.1001/archneurol.2010.248

Kim HJ, Cho KR, Jang H. et al. Intracerebroventricular injection of human umbilical cord blood mesenchymal stem cells in patients with Alzheimer's disease dementia: A phase I clinical trial. *Alzheimer's Res Ther*. 2021;13:154. doi:10.1186/s13195-021-00897-2

Kim HJ, Kim SU, Jung JK. et al. Stereotactic intraventricular implantation of mesenchymal stem cells in patients with Alzheimer's disease dementia: A phase I clinical trial. *Alzheimer's Res Ther*. 2015;7(1):59. doi:10.1186/s13195-015-0140-8

Kim J, Lee Y, Lee S, Kim K, Song M, Lee J. Mesenchymal stem cell therapy and Alzheimer's disease: Current status and future perspectives. *J Alzheimers Dis*. 2020;77(1):1–14. doi:10.3233/JAD-200219

Langer-Gould A, Cree CE, Williams BR, et al. Evolution of multiple sclerosis therapeutics: Potential lessons for the path forward in progressive multiple sclerosis. *Neurology*. 2023;100(3):119–33. doi:10.1212/WNL.0000000000201536

Laso-García F, Otero-Ortega ME, Colmenárez I, et al. Systemic mesenchymal stem cell administration enhances oligodendrocyte maturation and promotes functional recovery after oral ethidium bromide-induced demyelination in rats. *Stem Cell Res Ther*. 2018;9:133. doi:10.1186/s13287-018-0875-x

Leijs MJC, Villafuertes E, Haeck JC, Koevoet WJLM, Fernandez-Gutierrez B, Hoogduijn MJ, Verhaar JAN, Bernsen MR, van Buul GM, van Osch GJVM. Encapsulation of allogeneic mesenchymal stem cells in alginate extends local presence and therapeutic function. *Eur Cells Mater*. 2017;33:43–58. https://doi.org/10.22203/eCM.v033a04

Li J, Li N, Wei J, et al. Genetically engineered mesenchymal stem cells with dopamine synthesis for Parkinson's disease in animal models. *npj Parkinsons Dis*. 2022;8:175. doi:10.1038/s41531-022-00440-6

Li Z, Liu F, He X, Yang X, Shan F, Feng J. Exosomes derived from mesenchymal stem cells attenuate inflammation and demyelination of the central nervous system in EAE rats by regulating the polarization of microglia. *Int Immunopharmacol*. 2019;67:268–80. https://doi.org/10.1016/j.intimp.2018.12.001

Losurdo M, Pedrazzoli M, D'Agostino C, Elia CA, Massenzio F, Lonati E, Mauri M, Rizzi L, Molteni L, Bresciani E, Dander E, D'Amico G, Bulbarelli A, Torsello A, Matteoli M, Buffelli M, Coco S. Intranasal delivery of mesenchymal stem cell-derived extracellular vesicles exerts immunomodulatory and neuroprotective effects in a 3xTg model of Alzheimer's disease. *Stem Cells Transl Med.* 2020;9(9):1068–84. doi:10.1002/sctm.19-0327

Lublin FD, Reingold SC, Cohen JA, Cutter GR, Sørensen PS, Thompson AJ, Wolinsky JS, Balcer LJ, Banwell B, Barkhof F, Bebo B Jr, Calabresi PA, Clanet M, Comi G, Fox RJ, Freedman MS, Goodman AD, Inglese M, Kappos L, Kieseier BC, Lincoln JA, Lubetzki C, Miller AE, Montalban X, O'Connor PW, Petkau J, Pozzilli C, Rudick RA, Sormani MP, Stüve O, Waubant E, Polman CH. Defining the clinical course of multiple sclerosis: The 2013 revisions. *Neurology.* 2014;83(3):278–86. doi:10.1212/WNL.0000000000000560

Lukomska B, Stanaszek L, Zuba-Surma E, Legosz P, Sarzynska S, Drela K. Challenges and controversies in human mesenchymal stem cell therapy. *Stem Cells Int.* 2019;2019:9628536. doi:10.1155/2019/9628536

Mazzini L, Ferrero I, Luparello V, Rustichelli D, Gunetti M, Mareschi K, Testa L, Stecco A, Tarletti R, Miglioretti M, Fava E, Nasuelli N, Massara M, Cisari C, Fagioli F. Mesenchymal stem cell transplantation in amyotrophic lateral sclerosis: A phase I clinical trial. *Exp Neurol.* 2009;223(1):229–37. doi:10.1016/j.expneurol.2009.08.007

Mazzini L, et al. Stem cell treatment in amyotrophic lateral sclerosis. *J NeurolSci.* 2008;265(1–2):78–83. https://doi.org/10.1016/j.jns.2007.05.016

McCoy MK, Martinez SD, Botterman PR. Partial neurological recovery and neuroprotective effects induced by bone marrow-derived mesenchymal stem cell administration in a rat model of Parkinson's disease. *PLoS One.* 2008;3(7):e2698. doi:10.1371/journal.pone.0002698

Mendes AF, Rodrigues GB, Saldanha-Araujo FSM, Covas DT. Mesenchymal stromal cell-derived small extracellular vesicles in neurodegenerative disorders: Challenges and opportunities. *Front Cell Dev Biol.* 2019;7:200. doi:10.3389/fcell.2019.00200.

Mohyeddin Bonab M, Yazdanbakhsh K, Nezhad AL, et al. Does mesenchymal stem cell therapy help multiple sclerosis patients? Report of a pilot study. *Iran J Immunol.* 2007;4(1):50–7.

Moll G, Drzeniek E, Kamhieh-Milz SK, et al. Freshly isolated and culture-expanded mesenchymal stromal cells (MSCs) equally inhibit platelet activation in vitro and in vivo. *Stem Cell Res Ther.* 2020;11(1):94. doi:10.1186/s13287-020-1581-x

Mukhamedshina Y, et al. Mesenchymal stem cell therapy for spinal cord contusion: A comparative study on small and large animal models. *Biomolecules.* 2019;9(12):811. https://doi.org/10.3390/biom9120811

Müller L, Tunger A, Wobus M, von Bonin M, Towers R, Bornhäuser M, Dazzi F, Wehner R, Schmitz M. Immunomodulatory properties of mesenchymal stromal cells: An update. *Front Cell Dev Biol.* 2021;9:637725. doi:10.3389/fcell.2021.637725

Nagata T, Nakagawa Y, Hirota T, et al. Beneficial effect of intrathecal administration of fibroblast growth factor-1 in patients with amyotrophic lateral sclerosis. *Neurol Med Chir.* 2013;53(7):466–72. doi:10.2176/nmc.53.466

Neves AF, Camargo C, Premer C, Hare JM, Baumel BS, Pinto M. Intravenous administration of mesenchymal stem cells reduces Tau phosphorylation and inflammation in the 3xTg-AD mouse model of Alzheimer's disease. *Exp Neurol.* 2021;341:113706. doi:10.1016/j.expneurol.2021.113706

Pavlovic K, Tristán-Manzano M, Maldonado-Pérez N, Cortijo-Gutierrez M, Sánchez-Hernández S, Justicia-Lirio P, Carmona MD, Herrera C, Martin F, Benabdellah K. Using gene editing approaches to fine-tune the immune system. *Front Immunol.* 2020;11:570672. doi:10.3389/fimmu.2020.570672

Petrou P, Kassis I, Ginzberg A, Hallimi M, Karussis D. Effects of mesenchymal stem cell transplantation on cerebrospinal fluid biomarkers in progressive multiple sclerosis. *Stem Cells Transl Med.* 2022;11(1):55–8. doi:10.1093/stcltm/szab017

Petrou P, Levy YS, Ben Sira A, et al. Safety, tolerability, and preliminary efficacy of intrathecal administration of autologous mesenchymal stem cells in patients with amyotrophic lateral sclerosis: Results of a phase I/II trial. *JAMA Neurol.* 2016;73(3):337–44. doi:10.1001/jamaneurol.2015.4321

Phinney DG, Pittenger MF. Concise review: MSC-derived exosomes for cell-free therapy. *Stem Cells.* 2017;35(4):851–8. doi:10.1002/stem.2575. Epub 2017 Mar 10. Erratum in: Stem Cells. 2017 Sep;35(9):2103. doi:10.1002/stem.2626

Qu J, Zhang H. Roles of mesenchymal stem cells in spinal cord injury. *Stem Cells Int.* 2017;2017:5251313. doi:10.1155/2017/5251313

Rash BG, Ramdas KN, Oliva AA Jr. et al. Allogeneic MSC therapy in mild Alzheimer's disease: A randomized phase 2a trial. *Nat Med.* 2025;31:1257–66. doi:10.1038/s41591-025-03559-0

Razavi S, Ghasemi N, Mardani M, Salehi H. Remyelination improvement after neurotrophic factors secreting cells transplantation in rat spinal cord injury. *Iran J Basic Med Sci.* 2017;20(4):392–8. doi:10.22038/IJBMS.2017.8580

Sahraian MA, Bonab MM, Baghbanian SM, Owji M, Moghadasi AN. Therapeutic use of intrathecal mesenchymal stem cells in patients with multiple sclerosis: A pilot study with booster injection. *Immunol Invest.* 2019;48(2):160–8. https://doi.org/10.1080/08820139.2018.1504301

Salmikangas P, Carlsson B, Klumb C, Reimer T, Thirstrup S. Potency testing of cell and gene therapy products. *Front Med.* 2023;10:1190016. doi:10.3389/fmed.2023.1190016

Shin JY, Kim SH, Park YS, et al. Mesenchymal stem cells enhance autophagy and increase β-amyloid clearance in Alzheimer disease models. *Autophagy.* 2014;10(1):32–44. doi:10.4161/auto.26508

Sipp D, Robey PG, Turner L. Clear up this stem-cell mess. *Nature.* 2018 Sep;561(7724):455–7. doi:10.1038/d41586-018-06756-9

Siwek T, Jezierska-Woźniak K, Maksymowicz S, Barczewska M, Sowa M, Badowska W, Maksymowicz W. Repeat administration of bone marrow-derived mesenchymal stem cells for treatment of amyotrophic lateral sclerosis. *Med Sci Monit.* 2020;26:e927484. doi:10.12659/MSM.927484

Soares DM, Gonçalves RGJ, Vasques CL, et al. Current status of mesenchymal stem/stromal cells for treatment of neurological diseases. *Front Mol Neurosci.* 2022;15:913075. doi:10.3389/fnmol.2022.913075

Staff NP, Singer W, Froiland JO, et al. Safety of intrathecal autologous adipose-derived mesenchymal stromal cells in patients with ALS. *Neurology.* 2016;87(21):2230–34. doi:10.1212/WNL.0000000000003342

Sun H, Bénardais K, Stanslowsky N, Thau-Habermann N, Hensel N, Huang D, Claus P, Dengler R, Stangel M, Petri S. Therapeutic potential of mesenchymal stromal cells and MSC conditioned medium in amyotrophic lateral sclerosis (ALS) — In vitro evidence from primary motor neuron cultures, NSC-34 cells, astrocytes and microglia. *PLoS One.* 2013;8(9):e72926. doi:10.1371/journal.pone.0072926

Syková E, Rychmach J, Drahorádová P. et al. Transplantation of mesenchymal stromal cells in patients with amyotrophic lateral sclerosis: Results of phase I/IIa clinical trial. *Cell Transplant.* 2017;26(4):647–58. doi:10.3727/096368916X693498

Tseng N, Lambie SC, Huynh CQ, Sanford B, Patel M, Herson PS, Ormond DR. Mitochondrial transfer from mesenchymal stem cells improves neuronal metabolism after oxidant injury in vitro: The role of Miro1. *J Cereb Blood Flow Metab.* 2021;41(4):761–70. doi: 10.1177/0271678X20928147

Uccelli A, Laroni F, Freedman MS. Mesenchymal stem cells for the treatment of multiple sclerosis and other neurological diseases. *Lancet Neurol.* 2020;19(5):360–71. doi:10.1016/S1474-4422(19)30489-7

Uccelli A. et al. Safety, tolerability, and activity of mesenchymal stem cells versus placebo in multiple sclerosis (MESEMS): A phase 2, randomised, double-blind crossover trial. *Lancet Neurol.* 2021;20(11):917–29. doi:10.1016/S1474-4422(21)00301-X

Vaquero J, Zurita M, Bonilla C, Fernández-Mateos C, Rubio JJ. et al. Progressive increase in brain glucose metabolism after intrathecal administration of autologous mesenchymal stromal cells in patients with diffuse axonal injury. *Cytotherapy.* 2017;19(1):88–94. doi:10.1016/j.jcyt.2016.10.001

Venkataramana NK, Kumar SK, Balaraju S, Radhakrishnan RC, Bansal A, Dixit A, Rao DK, Das M, Jan M, Gupta PK, Totey SM. Open-labeled study of unilateral autologous bone-marrow-derived mesenchymal stem cell transplantation in Parkinson's disease. *Transl Res.* 2010;155(2):62–70. doi:10.1016/j.trsl.2009.07.006

Vercelli A, Mereuta F, Giunti S. et al. Human mesenchymal stem cell transplantation extends survival, improves motor performance and decreases neuroinflammation in mouse model of amyotrophic lateral sclerosis. *Neurobiol Dis.* 2008;31(3):395–405. doi:10.1016/j.nbd.2008.05.016.

Wang F, Yasuhara T, Shingo T, et al. Intravenous administration of mesenchymal stem cells exerts therapeutic effects on parkinsonian model of rats: Focusing on neuroprotective effects of stromal cell-derived factor-1α. *BMC Neurosci.* 2010;11:52. https://doi.org/10.1186/1471-2202-11-52

Wang J, Zhang SX, Chang JS, Cheng T, Jiang XJ, Su QY, Zhang JQ, Luo J, Li XF. Low-dose IL-2 improved clinical symptoms by restoring reduced regulatory T cells in patients with refractory rheumatoid arthritis: A randomized controlled trial. *Front Immunol.* 2022;13:947341. doi:10.3389/fimmu.2022.947341

Wei Z, Zhang P, Zhang Z. et al. Adipose stem cells-secreted neprilysin inhibits amyloid-β accumulation in a paracrine manner. *Stem Cells.* 2018;36(7):1062–73. doi:10.1002/stem.2806

Wang, et al. SDF-1/CXCR4 axis modulates bone marrow mesenchymal stem cell apoptosis, migration and cytokine secretion. *Protein Cell.* 2011;2(10):845–54. https://doi.org/10.1007/s13238-011-1097-z

Xu M, Shaw G, Murphy M, Barry F. Induced pluripotent stem cell-derived mesenchymal stromal cells are functionally and genetically different from bone marrow-derived mesenchymal stromal cells. *Stem Cells.* 2019;37(6):754–65. https://doi.org/10.1002/stem.2993

Yasuhara T, Kameda Y, Sasaki M. et al. Cell therapy for Parkinson's disease using a cultured human neural stem cell line: A pilot study. *J Neurosci Res.* 2017;95(9):1828–41. doi:10.1002/jnr.24005

Yu X, Chen D, Zhang Y, Wu X, Huang Z, Zhou H, Zhang Y, Zhang Z. Overexpression of CXCR4 in mesenchymal stem cells promotes migration, neuroprotection and angiogenesis in a rat model of stroke. *J Neurol Sci.* 2012;316(1–2):141–9. doi: 10.1016/j.jns.2012.01.001

Zhang J, Li Z, Li C. et al. Mesenchymal stem cells modify microglial phenotype and promote recovery via inhibition of methyl-CpG-binding protein 2 in TBI. *Mol Ther.* 2020;28(2):517–31. doi:10.1016/j.ymthe.2019.10.009

Zhao et al. Mesenchymal stem cell transplantation improves motor function and suppresses T-cell infiltration in the spinal cord of amyotrophic lateral sclerosis mice. *Stem Cells.* 2007;25(8):2067–73. doi:10.1634/stemcells.2006-0627

Zhu G, Zhao Q, Yang L. et al. Mesenchymal stem cell exosome therapy: Current research status in the treatment of neurodegenerative diseases and the possibility of reversing normal brain aging. *J Transl Med.* 2025;23(1):41. doi:10.1186/s12967-025-04126-1

5 Immunomodulatory Roles of Stem Cells in Neurodegenerative Diseases

5.1 INTRODUCTION

Neurodegenerative disorders such as Alzheimer's disease (AD), Parkinson's disease (PD), amyotrophic lateral sclerosis (ALS), and multiple sclerosis (MS) are characterized not only by progressive loss of specific neurons but also by chronic neuroinflammation. In these conditions, innate immune cells, microglia, and astrocytes in the central nervous system (CNS) become persistently activated, and in some cases, peripheral immune cells infiltrate the CNS, creating a sustained inflammatory milieu that contributes to neuronal damage (Chen et al. 2018). In AD and PD, chronic activation of microglia and astrocytes is believed to exacerbate pathology by releasing pro-inflammatory cytokines and toxic mediators, whereas in ALS and especially MS, dysfuntional immune responses, including T cell and macrophage infiltration play a direct role in disease progression (Chen et al. 2018). Modulating these immune responses has therefore emerged as a promising therapeutic strategy to slow neurodegeneration.

Stem cell-based therapies, originally pursued for their potential to replace lost neurons, are now recognized to also exert powerful immunomodulatory and neurotrophic effects that can modify disease environments. Two types of stem cells have attracted particular attention in this context: mesenchymal stem cells (MSCs) and neural stem cells (NSCs). MSCs are multipotent stromal cells obtained from adult tissues that are known for secreting anti-inflammatory and neuroprotective factors and interacting with immune cells to suppress pathological immune responses (Gao et al. 2016; Chen et al. 2018). NSCs, which reside in the developing and adult brain or can be derived in vitro, also secrete a cocktail of growth factors, cytokines, and extracellular vesicles (EVs) that can promote tissue repair and modulate inflammation without replacing the neurons (Willis et al. 2020; Genchi et al. 2023). These immunomodulatory properties position MSCs and NSCs as attractive candidates for treating neurodegenerative diseases where inflammation is a key component of pathology.

There has been an upsurge in preclinical studies, in vitro, animal models, and clinical trials exploring the use of MSCs and NSCs to ameliorate neurodegenerative disorders via immune modulation. Preclinical models have demonstrated that transplanted MSCs or NSCs can reduce neuroinflammation, protect neurons, and improve functional outcomes in models of AD, PD, ALS, and MS (Madhu et al. 2020). Early-phase clinical studies have also been initiated to test safety and efficacy in patients, and while results are mixed, they provide crucial insights into dosage, delivery, and biological activity of stem cell therapies in humans.

In this chapter, a comprehensive overview of the immunomodulatory roles of stem cells focusing on MSCs and NSCs in different neurodegenerative diseases is presented. The key findings encompassing both mechanistic studies and therapeutic applications of specific diseases is also

DOI: 10.1201/9781003582687-5

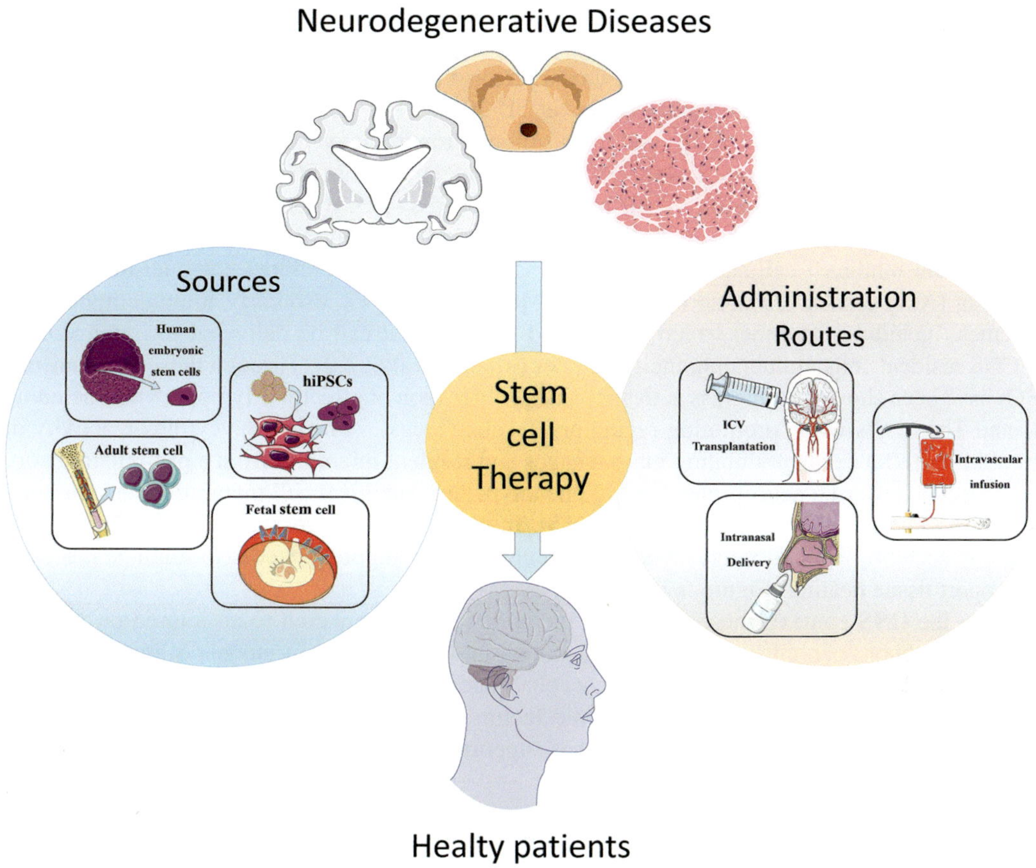

FIGURE 5.1 Stem Cell Sources and Delivery Routes for Neurodegenerative Therapy.

reviewed. The role of stem cell therapies in modulating immune responses in AD, PD, ALS, and MS is also discussed besides highlighting evidence from animal models and outcomes of clinical trials. The mechanistic basis of stem cell-mediated immunomodulation, including the strategies through which MSCs and NSCs interact with immune cells and secrete factors that alter the neuroinflammatory environment, is also discussed. A section on clinical trials examines the current state of translation, including recent trials, and their efficacy and safety. Finally, the chapter delineates the persisting challenges within the field, including achieving sustained engraftment, establishing standardized and reproducible cell preparations, and addressing ethical as well as regulatory complexities. It further outlines prospective directions encompassing advanced strategies such as genetic and functional cell engineering, along with innovative cell-free therapeutic modalities, particularly the utilization of stem cell–derived secretomes and EVs as promising alternatives to conventional cell-based interventions (Figure 5.1).

5.2 MECHANISMS OF IMMUNOMODULATION BY STEM CELLS IN THE CNS

MSCs have been characterized for their immunoregulatory capabilities. Unlike immune cells, MSCs are not inherently antigen-presenting and are least immune-evasive, enabling them to be transplanted with relatively low risk of immediate rejection (Gao et al. 2016). More importantly,

MSCs actively influence the behavior of multiple immune cell types through cell–cell contact and secretion of soluble factors. Upon sensing an inflammatory environment, for instance, high levels of interferon-γ (IFN-γ) or tumor necrosis factor-α (TNF-α), MSCs adopt a phenotype that produces anti-inflammatory mediators (Chen et al. 2018). The key soluble factors secreted by MSCs include interleukin-10 (IL-10) and transforming growth factor-β (TGF-β), which are potent suppressors of pro-inflammatory T cell responses, prostaglandin E2 (PGE2), and indoleamine 2,3-dioxygenase (IDO), which inhibit T cell proliferation and switch macrophages/microglia toward anti-inflammatory states, nitric oxide (NO) produced via inducible NO synthase, which at high local concentrations inhibits T cell and NK cell functions, and various chemokines that alter immune cell trafficking (Xiao et al. 2015; Gao et al. 2016). MSC-derived EVs also carry immunomodulatory molecules, including cytokines, microRNAs, and proteins that can be delivered to immune cells and CNS resident cells, influencing their behavior (Madhu et al. 2020). Through these mechanisms, MSCs have been shown to suppress activation and proliferation of effector T lymphocytes, including Th1 and Th17 cells while promoting regulatory T cells (Tregs), reducing the cytotoxic activity of natural killer (NK) cells, modulating macrophages and resident microglia from a pro-inflammatory M1 phenotype toward a neuroprotective M2 phenotype; and inhibit B cell maturation and antibody production (Chen et al. 2018; Xiao et al. 2015). In essence, MSCs respond to the inflammatory signals by secreting a broad range of factors that collectively suppress excessive immune reactions and support tissue healing (Figure 5.2).

Within the CNS context, these properties mean that MSCs delivered to an injured or diseased brain or spinal cord can mitigate chronic inflammation. For example, in models of injected MSCs secrete IFN-γ-induced factors that reduce encephalitogenic T cell responses and macrophage activation, while simultaneously releasing trophic factors like hepatocyte growth factor (HGF) and brain-derived neurotrophic factor (BDNF) that support oligodendrocyte survival and neurorepair

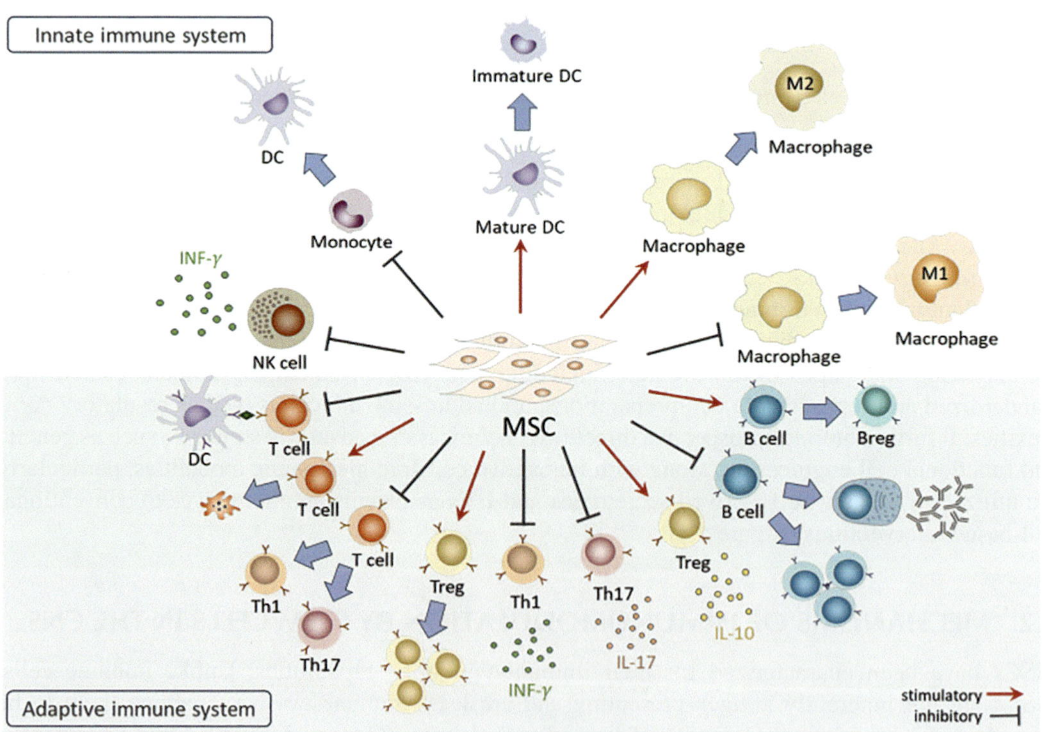

FIGURE 5.2 Immunomodulatory Actions of MSCs on Innate and Adaptive Immunity.

(Xiao et al. 2015). The net effect observed in experimental autoimmune encephalomyelitis is a reduction in inflammatory lesions and clinical severity, an effect strongly tied to MSCs' modulation of T cells and myeloid cells (Xiao et al. 2015). Similarly, in neurodegenerative models not driven primarily by autoimmunity such as ALS or PD, MSCs still exert beneficial effects largely through immunomodulation. They attenuate the reactivity of microglia and astrocytes, lower levels of pro-inflammatory cytokines in the CNS, and create a more permissive environment for neurons to survive or regenerate (Laroni et al. 2015; Madhu et al. 2020). It is noteworthy that MSCs rarely differentiate into neural cells in vivo and typically have poor long-term engraftment in the CNS. Thus, the therapeutic effects of MSCs in neurodegeneration are thought to arise predominantly from their paracrine actions and immune reprogramming rather than cell replacement (Kim et al. 2021). This paradigm is often referred to as the bystander effect of MSCs.

NSCs and their derivative neural progenitor cells (NPCs) exert notable immunomodulatory effects within the CNS, in addition to their well-characterized capacity for neurogenesis. While their primary role involves differentiation into neural lineage cells, emerging evidence indicates that both endogenous and exogenously administered NSCs or NPCs can modulate immune responses in a context-dependent manner. Within specialized neurogenic niches such as the subventricular zone, NSCs contribute to tissue repair processes. However, transplanted NSCs and NPCs have demonstrated the ability to attenuate neuroinflammation by influencing the activity of microglia, astrocytes, and infiltrating immune cells, thereby supporting neuroprotection and regeneration (Table 5.1; Pluchino et al. 2009; Willis et al. 2020). Transplanted NSCs secrete a spectrum of factors that have been shown to modulate microglial activation, T cell function, and even peripheral immune cell infiltration. Notably, NSC secretome analyses reveal a mixture of growth factors e.g. BDNF, glial cell line-derived neurotrophic factor (GDNF), vascular endothelial growth factor, anti-inflammatory cytokines, chemokines, and EVs (Willis et al. 2020; Table 5.2). These factors can

TABLE 5.1
Immunomodulatory Features of MSCs and NSCs

Feature	MSCs	NSCs
Origin	Mesodermal (bone marrow, adipose)	Neuroectodermal (brain, iPSC-derived)
Delivery route	Intravenous/intrathecal	Intrathecal/intraparenchymal
Primary mechanism	Paracrine immunomodulation	Context-dependent neuroimmune interaction
Key secreted factors	IL-10, TGF-β, PGE2, IDO	BDNF, GDNF, IL-10, LIF
Target immune cells	T cells, macrophages, NK cells	Microglia, T cells, astrocytes

TABLE 5.2
Stem Cell Applications across ND Diseases

Disease	Stem Cell Type	Mechanism of Action	Clinical Trial Phase
Alzheimer's disease (AD)	MSCs, NSCs	Modulate microglia, reduce Aβ and tau	Phase I/II
Parkinson's disease (PD)	MSCs, NSCs	Protect dopaminergic neurons, reduce inflammation	Phase I
Amyotrophic lateral sclerosis (ALS)	MSCs, NSCs	Reduce neuroinflammation, support motor neurons	Phase II/III
Multiple sclerosis (MS)	MSCs, NSCs	Suppress autoimmunity, promote remyelination	Phase II

suppress pro-inflammatory pathways and promote an environment conducive to regeneration. For instance, NSCs secrete IL-10 and TGF-β which can counteract inflammatory cytokines and also release molecules like leukemia inhibitory factor (LIF) and osteopontin that have been implicated in driving microglia towards a tissue-reparative phenotype (Einstein et al. 2016). In models of inflammatory CNS damage such as EAE or CNS injury, transplanted NPCs have been observed to reduce the infiltration of peripheral immune cells and shift the balance of CNS-resident microglia or macrophages toward anti-inflammatory states (Pluchino et al. 2009). A classic example comes from early studies by Pluchino and colleagues, who showed that NPC transplantation in EAE mice led to functional improvement not by replacing oligodendrocytes but by delivering T cell-suppressive and neurotrophic factors that reduced demyelination and axonal loss (Pluchino et al. 2005). NSCs can also induce the apoptotic elimination of infiltrating T cells in the CNS and promote expansion of Tregs, thereby decreasing autoimmune inflammation (Coyle 1996).

A distinctive feature of NSCs is their capacity to dynamically sense and respond to biochemical and cellular responses within the CNS microenvironment. This enables NSCs to adapt their secretory profile, releasing a context-specific array of cytokines, growth factors, and EVs that can influence neuroinflammation, support neuronal survival, and promote tissue regeneration. Such plasticity underscores their dual role in both neuro-regeneration and immunomodulation. When transplanted NSCs localize to sites of inflammation within the CNS, they respond by upregulating the expression and secretion of anti-inflammatory cytokines and antioxidant molecules, thereby attenuating immune activation and oxidative stress. In contrast, within neurodegenerative environments characterized by cellular loss but limited inflammation, NSCs preferentially shift toward a trophic support role, secreting neurotrophic and oligotrophic factors that promote the survival, differentiation, and repair of neurons and oligodendrocytes (Willis et al. 2020; Genchi et al. 2023). This responsiveness allows NSCs to act as localized regulators of the immune environment. Additionally, NSCs closely interact with resident glial cells. Studies indicate NSCs can directly communicate with microglia via gap junctions or ligand–receptor interactions, leading microglia to adopt phenotypes that facilitate debris clearance and release growth factors (Mathiasen and Kusuma 2022). NSCs can modulate astrocyte function by secreting paracrine factors that dampen astrocyte-derived pro-inflammatory mediators. This ability underscores the emerging field of regenerative neuroimmunology, which exploits NSCs and other stem cell populations to reprogram the postinjury immune milieu of the CNS from a neurotoxic to a pro-regenerative state (Willis et al. 2020).

While MSCs and NSCs operate in different contexts, MSCs are typically administered peripherally or into cerebrospinal fluid. While NSCs are often transplanted directly into CNS tissue, they share some common immunomodulatory mechanisms. Both cell types produce overlapping anti-inflammatory cytokines and growth factors and induce the generation of regulatory immune cells, including Tregs or anti-inflammatory myeloid cells (Chen et al. 2018; Pluchino et al. 2009). Both have demonstrated the ability to reduce levels of classic pro-inflammatory cytokines like TNF-α, IL-1β, and IL-6 in various neurodegenerative models. However, important distinctions exist between NSCs and MSCs in their immunomodulatory capacities. MSCs, derived from mesodermal lineages, exhibit robust interactions with peripheral immune components, including the suppression of systemic T cell responses, modulation of dendritic cell maturation, and alteration of lymphocyte trafficking. Owing to these properties, MSCs are more extensively studied in the context of systemic or peripheral immune regulation, particularly in inflammatory and autoimmune disorders, whereas NSCs are more specialized in modulating the neuroimmune milieu within the CNS. In contrast, NSCs operate within the CNS niche and provide local immunomodulation coupled with structural support, e.g., NSCs can differentiate and integrate into host tissue, possibly replacing some lost glial cells and thereby stabilizing the environment. NSCs also inherently address CNS-specific regenerative pathways, e.g., remyelination in MS models via oligodendrocyte differentiation that MSCs typically do not.

An emerging mechanism of interest in both MSC and NSC therapy is the role of EVs such as exosomes. These nano-sized vesicles carry proteins, mRNAs, and microRNAs from the parent stem cells and can distribute immunomodulatory signals broadly in the tissue. For instance, MSC-derived exosomes have been shown to inhibit activation of microglia and reduce secretion of TNF-α and IL-6 in models of ALS and PD, thereby protecting neurons (Pinho et al. 2020). Similarly, NSC-derived exosomes can transfer microRNAs that suppress pro-apoptotic and pro-inflammatory gene expression in target cells, contributing to neuroprotection (Nakano et al. 2020). Because EVs can cross biological barriers, they offer a cell-free means to recapitulate the immunomodulatory effects of stem cells. Research from the past few years suggests that the beneficial effects of stem cell transplants in neurodegeneration might be largely mediated by their secreted vesicles and factors thus focusing on the secretome could allow therapeutic exploitation of these effects without transplanting live cells (Willis et al. 2020; Madhu et al. 2020).

Thus, both MSCs and NSCs serve as modulators of the neuroimmune environment. They secrete overlapping sets of anti-inflammatory and trophic factors and interact with immune cells to suppress inflammation while promoting repairs. These actions underpin the therapeutic applications of stem cells in neurodegenerative diseases. It is important to understand the precise mechanisms by which immune pathways are affected and how critical it is for optimizing stem cell therapies, such as deciding which cell type or secretome components are best suited for a given disease, or how to engineer cells to enhance their immunomodulatory functions.

5.3 APPLICATION OF MLS AND NSC IN DIFFERENT ND DISEASES

5.3.1 Alzheimer's Disease

AD is characterized by extracellular amyloid-β (Aβ) plaques, intracellular tau neurofibrillary tangles, and a progressive loss of neurons, particularly in the hippocampus and cortex. In recent years, neuroinflammation has been recognized as a critical component of AD pathogenesis (Heneka et al. 2015). Microglia surrounding Aβ plaques become activated and release cytokines and chemokines that can contribute to synaptic dysfunction and neuron damage. Moreover, genetic studies have identified several AD risk genes that are expressed in microglia or immune-related pathways, underscoring the role of immune responses in disease progression. Given this, immunomodulatory therapies aiming to shift microglia and astrocytes from a pro-inflammatory, neurotoxic state to a more supportive are being explored to slow AD progression (Marsh et al. 2016). Stem cell-based strategies were initially introduced into AD research with the primary aim of delivering neurotrophic support and replacing lost or dysfunctional neurons (Figure 5.3). However, subsequent findings revealed that specific stem cell types, particularly MSCs, possess broader therapeutic potential. Beyond neuroregeneration, MSCs have demonstrated the capacity to modulate neuroinflammation and alter the progression of AD pathology by influencing amyloid-β (Aβ) accumulation and tau phosphorylation through the secretion of anti-inflammatory cytokines, growth factors, and EVs (Table 5.3). Multiple studies in AD transgenic mice overproducing Aβ have demonstrated beneficial effects of MSC administration. For example, intravenous or intrahippocampal transplantation of human MSCs in APP/PS1 transgenic AD mice has been shown to reduce microglial activation around plaques and shift the cytokine milieu towards anti-inflammatory profiles (Kim et al. 2015). Behavioral improvements such as better performance in memory tasks often accompany these molecular changes. Mechanistically, MSCs appear to help clear Aβ pathology and protect neurons via secreted factors. It has been indicated that human umbilical cord blood-derived MSCs secrete soluble intracellular adhesion molecule-1 (sICAM-1) and agouti-related protein (AgRP), which in AD models were found to reduce Aβ levels, possibly by enhancing microglial phagocytosis of Aβ (Kim et al. 2021). The same group showed that MSC-derived galectin-3 can bind to Aβ and decrease its neurotoxic effects on neurons (Kim et al. 2021). Additionally, MSCs were found to boost endogenous repair mechanisms in AD models. They increased adult hippocampal neurogenesis and

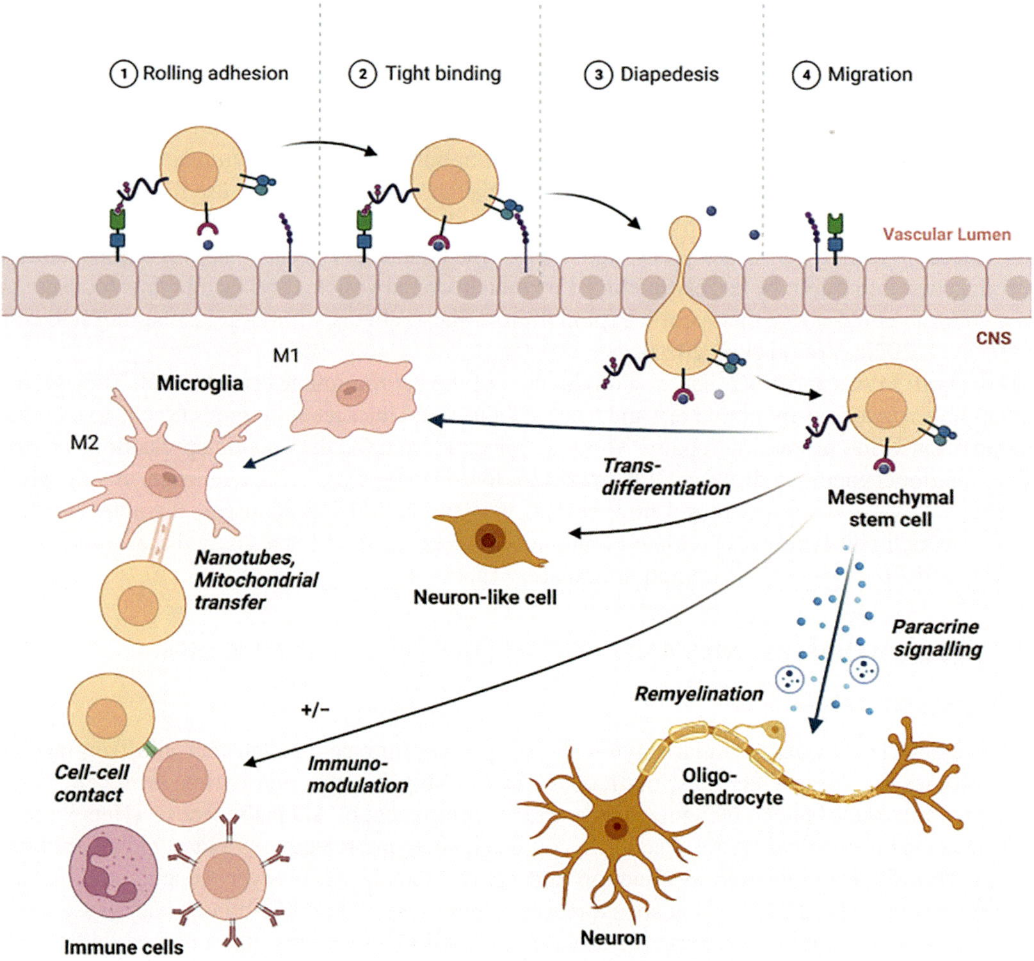

FIGURE 5.3 MSC Migration, Paracrine Signalling, and Repair Mechanisms in the CNS.

TABLE 5.3
Secretome Components of MSCs and NSCs

Molecule	Secreted By	Function
IL-10	MSCs/NSCs	Anti-inflammatory cytokine
TGF-β	MSCs/NSCs	Suppress immune activation
PGE2	MSCs	Modulates macrophages
BDNF	NSCs	Supports neuronal survival
LIF	NSCs	Promotes Treg and glial repair
GDNF	NSCs	Neurotrophic support
IDO	MSCs	T cell inhibition
NO	MSCs	Inhibits T/NK cell activity

synaptic density, an effect attributed to MSC secretion of growth and differentiation factor-15 (GDF-15) and Activin A, factors known to support neuronal growth (Kim et al. 2021). This preclinical data suggests that MSC therapy in AD provides a multifaceted benefit, reducing toxic protein accumulation and inflammation while enhancing neuron survival and regeneration, all mediated through paracrine signaling rather than neuronal differentiation.

An alternative strategy explored in preclinical models involves the administration of concentrated MSC-derived secretome or isolated exosomes, rather than whole-cell transplantation. This cell-free approach aims to harness the immunomodulatory and neuroprotective properties of MSCs, primarily mediated by their bioactive secretions while avoiding potential risks associated with cell engraftment, such as immune rejection or uncontrolled differentiation. These secreted vesicles carry a complex cargo of proteins, lipids, and nucleic acids capable of modulating inflammatory pathways, enhancing neuronal survival, and potentially influencing the hallmark of AD pathologies. In AD mouse models, the administration of MSC-conditioned media or purified exosomes has recapitulated several benefits of the cells themselves. For instance, exosomes from human MSCs reduced neuroinflammatory cytokine levels (TNF-α, IL-1β) and protected cognitive function when injected into AD mice, supporting the idea that MSC-derived vesicles carry critical immunomodulatory molecules (Reza-Zaldivar et al. 2018). These findings provide a rationale for cell-free therapies in the future, but so far, the majority of translational effort in AD has focused on cell-based interventions.

NSC transplantation has also been explored in AD models, though less extensively than MSCs. NSCs transplanted into the brains of AD transgenic mice have shown the ability to migrate towards sites of pathology, e.g., cortex and hippocampus with high Aβ burden and exert beneficial effects. A study by Blurton-Jones et al. (2009) demonstrated that NSC grafts improved cognitive function in APP/PS1 mice partly by releasing brain-derived neurotrophic factor, which enhanced synaptic density and neuronal survival. While that effect was more neurotrophic, subsequent research has highlighted immunomodulatory contributions as well. NSCs can produce modulators of microglial activity; for example, some transplanted NSCs secrete IL-4, a cytokine that can induce a supportive M2 microglial phenotype that aids in Aβ clearance (Brough et al. 2020). In a 2018 study, human NSCs introduced into an AD mouse model led to improved cognition and reduced plaque load, accompanied by evidence of microglial phenotype switching away from pro-inflammatory states. These findings suggest that NSCs modulate the activity of the brain's innate immune cells, enhancing their capacity to clear pathological protein aggregates while simultaneously limiting bystander or collateral neuronal injury. NSC transplantation in AD models has also been found to reduce tau hyperphosphorylation and aggregation in some cases (Oh et al. 2016), potentially through secreted factors that influence kinase and phosphatase activity in neurons or decrease inflammation-driven tau pathology.

Building on encouraging preclinical data, early-phase clinical trials have been initiated to assess stem cell therapy in AD patients, particularly with MSCs (Figure 5.4). One of the pioneering studies was a Phase I trial that involved stereotactic injection of autologous human umbilical cord blood-derived MSCs into the hippocampus and precuneus of nine patients with mild-to-moderate AD dementia (Hee et al. 2015). This investigation was primarily designed as a safety study, and notably, it reported no serious adverse events attributable to either the cell product or the surgical procedure. The method of delivery, direct intracerebral injection via drilled burr holes, was deemed technically feasible, albeit invasive. These findings support the procedural tolerability of intracerebral stem cell transplantation in a clinical context, laying the groundwork for subsequent efficacy-focused trials. Patients were monitored over a 1-year period following transplantation. Although the trial was not statistically powered to evaluate therapeutic efficacy, preliminary observations suggested a potential stabilization of cognitive decline in a subset of participants. However, these findings were anecdotal and insufficient to support definitive conclusions regarding clinical benefit, reinforcing that the primary outcome of the study was safety.

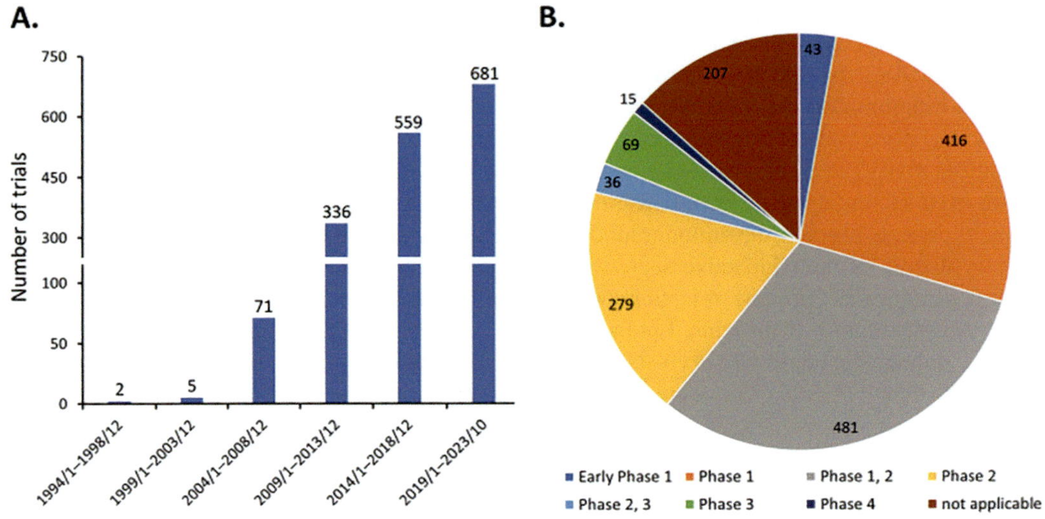

FIGURE 5.4 Clinical Trial Trends in Stem Cell Therapy for Neurodegenerative Disorders.

A subsequent Phase I/IIa trial by the same group (Kim et al. 2021) employed a less invasive delivery method, an Ommaya reservoir was implanted to allow repeated intracerebroventricular (ICV) infusions of cord blood MSCs. Nine AD patients received three monthly ICV injections of MSCs and were followed for safety and some exploratory efficacy endpoints. The results confirmed that repeated ICV administration was relatively safe and well-tolerated, transient fevers and headaches were common postinfusion likely reflecting immune activation in response to cells but resolved within 1–2 days, and no long-term adverse effects were seen. Notably, all patients remained immunologically stable with no signs of allo-sensitization to the donor MSCs (Kim et al. 2021). In terms of clinical outcomes, this small open-label study provided preliminary signals suggesting a potential slowing of cognitive decline. A few patients exhibited less deterioration in memory assessments at one year than would typically be expected in the natural course of AD. However, in the absence of a control group, these observations remain speculative, and no relationship to the stem cell intervention can be definitively established. The researchers also performed biomarker analyses, while CSF Aβ and tau levels did not change significantly, there was evidence of reduced inflammatory markers in the CSF such as lower IL-6 posttreatment in several patients (Kim et al. 2021). These findings, albeit preliminary, provided the first proof-of-concept that MSC infusions into the brain could modulate the neuroinflammatory environment in human AD.

More recently, a notable randomized controlled trial was completed that tested laromestrocel-L (also known as Lomecel-B), an allogeneic bone marrow-derived MSC product, in mild AD. In this Phase IIa trial (Rash et al. 2025), 50 patients were randomized to receive either placebo or varying doses of MSC infusions intravenously over a period of four months with some groups getting multiple monthly infusions. The primary endpoint was safety, and indeed, the MSC infusions were well tolerated with no significant differences in serious adverse events between treated and placebo groups (Rash et al. 2025). Importantly, this trial also employed a battery of clinical and imaging endpoints to seek efficacy signals. At 9 months, approximately 5 months post final infusion, patients who received the highest dose of MSCs showed slower cognitive decline on a composite AD cognitive measure compared to placebo, although the difference was modest. MRI analysis revealed that MSC-treated groups had a significantly lower rate of brain atrophy, whole brain volume loss over 9 months was roughly half that of the placebo group, and hippocampal volume loss was similarly reduced (Rash et al. 2025). Additionally, advanced MRI diffusion imaging suggested a reduction in neuroinflammation in treated patients, consistent with the idea that MSCs had a protective, anti-inflammatory effect on the brain. While the cognitive benefits did not reach robust statistical significance in all measures, the convergence of imaging and clinical trends was encouraging. This trial therefore provided the strongest evidence to date that MSC therapy might modify AD progression. On the strength of these results, larger Phase II/III trials are being planned to more definitively test efficacy with clinical outcomes.

It is worth noting that unlike other fields, NSC or NPC clinical trials in AD have not yet been prominent. Ethical constraints and substantial technical hurdles, including the diffuse, multifocal pathology typical of an elderly AD population have markedly curtailed attempts to transplant fetal-derived or induced NSCs into the AD brain. Most focus has remained on MSCs or similar mesenchymal-like cells for AD. MSC therapy for AD remains experimental; however, the accumulated evidence suggests it can modulate neuroinflammation and possibly yield neuroprotective effects. Future research should address optimal dosing, frequency of administration, delivery route (IV, ICV, or intranasal), and patient selection, as therapeutic response may vary with disease stage and inflammatory profile.

5.3.2 Parkinson's Disease

PD is primarily a movement disorder caused by degeneration of dopaminergic neurons in the substantia nigra pars compacta, leading to dopamine deficiency in the striatum. The classical pathology involves Lewy bodies, that are aggregates of α-synuclein in neurons. Neuroinflammation in PD is not as dominant a driver as in MS, but it is increasingly recognized that activated microglia and peripheral immune factors contribute to the progressive neuron loss in PD (Tansey et al. 2022). Postmortem studies show elevated cytokines like TNF-α and IL-1β and active microglia in the substantia nigra of PD brains (Table 5.3). α-Synuclein aggregates can activate microglia, and there is evidence that peripheral T cells infiltrate the brain in PD and may attack neurons presenting α-synuclein peptides. Therefore, therapies that reduce inflammation or protect neurons from inflammation-induced damage could be disease-modifying in PD.

MSCs have been tested in various PD animal models, including the neurotoxin models, MPTP in mice or 6-hydroxydopamine in rats that mimic dopaminergic neuron loss. MSCs have consistently shown neuroprotective effects in these models. For instance, in MPTP-lesioned mice, systemic or intranasal administration of human MSCs significantly preserved tyrosine hydroxylase-positive dopaminergic neurons and improved motor function (Park et al. 2017; Boruch et al. 2019). These improvements correlated with reduced neuroinflammation, MSC-treated PD model animals exhibit fewer activated microglia in the substantia nigra and lower expression of inflammatory mediators. A study by Dr. J. Oh's team demonstrated that MSC transplantation inhibited the elevation of TNF-α and IL-6 in the nigrostriatal system after MPTP injury, likely through secretion of anti-inflammatory factors like TGF-β and PGE2 that suppressed microglial activation (Oh et al. 2016).

Interestingly, MSCs may also influence the gut–brain axis in PD. Growing evidence indicates that PD pathology extends to the gastrointestinal tract, as reflected by early nonmotor symptoms such as constipation and the observation that α-synuclein aggregates may originate in the enteric nervous system before spreading to the brain.

A recent mouse study found that intranasally delivered human umbilical cord MSCs not only reduced neuroinflammation in the brain but also modulated the gut microbiota composition and reduced colonic inflammation in MPTP-treated mice (Li et al. 2022). In MSC-treated PD mice, gut microbiota composition was normalized, and intestinal barrier integrity was preserved, which corresponded with reduced systemic inflammatory responses (Li et al. 2022). These findings suggest that MSCs may exert broad immunomodulatory effects in PD, attenuating inflammatory responses not only within the CNS but also in peripheral immune compartments implicated in PD pathogenesis.

This dual-site modulation highlights the potential of MSCs to influence the neuroimmune axis and alter disease-relevant inflammatory cascades systemically.

Beyond immunomodulation, MSCs also secrete growth factors such as glial-derived neurotrophic factor, GDNF, and BDNF that support dopaminergic neuron survival and spur endogenous repair. Some preclinical experiments have used genetically engineered MSCs for enhanced effects, e.g., MSCs overexpressing Nurr1, a transcription factor that supports dopaminergic neuron health were transplanted into a 6-OHDA rat PD model and showed improved survival of host dopamine neurons and reduced inflammatory responses compared to unmodified MSCs (Tsai et al. 2015). This indicates that augmenting MSCs' neurotrophic or immunomodulatory capacity can further benefit outcomes.

NSC transplantation in PD models has largely been pursued for cell replacement, i.e., to differentiate into new dopamine-producing neurons. Fetal neural progenitors and dopaminergic precursors derived from pluripotent stem cells have demonstrated the ability to survive following transplantation into the striatum and partially restore dopaminergic tone, offering functional benefits in preclinical models. This cell replacement strategy remains a subject of active investigation, particularly in the context of PD, as researchers seek to optimize graft survival, integration, and long-term efficacy (Kikuchi et al. 2017). While the primary therapeutic objective of NSC transplantation in PD has traditionally focused on neuronal replacement, particularly of dopaminergic neurons, the concept of employing NSCs purely for their immunomodulatory properties remains less commonly pursued. Nevertheless, emerging evidence suggests that some of the functional improvements observed following NSC grafts may be attributed to their bystander effects, namely, the secretion of anti-inflammatory and neurotrophic factors that modulate the local microenvironment, reduce neuroinflammation, and support host neuronal function. For example, NSCs grafted into the striatum of parkinsonian rats have been observed to reduce astrocyte and microglial reactivity in addition to forming some new neurons (Vilà et al. 2008). NSC secreted factors like GDNF are highly relevant for PD, indeed, some NSC-based trials involve NSCs engineered to secrete GDNF or other neurotrophic factors, thereby combining regenerative and immunomodulatory support.

Clinical translation of stem cell therapy in PD has followed two paths: (1) the transplantation of dopamine-precursor cells from fetal tissue, embryonic stem cells, or induced pluripotent stem cells to replace lost neurons, a strategy aimed at symptomatic relief by restoring dopamine (Barker et al. 2017); and (2) the use of MSCs to provide disease-modifying effects through neuroprotection and immunomodulation. The first approach has achieved notable recent advancements, exemplified by ongoing clinical trials employing induced pluripotent stem cell-derived dopaminergic neuron grafts in Japan and the United States. Several early-phase studies have examined MSC administration in PD patients, primarily to assess safety. A Phase I trial (Schiess et al. 2021) tested intravenous allogeneic bone marrow MSCs in 20 mild-to-moderate PD patients. Patients received a single infusion at one of four dose levels (1, 3, 6, or 10 million cells per kg body weight) and were followed for 1 year. The results were encouraging, with no serious adverse events or immunological reactions attributable to MSC infusion, and the procedure was considered safe across all administered dose levels (Schiess et al. 2021). Importantly, even though it was primarily a safety trial, exploratory efficacy observations were made. Patients at the highest dose (10×10^6 cells/kg) showed trends toward clinical improvement, their motor symptoms measured by UPDRS scores improved modestly at 6 and 12 months compared to baseline, whereas one would expect gradual worsening in PD over that period (Schiess et al. 2021). Although the study was uncontrolled, the findings suggested a potential disease-modifying or symptomatic benefit. Supporting this observation, analysis of blood biomarkers in the high-dose cohort showed a significant reduction in inflammatory cytokines, most notably, TNF-α levels at 52 weeks post-infusion compared to baseline and a concomitant increase in brain-derived neurotrophic factor levels (Schiess et al. 2021). The investigators also performed advanced imaging and noted improved cerebral blood flow in specific brain regions of treated patients, suggesting a neurovascular or metabolic enhancement from MSC therapy. While these results must be interpreted with caution, they provided the first evidence in PD patients that MSC infusion can be safe and might confer multidimensional benefits.

Another approach has been intrathecal administration of MSCs in PD. A small pilot trial in 2015 delivered autologous adipose-derived MSCs intrathecally to 7 PD patients and followed them for a year (Emadi et al. 2015). It reported no major safety issues and some improvements in nonmotor symptoms, although the study design and size limited conclusions. The emerging consensus from early-phase clinical studies is that MSC therapy in PD is both feasible and safe. Unlike conditions such as MS or ALS, where inflammation is a central pathogenic driver and thus a clear therapeutic target, the rationale for MSC use in PD is more limited. In PD, MSCs are not expected to directly modify the core pathology of alpha-synuclein aggregation, but rather to exert beneficial bystander effects, modulating neuroinflammation, secreting neurotrophic factors, and fostering a more supportive microenvironment for neuronal survival. This indirect mode of action complicates the evaluation of therapeutic efficacy, as conventional biomarkers may not adequately capture treatment effects. Consequently, endpoints such as disease progression rate, functional performance, and quality of life will serve as more meaningful indicators, underscoring the need for larger, long-term clinical trials to establish definitive outcomes.

It is noteworthy that there has been interest in MSCs as a disease-modifying therapy for PD. For example, a Phase II trial (STEM-PD) was launched to test repeated IV MSC infusions in early PD with a placebo control, aiming to see if MSCs can slow the worsening of motor symptoms over 1–2 years. Additionally, as previously noted, gut microbiome studies indicate that MSC therapy may target peripheral contributors to PD, such as intestinal inflammation which is not addressed by conventional dopamine replacement treatments. In parallel, combination strategies are being explored, such as employing MSCs as adjunctive therapy alongside dopaminergic cell replacement. Given that grafted neurons require a supportive, anti-inflammatory microenvironment for optimal survival and integration, such combinatorial approaches may exert synergistic therapeutic effects

5.3.3 Amyotrophic Lateral Sclerosis

ALS is a fatal neurodegenerative disease characterized by progressive loss of motor neurons in the spinal cord, brainstem, and motor cortex, leading to muscle weakness and paralysis. Neuroinflammation is a hallmark of ALS pathology, characterized by activated microglia and reactive astrocytes surrounding degenerating motor neurons. Moreover, studies have demonstrated infiltration of peripheral immune cells, including T lymphocytes and monocytes, into the spinal cord of ALS patients (Philips and Rothstein 2015). These immune cells release cytotoxic mediators, including proinflammatory cytokines and reactive oxygen species, which exacerbate motor neuron degeneration. Interestingly, however, specific immune subsets such as regulatory T cells and M2-polarized microglia exert neuroprotective effects and are associated with a slower rate of disease progression.

Thus, the immune system in ALS exhibits a dual role. During early disease stages, microglia may exert protective effects, however, as the disease progresses, they transition toward a proinflammatory and neurotoxic phenotype (Zondler et al. 2016). Therapeutic strategies aimed at enhancing protective immune responses or suppressing deleterious inflammation may promote motor neuron survival in ALS. Stem cell–based interventions have been investigated with two principal objectives, delivering neurotrophic support to degenerating motor neurons and restoring a more balanced, neuroprotective immune milieu within the CNS (Table 5.3). Both MSCs and NSCs have been investigated extensively in preclinical ALS models and progressively in clinical trials.

MSCs have shown remarkable protective effects in ALS mouse models. Early have found that MSC injections either, intravenous or into spinal cord resulted in prolonged survival of ALS mice and preserved motor neurons, which were initially attributed to neurotrophic factors e.g., MSCs secrete Recent studies have highlighted immunomodulation as a key mechanism. For instance,

MSC transplantation in ALS mice reduces the pro-inflammatory cytokines like TNF-α and IL-1β in the spinal cord and increases anti-inflammatory cytokines like IL-10 (Sharma et al. 2020). MSCs also induce local microglia to shift toward an M2 phenotype that expresses neuroprotective factors (Zhao et al. 2019). As a result, there is less glial-mediated damage to motor neurons. Nam et al. (2023) has found that repeated intrathecal delivery of an autologous bone marrow MSC product (Neuronata-R) in ALS mice not only modulated inflammatory mediators but also promoted the clearance of misfolded SOD1 protein from the CNS (Nam et al. 2023). The MSCs somehow enhanced proteostatic mechanisms or microglial phagocytosis to reduce protein aggregates, linking immunomodulation with a direct impact on toxic protein burden.

Building on encouraging preclinical data, MSC secretome therapy has gained increasing attention. A 2023 study demonstrated that intranasal administration of MSC-derived exosomes in ALS mouse models produced outcomes comparable to whole-cell transplantation, including improved motor performance, reduced microglial activation, and preservation of motor neuron populations relative to untreated controls (Tang et al. 2023). Notably, MRI tracking in this study confirmed that intranasally administered MSC-derived exosomes successfully reached both the brain and spinal cord, with preferential accumulation in lesion sites (Tang et al. 2023). This finding suggests that intranasal administration may serve as a noninvasive and efficient route for delivering the immunomodulatory payload of MSCs across the neuraxis.

NSC transplantation in ALS models has been designed to both replenish supportive glial populations and modulate the pathological microenvironment toward a more neuroprotective state. Work by Glass and colleagues (Glass et al. 2014) involved transplanting human fetal spinal cord neural progenitors into the spinal cord of SOD1 mutant ALS rats. These grafts did not replace lost motor neurons per se, but many of the transplanted cells became astrocyte-like or neurotrophic factor-secreting cells. The result was prolonged survival and improved motor performance in treated rats. The underlying mechanism appeared predominantly paracrine, with NSCs secreting glial cell line-derived neurotrophic factors and other trophic molecules that conferred protection to host motor neurons. Furthermore, local expression of proinflammatory genes was markedly reduced in regions adjacent to the graft sites (Glass et al. 2014). A related approach involves utilizing NSCs or NPCs is biological mini pumps for sustained delivery of neuroprotective factors. For instance, NSCs genetically engineered to overexpress GDNF have demonstrated enhanced therapeutic efficacy in ALS models, effectively integrating principles of cell therapy with gene therapy (Suzuki et al. 2007).

A defining feature of ALS pathophysiology is the coexistence of cell autonomous motor neuron degeneration and noncell autonomous mechanisms, wherein surrounding glial and immune cells actively contribute to neuronal loss. NSC transplantation primarily targets these non-neuronal components by modulating the local microenvironment. Experimental studies have demonstrated that transplanted NPCs integrate within the spinal cord and attenuate the inflammatory activation of resident astrocytes and microglia, partly through the secretion of anti-inflammatory cytokines such as IL-10 and IL-27 (Cusimano et al. 2018). Therefore, in the context of ALS, NSCs primarily act as neuroprotective modulators rather than direct neuronal replacements, exerting their therapeutic effects by reshaping the surrounding microenvironment to promote motor neuron survival and reduce neuroinflammation. By mitigating the neurotoxic activity of reactive astrocytes and microglia, NSCs suppress local inflammatory cascades and oxidative stress, thereby preserving neuronal integrity and functional viability within the affected spinal cord.

ALS has been a central focus of stem cell–based clinical research, owing to the absence of effective disease-modifying therapies, as currently, only riluzole and edaravone are approved, both offering limited clinical benefit. One of the most prominent clinical programs is led by BrainStorm Cell Therapeutics, which developed the product NurOwn. This therapy utilizes autologous bone marrow–derived MSCs isolated from each patient, expanded ex vivo, and subsequently conditioned under specific growth factor–rich culture environments to enhance the secretion of neurotrophic factor. These cells are then transplanted into the patient, typically via intrathecal injection into the cerebrospinal fluid. Early-phase trials (Phase I/II) indicated that the procedure was generally safe and that some patients showed slowing of ALS progression or slight improvements in muscle function (Petrou et al. 2016). In a randomized Phase II trial, although the primary endpoint was not statistically met, post hoc analyses suggested that a subgroup of patients with less advanced ALS had a better outcome on NurOwn than placebo, with slower decline in respiratory function and muscle strength (Berry et al. 2019). This led to a larger Phase III trial which was completed in 2020. The Phase III results were mixed, the trial overall did not meet its primary endpoint, an dproportion of patients with a significant slowdown in decline was not different between NurOwn and placebo. However, consistent with earlier findings, a subgroup analysis, excluding the most severe patients who likely had little capacity for improvement showed a trend that more NurOwn-treated patients experienced stabilization or slight improvement in ALS Functional Rating Scale (ALSFRS) scores compared to placebo (Berry et al. 2022). Additionally, patients receiving NurOwn had increased levels of neurotrophic factors, like VEGF and decreased inflammatory markers, like MCP-1, a chemokine in their CSF after treatment, indicating a biological effect (Berry et al. 2022). Although the Phase III trial did not demonstrate definitive efficacy, it yielded valuable insights regarding dosing parameters and emphasized the importance of treatment timing in ALS. Early intervention appears critical, as therapeutic benefit is more likely before substantial motor neuron loss occurs as damage is irreversible once neurons have degenerated. Apart from BrainStorm's program, other MSC trials have been conducted. For example, a Phase I/II trial (Mazzini et al. 2012) delivered autologous bone marrow MSCs directly into the spinal cord of ALS patients. This invasive approach was designed to deliver cells directly into regions harboring motor neurons. It was found to be safe in the limited patient cohort, with some individuals exhibiting a slower rate of functional decline following transplantation. However, direct intraspinal injection entails notable surgical risks and, consequently, has not been widely implemented beyond early-phase pilot studies.

More recently, in another study, the MSC product, Neuronata-R, was tested in mice as. A long-term observational study (Nam et al. 2023) compared ALS patients who received repeated intrathecal MSC injections to matched controls. The MSC-treated group showed a modest but statistically significant extension of survival and a slower rate of ALSFRS decline over 1–2 years (Nam et al. 2023). No major safety concerns were identified, supporting the feasibility of multiple dosing. These findings suggest that sustained and repeated modulation of the immune microenvironment through successive MSC administrations may be necessary in ALS, as a single dose is unlikely to produce durable therapeutic effects in such a rapidly progressive disease.

A landmark clinical trial conducted by Neuralstem, Inc. investigated the use of a fetal-derived spinal cord NSC line (NSI-566) in ALS. In the Phase I study (Feldman et al. 2014), 18 patients with ALS received unilateral or bilateral intraspinal injections of NSCs into the lumbar region, with a subset additionally administered cells into the cervical spinal cord to assess safety. The results indicated that surgical transplantation of NSCs was generally safe, though not entirely without risk; a few participants experienced transient pain or localized inflammation. Importantly, a subgroup of patients who received both lumbar and cervical injections exhibited signs of slower disease progression in the injected leg muscles compared to the contralateral side or the expected rate of decline, suggesting a potential localized therapeutic effect (Feldman et al. 2014). While the uncontrolled design precluded definitive conclusions, the study established the feasibility of the procedure. A subsequent Phase II dose-escalation trial further confirmed safety, showing that administration of up to 16 million cells in the cervical and lumbar spinal regions was well tolerated under immunosuppressive therapy to prevent rejection of allogeneic NSCs (Glass et al. 2016). Some participants in the Phase II trial demonstrated slower functional decline over a nine-month period relative to historical ALS controls, although the sample size remained limited.

A more refined scientific version could read as follows:

A recent Phase I/IIa clinical study at Cedars-Sinai (Baloh et al. 2022) evaluated human embryonic stem cell–derived NPCs engineered to secrete glial cell line–derived neurotrophic factor, integrating trophic and immunomodulatory mechanisms. These NPCs were transplanted into the lumbar spinal cord of ALS patients. The procedure was well tolerated, and analyses of cerebrospinal fluid and post-mortem tissue from one participant confirmed that the grafted cells survived and secreted GDNF for over one year. Although the trial did not demonstrate clear clinical improvement, a subset of patients exhibited slower functional decline.

Importantly, high GDNF concentrations were detected in CSF, indicating sustained cellular activity. The findings highlight the feasibility of NSC-based localized drug delivery within the CNS and the potential for concurrent modulation of neuroinflammation, as GDNF exerts anti-inflammatory effects on microglia. Supporting this, exploratory analyses revealed reduced inflammatory cytokine levels in CSF following transplantation, though the small sample size precluded definitive conclusions.

Overall, ALS clinical trials with stem cells indicate: (1) Safety is attainable with both MSC and NSC approaches, albeit NSC transplants require invasive procedures and immunosuppression. (2) There are biologic signs of efficacy such as increased neurotrophic factors and reduced inflammatory markers in patients' CSF after treatment, consistent with the proposed mechanisms. (3) Clinical efficacy signals have been inconsistent, but there are glimmers of positive effects, especially in faster-progressing diseases like ALS where any plateau or slowing in decline is notable. The consensus emerging is that a multi-dose regimen may be necessary, one-time cell infusion may only transiently alter the immune landscape (Petrou et al. 2021). Repeated doses or sustained delivery via implanted cells might yield better outcomes. Additionally, patient selection could enhance the observable benefits. In fact, an interesting finding is that ALS patients with higher baseline levels of inflammation, e.g., elevated CSF cytokines might respond more to MSC therapy than those without, a hypothesis that needs testing but suggests that future trials may stratify or select patients by inflammatory biomarkers.

Thus, stem cell-based therapies for ALS, particularly those utilizing MSCs, are designed to shift the neuroimmune milieu from a pro-inflammatory, neurotoxic state toward a neuroprotective phenotype, while simultaneously providing trophic support to vulnerable motor neurons. Although these approaches have yet to demonstrate clinical efficacy, significant progress over the past decade in mechanistic understanding and clinical trial methodology has laid the groundwork for larger, more rigorous investigations. Given the progressive and currently untreatable nature of ALS, even modest reductions in disease progression are of considerable clinical interest. As biologically active, immunomodulatory platforms, stem cells offer a promising adjunct within a future multimodal therapeutic framework that may also include gene-targeting strategies such as antisense oligonucleotides and gene therapy for specific ALS-associated mutations.

5.3.4 Multiple Sclerosis

MS is an immune-mediated demyelinating disease of the CNS, where autoreactive lymphocytes attack the myelin sheath of neurons leading to inflammation and neurodegeneration. MS is characterized by episodes of neurologic deficits and, in many patients, progressive accumulation of disability. The central driver in MS is the immune system, peripheral immune cells infiltrate the brain and spinal cord, and along with activated resident microglia and macrophages, which in turn leads to demyelination and axonal injury. Current approved therapies for MS are mostly immunosuppressive or immunomodulatory drugs that reduce inflammation and relapses (Table 5.3). However, advanced progressive MS has less overt inflammation and more neurodegeneration for which current drugs are inadequate. Stem cell therapy in MS has been explored in two distinct forms, hematopoietic stem cell transplantation (HSCT) to essentially reboot the immune system, and mesenchymal or NSC therapy to locally modulate immune responses and potentially promote repair. MSCs are particularly attractive in the context of MS due to their dual therapeutic potential, they can modulate the aberrant immune response underlying the autoimmune attack and simultaneously secrete trophic factors that promote remyelination and neuroprotection. This ability to target both immunological dysregulation and neurodegeneration makes MSCs a promising candidate for addressing the multifaceted pathology of MS (Uccelli et al. 2019).

In the standard MS animal model, EAE, induced by immunization with myelin antigens, MSCs have repeatedly shown strong therapeutic effects. As early as 2005, it was reported that MSC infusion ameliorated EAE, attributed to induction of T cell anergy (Zappia et al. 2005). Subsequent studies have extensively characterized the immunomodulatory effects of MSCs in the experimental autoimmune encephalomyelitis model of MS. MSCs have been shown to suppress the activity of pathogenic Th17 and Th1 cells, the key drivers of MS relapses while promoting the expansion of regulatory T cells (Tregs). Additionally, MSC treatment reduces demyelinating lesions and has been associated with partial remyelination, suggesting a capacity not only to modulate the immune cascade but also to contribute to tissue repair processes within the CNS (Xiao et al. 2015). Mechanistically, one key action is the MSC secretion of IFN-γ-stimulated IDO and PGE2, which together lead to suppressed proliferation of T cells and reduced activation of microglia/macrophages (Xiao et al. 2015). MSCs also produce chemokines that attract immune cells to themselves, essentially acting as sinks to divert immune cells away from attacking myelin. This was shown in studies where MSCs lured autoreactive T cells and then rendered them apoptotic or inactive through NO and FasL pathways (Garg et al. 2017).

The overall immunological outcome in the EAE model is a marked attenuation of neuroinflammation following MSC treatment. MSC-treated mice exhibit reduced inflammatory cell infiltration within the CNS and decreased levels of proinflammatory cytokines such as IFN-γ and IL-17, accompanied by increased expression of anti-inflammatory cytokines including IL-10 and IL-4. Consequently, disease severity is mitigated, and in some instances, onset is delayed or entirely prevented.

Beyond their immunosuppressive functions, MSCs also secrete a repertoire of trophic factors, such as brain-derived neurotrophic factor, insulin-like growth factor-1, and interleukin-6 that promote oligodendrocyte survival and may stimulate endogenous precursor cells to differentiate into myelinating oligodendrocytes, thereby facilitating remyelination (Uccelli et al. 2011). Supporting this reparative capacity, several studies have demonstrated the presence of remyelinated axons and improved axonal integrity in MSC-treated EAE animals (Jaramillo-Merchán et al. 2013).

Neural stem and precursor cells (NSCs/NPCs) have also been extensively evaluated in the experimental autoimmune encephalomyelitis model, where they demonstrate significant therapeutic benefits mediated primarily through immunomodulation rather than direct oligodendrocyte replacement. In a landmark study, Pluchino et al. (2003) showed that transplantation of neural precursors into EAE mice markedly reduced neurological deficits despite limited differentiation into myelin-forming cells. Follow-up work confirmed that these transplanted NSCs migrate preferentially to inflamed CNS regions, where they secrete anti-inflammatory and neuroprotective factors that foster a permissive microenvironment for endogenous repair (Pluchino et al. 2005).

NSC administration in EAE results in fewer inflammatory lesions within the CNS and enhanced oligodendrogenesis, likely through preservation and stimulation of the host's own oligodendrocyte precursor cells. One critical mediator of these effects is LIF, secreted by NSCs, which exerts potent immunosuppressive activity on encephalitogenic T cells while simultaneously supporting oligodendrocyte survival. Experimental blockade of LIF negates the therapeutic benefits of NSC transplantation, underscoring its central role in this process (Yang et al. 2012). In addition, NSCs produce molecules such as osteopontin and matrix metalloproteinases, which contribute to extracellular matrix remodeling and recruitment of reparative immune cells to sites of damage.

Overall, NSCs in MS models exhibit a distinctive and therapeutically valuable ability to sense and migrate toward regions of active demyelination. This targeted homing is directed by chemotactic signals from inflamed tissue, allowing NSCs to localize their effects precisely within zones of neuroinflammatory injury. Once there, they exert context-specific immunomodulatory actions, including the induction of IL-10–producing regulatory B cells and a broader shift of the immune milieu away from pathogenic autoimmunity toward a neuroprotective state (Rossi et al. 2010). Among neurodegenerative disorders, MS has been one of the most extensively explored conditions for MSC-based therapy. The initial wave of small, open-label trials conducted over a decade ago consistently demonstrated that autologous MSC infusions administered either intravenously or intrathecally were generally safe for MS patients, with only mild, transient adverse effects such as headaches or low-grade fever and no evidence of disease exacerbation (Mohyeddin Bonab et al. 2007). Several of these early studies also reported anecdotal observations of reduced relapse frequency or improved visual function in cases of optic neuritis following MSC therapy. However, in the absence of control groups, these findings could not be distinguished from placebo effects or natural disease variability.

To obtain more definitive evidence, the international MESEMS trial, a multicenter, Phase II, randomized, placebo-controlled studywas undertaken (Uccelli et al. 2020). This trial enrolled 144 patients with active MS, predominantly relapsing–remitting and a subset with active progressive forms, who were randomized to receive intravenous infusions of autologous bone marrow–derived MSCs or placebo in a crossover design. The primary endpoint was the change in new MRI lesion count at 24 weeks. The results demonstrated that MSC therapy was safe and well tolerated, with no significant differences in adverse events between treatment and placebo groups, and no induction of MS relapses following infusion.

However, with respect to efficacy, the study did not meet its primary endpoint. MSC treatment did not significantly reduce new MRI lesion formation or relapse rates compared with placebo during the study period (Uccelli et al. 2020). While there was a nonsignificant trend toward fewer lesions in the MSC-treated cohort, the results indicated that, as an anti-inflammatory therapy for active relapsing MS, MSCs did not outperform placebo under rigorous randomized conditions. These findings moderated some of the initial optimism around MSC therapy in MS, indicating that MSC infusion may not exert the same rapid or robust anti-inflammatory effects as conventional immunosuppressive or disease-modifying therapies, e.g., high-dose corticosteroids or IFN-beta. However, the trial did find some intriguing secondary outcomes, patients who received MSCs showed a reduction in certain inflammatory cytokines in blood and an increase in neuroprotective growth factors, indicating a biological effect (Uccelli et al. 2020). Additionally, a subset analysis involving patients with progressive MS, although limited in sample size, revealed that those treated with MSCs demonstrated modestly improved functional outcomes compared to placebo. This observation raises the possibility that MSCs may confer greater benefit in progressive MS, a disease stage characterized more by chronic neurodegeneration than by acute inflammatory relapses.

In parallel with the MESEMS trial, additional investigations have focused on intrathecal administration and repeated dosing of MSCs in MS. In an open-label study, Petrou et al. (2021) administered multiple intrathecal infusions up to eight over four years of autologous MSCs to 24 patients with progressive MS. The regimen was well tolerated, with no major adverse events. Remarkably, 22 of 24 participants exhibited clinical stabilization or improvement rather than the expected progressive decline, and 10 demonstrated a reduction of more than 0.5 points in Expanded Disability Status Scale score, an outcome rarely observed in this disease stage. Patients receiving more than two infusions were more likely to show benefit, suggesting a cumulative or dose-dependent effect. Immunologically, transient increases in circulating regulatory T cells and decreased T-cell proliferative responses were observed following each administration, consistent with MSC-mediated immunosuppression. Although uncontrolled, these findings indicate that sustained MSC exposure through repeated dosing may be required to achieve measurable therapeutic benefit in progressive MS, potentially through gradual modulation of chronic neuroinflammation rather than acute relapse suppression.

NSC-based therapy has more recently advanced to early clinical evaluation. The STEMS trial, a Phase I open-label study (Genchi et al. 2023), investigated the intrathecal delivery of human fetal-derived NSCs in 12 patients with progressive MS. Participants received escalating doses up to 50 million cells via lumbar puncture and were followed for two years to assess safety, persistence, and biological activity. The procedure was generally well tolerated as no serious adverse events related to the transplanted cells occurred. Transient postprocedural headaches and mild meningitic symptoms were observed in a few participants and resolved spontaneously, fulfilling the primary safety endpoint.

Exploratory analyses demonstrated that recipients of the highest NSC dose exhibited a slower rate of brain atrophy on MRI relative to the expected trajectory of progressive MS. Cerebrospinal fluid studies showed increased concentrations of anti-inflammatory cytokines such as IL-10 and elevated neurotrophic factors including brain-derived neurotrophic factor, suggesting active paracrine signaling from engrafted NSCs. Although no significant improvements in clinical disability were recorded within the observation period, several patients reported subjective stabilization. Collectively, these findings provide initial clinical evidence that NSCs can be safely administered to the CNS via intrathecal delivery and remain biologically active, exerting immunomodulatory and neuroprotective effects. The study establishes a foundation for subsequent controlled trials aimed at assessing efficacy in progressive MS, representing a regenerative approach distinct from immune-ablative strategies such as HSCT.

MSCs and NSCs exhibit immunomodulatory and reparative effects in MS models, primarily by suppressing autoimmunity and promoting neural repair. Clinical trials confirm safety, though efficacy remains unproven. Future strategies may combine cell therapy with disease-modifying treatments or focus on refractory and progressive MS cases. Additionally, the immune-privileged and modulatory properties of these cells could be utilized for targeted delivery of therapeutic agents, such as enzymes or antibodies, within the CNS.

5.4 CURRENT STATUS AND OUTCOMES OF CLINICAL TRIALS

Stem cell-based therapies for neurodegenerative diseases have progressed to clinical evaluation across all four major disorders (AD, PD, ALS, MS), each employing distinct strategies and at varying stages of development.

Across neurodegenerative disease trials, MSCs have demonstrated a consistently favorable safety profile when appropriately prepared and administered. Intravenous MSC infusions tested in PD, MS, and AD have produced only transient, mild adverse effects such as low-grade fever, headache, or localized infusion-site discomfort, typically resolving within 24–48 hours (Schiess et al. 2021; Uccelli et al. 2020). Importantly, even allogeneic MSCs have not elicited acute immune rejection or hypersensitivity responses. In the PD trial by Schiess et al. (2021), none of the 20 participants developed donor-specific antibodies or serum sickness, suggesting that MSCs are inherently low in immunogenicity, likely due to their minimal expression of MHC class II and costimulatory molecules, as well as their capacity to induce host immune tolerance.

Furthermore, intrathecal administration of MSCs tends to be more reactogenic, with some patients developing transient, mild meningitis-like symptoms, likely reflecting immune responses within the cerebrospinal fluid to either the DMSO preservative or cellular components. These effects are self-limiting and readily managed with standard supportive care (Kim et al. 2021; Petrou et al. 2021). Direct intraparenchymal delivery, employed in selected NSC and MSC trials for Alzheimer's and PD via stereotactic injection, involves procedural risks inherent to neurosurgical intervention, however, no serious adverse events have been attributed to the transplanted cells. Importantly, tumorigenicity remains a primary safety consideration, and to date, no clinical trial has reported any evidence of neoplastic transformation or ectopic tissue formation following MSC or NSC transplantation.

In AD, MRI findings of reduced brain atrophy and modest cognitive improvement have been reported following MSC therapy (Rash et al. 2025). In PD, small UPDRS score improvements were observed in open-label studies without controlled evidence of disease modification (Schiess et al. 2021). In ALS, repeated MSC dosing showed slower functional decline (Petrou et al. 2016) and extended survival in matched analyses (Nam et al. 2023), though a Phase III trial failed to meet primary endpoints (Berry et al. 2022). In MS, the largest controlled trial was negative (Uccelli et al. 2020), while smaller studies suggested stabilization or modest improvement in progressive cases (Petrou et al. 2021).

A consistent pattern emerges across trials as in more advanced or progressive conditions such as ALS and progressive MS, reversal of established damage is difficult to achieve, though slowing of disease progression may be observed if the therapy effectively targets ongoing neurodegenerative processes. In contrast, disorders with relapsing or variable courses, such as relapsing-remitting MS or early-stage PD, pose challenges for detecting treatment effects over short observation periods due to inherent variability. Consequently, trial design must be disease-specific, ALS studies typically assess the slope of ALSFRS decline, MS trials emphasize MRI lesion burden or mobility measures, and AD studies rely on cognitive scales. The heterogeneity of clinical endpoints across diseases complicates direct comparisons of therapeutic efficacy.

Across studies, a consistent observation is that stem cell therapy induces measurable biochemical changes indicative of immunomodulation. In MS and ALS trials, treated patients frequently exhibit increased circulating regulatory T cells and reduced proliferation or cytokine secretion by proinflammatory T cells (Petrou et al. 2021; Mazzini et al. 2018). Cerebrospinal fluid analyses further confirm central effects, in the NurOwn ALS trial, CSF concentrations of neurotrophic factors such as VEGF and HGF increased significantly, while inflammatory mediators including MCP-1 and IL-12 decreased in MSC-treated patients compared with placebo (Berry et al. 2022). Similarly, in the AD laromestrocel trial, MRI diffusion metrics suggested reduced neuroinflammatory activity (Rash et al. 2025). These biomarker changes support active immunological engagement by cell therapy and may inform future trial design by identifying responder subgroups through biomarker-based stratification.

Clinical findings increasingly underscore that the route of stem cell administration critically influences therapeutic efficacy by determining the extent and nature of target engagement. Intravenous infusion of MSCs is technically straightforward and achieves systemic distribution. However, a large proportion of infused cells are retained within peripheral organs such as the lungs. Despite limited CNS trafficking, intravenously administered MSCs can mediate indirect neuroprotective effects through systemic immunomodulation, attenuation of peripheral inflammatory activity, and release of EVs capable of crossing into the CNS.

In contrast, intrathecal administration places stem cells in direct contact with cerebrospinal fluid, allowing interaction with meningeal and perivascular compartments and enabling limited parenchymal access. This anatomical proximity enhances localized immunomodulatory and neuroprotective actions, as supported by ALS studies where intrathecal MSC administration produced modest efficacy signals, likely reflecting improved engagement with the CNS microenvironment. Direct intraparenchymal delivery, such as intranigral transplantation in PD, intrahippocampal infusion in AD, or spinal cord injection in ALS ensures precise targeting but remains limited by surgical invasiveness and patient eligibility constraints. Consequently, current clinical strategies increasingly favor the intrathecal route as an optimal balance between safety and CNS accessibility. This approach is now being prioritized in upcoming trials for progressive MS, aiming to enhance CNS exposure following the limited efficacy observed with intravenous administration in the MESEMS trial.

Another key factor influencing therapeutic outcomes is dosing frequency. Earlier clinical trials typically employed single-dose regimens, whereas more recent studies favor repeated administrations, often delivered monthly over several months. The rationale stems from the limited in vivo persistence of MSCs, which remain viable and functionally active for only a few weeks. Consequently, maintaining sustained immunomodulatory activity likely requires repeated dosing, analogous to the chronic administration of immunotherapies in other inflammatory disorders.

The study by Petrou et al. (2021) in MS demonstrated that patients receiving long-term, repeated MSC infusions exhibited superior clinical stabilization compared with those who received only one or two doses, suggesting a cumulative or maintenance-dependent therapeutic effect. Similarly, the ALS NurOwn trial incorporated multiple dosing in its extension phase, and ongoing AD trials have adopted comparable approaches. Collectively, these findings indicate a paradigm shift, stem cell–based therapy is increasingly viewed not as a single curative intervention but as a periodic, adjunctive treatment requiring sustained administration to achieve lasting immunomodulatory and neuroprotective effects.

Stem cell therapy represents a complementary modality within the evolving landscape of neurotherapeutics. In MS, autologous hematopoietic stem cell transplantation achieves durable immune reconstitution with high efficacy but substantial procedural risk, whereas MSC therapy provides a safer, immunomodulatory alternative with potential reparative benefit following immune ablation. In ALS, antisense oligonucleotides targeting SOD1 and C9orf72 mutations offer gene-specific interventions, while MSC-based approaches act through mutation-independent neurotrophic and anti-inflammatory mechanisms, supporting their potential use in combination regimens. In AD, antiamyloid monoclonal antibodies such as lecanemab demonstrate disease-modifying activity but are limited by ARIA-related neuroinflammation, an area where stem cell–based therapies may confer adjunctive neuroprotective and immunomodulatory advantages.

As of 2025, no stem cell–based therapy for any neurodegenerative disorder has received full regulatory approval in either the United States or the European Union. Nonetheless, several candidates have advanced to late-stage clinical evaluation or obtained conditional authorization in select jurisdictions. In South Korea, Neuronata-R, an autologous bone marrow–derived MSC product for ALS, has held conditional approval since 2014 for restricted clinical use, representing one of the earliest commercialized MSC-based interventions for a neurodegenerative indication (Oh et al. 2016). BrainStorm's NurOwn, another MSC-derived product for ALS, remains under review by the U.S. Food and Drug Administration. Although its Phase III trial did not achieve statistical significance for efficacy, post hoc analyses and plans for further studies in early-stage ALS are ongoing. The FDA has granted NurOwn Fast Track designation, reflecting regulatory interest in its therapeutic potential.

In MS, no MSC-based product is currently near regulatory submission owing to insufficient efficacy evidence, although a large-scale Phase III trial may become feasible once standardized endpoints for repair and functional benefit are established. In AD and PD, stem cell programs remain in early-phase (I/II) clinical development, with laromestrocel in AD appearing to be the most advanced candidate approaching potential expansion into late-stage trials.

These clinical investigations have underscored several scientific and operational challenges, foremost among them being product variability. Autologous MSC preparations display substantial inter-patient heterogeneity in proliferative capacity and immunomodulatory potency, while allogeneic MSCs can vary according to donor characteristics, tissue source, and culture conditions, complicating standardization and reproducibility. As emphasized by Freedman (2021), two MSC preparations are not necessarily equivalent, and such variability may have contributed to outcome differences across sites in multicenter trials such as MESEMS.

To address this, current strategies emphasize centralized manufacturing, implementation of Good Manufacturing Practice (GMP) standards, and the use of well-characterized allogeneic MSC lines to ensure batch consistency. Another persistent challenge is clinical heterogeneity within trial populations. Early studies frequently enrolled mixed cohorts such as both relapsing and progressive MS patients, or ALS patients spanning early to late disease stages thereby diluting measurable efficacy signals. Future trial designs are increasingly focused on refining patient selection and stratification based on disease stage, progression rate, or biological markers such as baseline inflammatory activity, to enhance the likelihood of detecting meaningful therapeutic effects.

5.5 CHALLENGES AND FUTURE DIRECTIONS

The application of stem cell-based interventions for neurodegenerative diseases represents a rapidly advancing area of translational research. Despite substantial mechanistic progress, multiple scientific and clinical challenges continue to limit the realization of their full therapeutic potential. Concurrently, innovative strategies are being developed to address these obstacles and refine the precision of treatment approaches. This section summarizes the principal challenges and outlines future directions likely to define the next phase of development in the field (Figure 5.5).

1) Heterogeneity of Disease and Timing of Intervention

Neurodegenerative diseases exhibit pronounced heterogeneity in their underlying pathology, rate of progression, and patient-specific clinical trajectories. This variability undermines the applicability of a uniform stem cell treatment paradigm. For example, in ALS, pathology may predominate in upper or lower motor neurons, while in AD, some patients display extensive amyloid pathology with robust inflammation, whereas others exhibit predominantly tau-driven neurodegeneration. Similarly, the immunological landscape evolves with disease stage, early AD may feature compensatory immune activation, whereas late-stage disease is often characterized by chronic, maladaptive inflammation. Stem cell-based interventions are most effective when administered during a therapeutic window in which inflammatory and degenerative processes are active but before extensive neuronal loss and gliosis have occurred. Treatment at late disease stages, where irreversible structural and cellular damage predominates, has shown limited efficacy, as reflected in clinical trials where patients with advanced pathology demonstrated minimal responsiveness (Berry et al. 2022). Conversely, premature interventions such as in very mild or preclinical disease may yield negligible

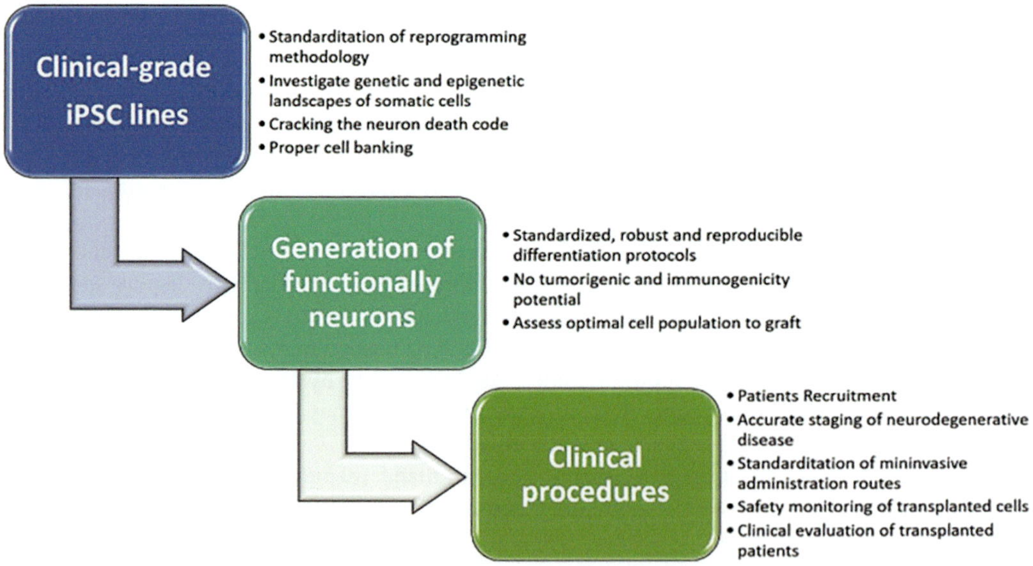

FIGURE 5.5 Translational Pipeline for iPSC-Derived Neuronal Therapies.

benefit or disrupt transiently protective immune mechanisms, such as early microglial clearance of amyloid in AD.

Future strategies will depend on improved biomarker-based staging to optimize treatment timing. Quantitative imaging of neuroinflammation and cerebrospinal fluid cytokine profiling may facilitate stratification of patients according to inflammatory burden and disease activity, enabling individualized intervention schedules. Furthermore, rational combination therapies may help address temporal limitations, for instance, deploying conventional immunosuppressive agents in MS to suppress acute relapses, followed by stem cell administration during the recovery phase to promote repair and sustained immunomodulation.

B) Delivery to CNS and Cell Survival

The CNS constitutes a highly protected compartment that, while effectively shielding neural tissue from harmful agents, simultaneously poses significant challenges for therapeutic delivery. Following intravenous administration, only a minimal proportion of MSCs transiently reach the CNS. For NSCs or MSCs to exert therapeutic impact within the brain or spinal cord, targeted delivery routes such as intra-arterial infusion with blood–brain barrier modulation, intrathecal injection, or direct parenchymal administration are necessary. Each method presents inherent limitations, for example, intrathecal delivery is less invasive but confines cells largely to cerebrospinal fluid spaces, whereas direct parenchymal injection enables localized delivery but necessitates surgical intervention. Future advancements aim to optimize cellular delivery and biodistribution. Strategies include transient BBB opening via focused ultrasound to permit localized entry of systemically administered cells or exosomes (Khan et al. 2021), use of injectable scaffolds or hydrogels to enhance cellular retention and migration (Assunção-Silva et al. 2015), and genetic modification of MSCs to express chemokine receptors such as CXCR4, improving homing to inflamed neural tissue. Posttransplantation survival remains a key limitation due to inflammatory and oxidative stressors within the host environment. Preconditioning approaches, including hypoxic culture or pharmacological priming, have been shown to enhance MSC resilience and secretory profiles (Moll et al. 2020). For example, IFN-γ–preconditioned MSCs exhibit elevated indoleamine 2,3-dioxygenase (IDO) production and superior immunomodulatory capacity (Chen et al. 2018).

C) Consistency and Potency of Cell Products

A key obstacle to large-scale clinical translation of MSC therapy is maintaining consistent batch quality and therapeutic potency. Autologous MSCs exhibit donor-dependent variability influenced by factors such as age, disease state, and culture conditions, often resulting in divergent cytokine expression profiles. In contrast, allogeneic MSCs derived from young, healthy donors offer greater uniformity and scalability but pose potential risks of immune sensitization with repeated administrations, despite their low inherent immunogenicity. Future strategies focus on developing standardized, off-the-shelf allogeneic cell lines with defined characteristics. Emerging platforms employ clonal MSC lines or mesenchymal-like cells differentiated from induced pluripotent stem cells, engineered to minimize antigenicity and enhance immunomodulatory properties, thereby reducing batch variability and ensuring predictable therapeutic potency (Mendicino et al. 2019).

D) Scaling Up and Manufacturing

For large-scale clinical application, the scalability and cost-effectiveness of cell therapies remain critical challenges. The production of viable cellular therapeutics is considerably more intricate than the synthesis of small molecules or monoclonal antibodies, as it necessitates stringent aseptic conditions and the expansion of vast quantities of cells, often ranging from tens to hundreds of millions per dose. To address these demands, biomanufacturing is increasingly shifting toward automated bioreactor systems that enable controlled, reproducible, and high-throughput cell culture processes (Watson et al. 2022). Encapsulation technology represents another promising direction; wherein therapeutic cells are enclosed within biocompatible semipermeable capsules. This approach can permit implantation without systemic immunosuppression while enhancing cellular viability by providing localized nutrient and trophic support. Although encapsulation has been primarily explored with pancreatic islet cells and engineered cell lines, its extension to NSCs or other therapeutic cell types holds significant potential. For instance, encapsulated NSCs designed to secrete neuroprotective factors could be implanted into the spinal cord of patients with ALS, offering prolonged immunomodulatory and neurotrophic benefits, with the possibility of retrieval or replacement as required.

E) Monitoring and Controlling the Effects

Unlike conventional pharmacological agents that can be quantified through blood concentration measurements, living cell therapeutics present unique challenges due to their capacity to migrate, differentiate, and dynamically alter behavior after administration. Consequently, robust noninvasive monitoring strategies are essential to assess their biodistribution and persistence in vivo. Current developments include labeling transplanted cells with superparamagnetic iron oxide nanoparticles or similar tracers, enabling MRI to verify early post-transplant localization, as demonstrated in several MSC-based clinical trials (Janowski et al. 2014). Additionally, reporter gene imaging techniques, such as PET using genetically encoded reporter constructs offer the potential to longitudinally visualize cell survival, proliferation, and functional integration over extended periods.

F) Ethical and Regulatory Issues

The use of certain stem cell sources, particularly fetal-derived NSCs, raises significant ethical and societal concerns that may limit their clinical acceptability. Consequently, future strategies are expected to increasingly favor induced pluripotent stem cell-derived neural cells, which mitigate ethical issues and offer the possibility of generating autologous patient-specific therapies. However, individualized iPSC derivation remains costly and time intensive. An emerging alternative involves the establishment of iPSC banks comprising immunologically universal lines, either selected for common HLA haplotypes or genetically engineered to evade immune recognition, thereby providing a practical balance between autologous and allogeneic approaches (Deuse et al. 2019). In parallel, regulatory agencies continue to refine oversight frameworks for regenerative medicine. While generally supportive of innovation, they emphasize stringent safety evaluations, including long-term posttreatment monitoring often extending five to ten years to detect delayed adverse effects such as tumorigenicity or ectopic tissue formation.

5.6 FUTURE DIRECTION

A key emerging strategy involves the integration of stem cell therapy with complementary treatment modalities to achieve synergistic therapeutic outcomes. Combining stem cells with antibody or gene-based interventions enables simultaneous targeting of distinct disease mechanisms. For instance, in AD, antiamyloid antibodies may mitigate toxic plaque accumulation while MSCs attenuate secondary inflammation and promote synaptic repair. Similarly, in MS, coadministration of an anti-inflammatory agent could suppress new inflammatory episodes, allowing MSCs to facilitate remyelination and neuroprotection. Another promising paradigm is the use of drug-enhanced stem cell therapies, wherein pharmacological agents augment the efficacy of transplanted or endogenous stem cells, for example, small molecules that mobilize intrinsic progenitors or local injection of matrix metalloproteinases to improve cellular engraftment and migration.

Thus, the future trajectory of stem cell applications in neurodegenerative diseases is expected to emphasize precision, combinatorial, and adjunctive approaches rather than stand-alone monotherapies. The overarching goal is to transform the natural course of disorders such as AD, PD, and MS from progressive degeneration to controllable chronic conditions, ultimately achieving functional recovery and regeneration.

5.7 SUMMARY

This chapter provides a comprehensive examination of the immunomodulatory mechanisms by which stem cells, particularly MSCs and NSCs influence neurodegenerative disease progression. It provides with an overview of the contribution of neuroinflammation to the pathophysiology of AD, PD, ALS, and MS. MSCs are recognized for secreting anti-inflammatory mediators such as IL-10, TGF-β, and PGE_2, promoting regulatory T-cell differentiation while suppressing pro-inflammatory lymphocytes and macrophages. Similarly, NSCs modulate the CNS immune milieu through context-dependent signaling, secretion of trophic factors, and glial phenotype modulation, with the potential to integrate into damaged neural networks.

In AD, MSC transplantation has shown preclinical efficacy in reducing amyloid burden and neuroinflammation, with early-phase clinical trials confirming safety and suggesting biomarker modulation. In PD, MSCs exert neuroprotective effects via microglial deactivation and motor function improvement, while NSCs may contribute to dopaminergic neuron restoration alongside immunomodulation. In ALS, MSCs promote M2-type microglial polarization and motor neuron preservation, correlating with slower disease progression in some clinical studies. In MS, both MSCs and NSCs suppress autoreactive immune activity and facilitate remyelination, though large-scale trials such as MESEMS have yielded variable outcomes.

The chapter also highlights EVs, including exosomes, as emerging acellular alternatives capable of replicating many of the immunoregulatory and neurotrophic functions of their parent cells. It concludes by discussing translational challenges such as delivery optimization, immune compatibility, manufacturing standardization, and ethical considerations. Future directions emphasize personalized and combinatorial therapeutic strategies, integrating stem cells, EVs, and pharmacological interventions to enhance efficacy and durability of response in neurodegenerative disease management.

REFERENCES

Assunção-Silva RC, Gomes NG, Sousa N, Salgado AJ, Silva JO. Hydrogel-based strategies to advance therapies for chronic spinal cord injury. *Materials*. 2015;8(12):8030–42.

Baloh RH, Johnson JP, Avalos P, et al. Transplantation of human neural progenitor cells secreting GDNF into the spinal cord of patients with ALS: A phase 1/2a trial. *Nat Med.* 2022;28(10):1813–22. doi:10.1038/s41591-022-01956-3

Barker RA, Rosenblad C, Björklund A. Emerging paradigms in the cell and gene therapy of Parkinson's disease. *Prog Brain Res*. 2017;230:165–203.

Berry JD, Cudkowicz JD, Cudkowicz ME, et al. NurOwn, a phase 3 trial of autologous MSC-NTF cells in patients with ALS: Failed primary endpoint–a focal perspective. *Muscle Nerve*. 2022;66(4):418–26.

Berry JD, Cudkowicz ME, Windebank AJ, Staff NP, Owegi M, Nicholson K, McKenna-Yasek D, et al. NurOwn, phase 2, randomized, clinical trial in patients with ALS: Safety, clinical, and biomarker results. *Neurology*. 2019;93(24):e2294–e2305. doi:10.1212/WNL.0000000000008620

Blurton-Jones M, Kitazawa M, Martinez-Coria H, Castello NA, Müller FJ, Loring JF, Yamasaki TR, Poon WW, Green KN, LaFerla FM. Neural stem cells improve cognition via BDNF in a transgenic model of Alzheimer disease. *Proc Natl Acad Sci*. 2009;106(32):13594–9.

Boruch AV, Conover RS, Qian JW, Pan YS. Human umbilical cord mesenchymal stem cells and the treatment of Parkinson's disease: A systematic review. *Cell Transplant*. 2019;28(4):447–71.

Brough D, Rozemuller SJ, Lute PLM. Neural stem cell therapy for Alzheimer's disease: The role of neuroinflammation. *J Neurochem*. 2020;155(1):27–43.

Chen X, Wang S, Cao W. Mesenchymal stem cell-mediated immunomodulation in cell therapy of neurodegenerative diseases. *Cell Immunol*. 2018;326(April):8–14.

Coyle PK. The neuroimmunology of multiple sclerosis. *Adv Neuroimmunol.* 1996;6(2):143–54. doi:10.1016/0960-5428(96)00013-7

Cusimano M, Bagnato A, Gualdi C, et al. Transplanted neural stem/progenitor cells protect from chronic brain inflammation and progressive neurodegeneration in the QA model of Huntington's disease. *Cell Transplant.* 2018;27(1):117–29.

Deuse T, Hu S, Gravina W, et al. Hypoimmunogenic stem cells induce immunologic tolerance in hearts. *Nat Biotechnol.* 2019;37(8):955–62.

Einstein O, Friedman-Levi A, Grigoriadis L, Ben-Hur D. Transplanted neural precursors enhance alcohol breakdown and improve mitochondrial dynamics in the chronic ethanol exposure model. *J Neurosci.* 2016;36(44):11750–62.

Emadi N, Rezaei M, Keyhanian MH, et al. Intrathecal autologous mesenchymal stem cell therapy in patients with Parkinson's disease. *J Clin Neurosci.* 2015;22(3):421–5.

Feldman EL, Glass JD, Miller NM, et al. Phase I trial of intracerebral neural stem cell transplantation in ALS. *Ann Neurol.* 2014;75(3):363–73.

Freedman MS. Mesenchymal stem cell therapy for multiple sclerosis: Hype or hope? *Lancet Neurol.* 2021;20(11):870–1.

Gao F, Chiu SM, Motan DA, Zhang Z, Chen L, Ji HL, Tse HF, Fu QL, Lian Q. Mesenchymal stem cells and immunomodulation: Current status and future prospects. *Cell Mol Immunol.* 2016;13(1):23–30.

Garg T, Kant S, Singh JS, Kaur I. Mesenchymal stem cell-based therapy in experimental autoimmune encephalomyelitis: A meta-analysis. *Mult Scler Relat Disord.* 2017;18:47–55.

Genchi A, Brambilla E, Sangalli F, et al. Neural stem cell transplantation in patients with progressive multiple sclerosis: An open-label, phase 1 study. *Nat Med.* 2023;29(1):116–26.

Glass JD, Boulis NM, Feldman EL, et al. Intrathecal administration of neural stem cells in ALS: A phase I trial, safety, and challenges. *Amyotroph Lateral Scler Frontotemporal Degener.* 2014;15(1–2):70–1.

Glass JD, Boulis NM, Johe WG, et al. Lumbar intraspinal injection of neural stem cells in patients with ALS: Results of a phase I trial in 12 patients. *Stem Cells.* 2016;34(4):623–31.

Heneka MT, Carson MJ, El Khoury J, et al. Neuroinflammation in Alzheimer's disease. *Lancet Neurol.* 2015;14(4):388–405.

Janowski M, Lyczek P, Suszynski P, et al. Cell grafting for stroke: The quest for stroke-adjusted cell therapy. *Dev Neurobiol.* 2014;74(3):288–302.

Jaramillo-Merchán J, Jones J, Ivorra J, et al. Mesenchymal stromal-cell transplants induce oligodendrocyte progenitor migration and remyelination in a chronic demyelination model. *Cell Death Dis.* 2013;4:e779. doi: 10.1038/cddis.2013.304.

Khan AR, Anwar JJ, Kim HW. Focused ultrasound and microbubbles-mediated drug delivery to the brain. *J Control Release.* 2021;330:84–99.

Kikuchi T, Hatanaka N, Yamamoto AS, et al. Human iPS cell-derived dopaminergic neurons function in a primate Parkinson's disease model. *Nature.* 2017;548(7669):592–6.

Kim HJ, Cho KR, Jang H, et al. Intracerebroventricular injection of human umbilical cord blood mesenchymal stem cells in patients with Alzheimer's disease dementia: A phase I clinical trial. *Alzheimer's Res Ther.* 2021;13:154.

Kim HJ, Seo SW, Chang JW, et al. Stereotactic brain injection of human umbilical cord blood mesenchymal stem cells in patients with Alzheimer's disease dementia: A phase 1 clinical trial. *Alzheimer's Demen: Transl Res Clin Interv.* 2015;1(2):95–102.

Laroni A, de Rosbo NK, Uccelli A. Mesenchymal stem cells for the treatment of neurological diseases: Immunoregulation beyond neuroprotection. *Immunol Lett.* 2015;168 (2):183–90. https://doi.org/10.1016/j.imlet.2015.08.007

Li R, Liu J, Yu X, et al. Human umbilical cord mesenchymal stem cells improve locomotor function in Parkinson's disease mouse model through regulating intestinal microorganisms. *Front Cell Dev Biol.* 2022;9:808905.

Madhu LN, Raju G, Mathkar MS, et al. Extracellular vesicles from human-induced pluripotent stem cell-derived neural stem cells alleviate proinflammatory cascades within disease-associated microglia in Alzheimer's disease. *Stem Cells.* 38(9):1150–64.

Marsh SE, Blurton-Jones SE. Examining the mechanisms that link β-amyloid and microglial dysfunction in Alzheimer's disease. *Front Cell Neurosci.* 2016;10:153.

Mathiasen AB, Kusuma GS. Mesenchymal stromal cells in neuroinflammation: Mechanisms and therapeutic opportunities. *J Neuroinflamm.* 2022;19(1):210.

Mazzini L, Gelati M, Profico DC, Sgaravizzi G, Projetti Pensi M, Muzi G, et al. Results from phase I clinical trial with intraspinal injection of neural stem cells in ALS. *Stem Cells Transl Med.* 2018;7(6):433–2.

Mazzini L, Mareschi F, Ferrero I, et al. Autologous mesenchymal stem cells: Clinical applications in amyotrophic lateral sclerosis. *Neurol Res.* 2012;34(3):298–303.

Mendicino M, Bailey BL, Wonnacott B, Puri JS, Helms W. FDA perspective on preclinical development of cell-based regenerative medicine products. *Regen Med.* 2019;14(8):969–79.

Mohyeddin Bonab M, Yazdanbakhsh K, Lotfi A, et al. Does mesenchymal stem cell therapy help multiple sclerosis patients? Report of a pilot study. *Eur J Neurol.* 2007;14(12):1485–9.

Moll G, Ankrum E, et al. Improving mesenchymal stromal cell potency and efficacy: Overcoming hurdles to clinical application. *Stem Cells Transl Med.* 2020;9(1):17–29.

Nakano M, Nagaishi D, Konari Y, et al. Bone marrow-derived mesenchymal stem cells improve diabetes-induced cognitive impairment by exosome transfer into damaged neurons and astrocytes. *Sci Rep.* 2020;10(1):21166.

Nam JY, et al. Long-term survival benefits of intrathecal autologous bone marrow-derived mesenchymal stem cells (Neuronata-R®: Lenzumestrocel) treatment in ALS: Propensity-score-matched control, surveillance study. *Front Aging Neurosci.* 2023;15:1148444.

Oh SH, Kim HN, Park HJ, Shin JY, Bae E-J, Sunwoo MK, Lee S-J, Lee PH. Mesenchymal stem cells inhibit transmission of α-synuclein by modulating clathrin-mediated endocytosis in a parkinsonian model. *Cell Rep.* 2016;14:835–49.

Park HS, Lee YE, Kim CY, et al. Intranasal delivery of mesenchymal stem cells enhances therapeutic efficacy in a model of Parkinson disease. *Stem Cells.* 2017;35(7):1925–37.

Petrou P, Gothelf Y, Argov Z, et al. Safety and clinical effects of mesenchymal stem cells secreting neurotrophic factor transplantation in patients with amyotrophic lateral sclerosis: Results of phase 1/2 and 2a clinical trials. *JAMA Neurol.* 2016;73(3):337–44. doi: 10.1001/jamaneurol.2015.4321.

Petrou P, et al. Long-term clinical and immunological effects of repeated mesenchymal stem cell injections in patients with progressive forms of multiple sclerosis. *Front Neurol.* 2021;12:639315.

Philips T, Rothstein JD. Glial cells in amyotrophic lateral sclerosis. *Exp Neurol.* 2015;262(Pt B):111–20.

Pinho ACM, Silva RA, Rebelo C, et al. Therapeutic potential of mesenchymal stem cell-derived extracellular vesicles in Parkinson's disease: An in vitro study. *J Control Release.* 2020;323:88–97.

Pluchino S, Gritti A, Blezer E, et al. Human neural stem cells ameliorate autoimmune encephalomyelitis in non-human primates. *Ann Neurol.* 2009;66(3):343–54. doi: 10.1002/ana.21745.

Pluchino S, Martino G, et al. Injection of adult neurospheres induces recovery in a chronic model of multiple sclerosis. *Nature.* 2003;422(6933):688–94.

Pluchino S, et al. Neurosphere-derived multipotent precursors promote neuroprotection by an immunomodulatory mechanism. *Nature.* 2005;436(7048):266–71.

Rash BG, Ramdas KN, Agafonova N, et al. Allogeneic mesenchymal stem cell therapy with laromestrocel in mild Alzheimer's disease: A randomized controlled phase 2a trial. *Nat Med.* 2025;31(4):711–21.

Reza-Zaldivar EE, Hernández-Sapiéns KA, Minjarez R, et al. Potential effects of MSC-derived exosomes in neuroplasticity in Alzheimer's disease. *Front Cell Neurosci.* 2018;12:317.

Rossi S, Martorana MA, Brambilla M, et al. Potential therapeutic action of neural stem cells in multiple sclerosis: Protection of oligodendrocytes and induction of immunomodulation. *Ann Neurol.* 2010;68(3):369–80.

Schiess MC, Suescun J, Green C, Tharp E, Chandra S, Adams C, Shahnawaz M, Rodarte E, Thyne V, Ellmore TM. Allogeneic bone marrow–derived mesenchymal stem cell safety in idiopathic Parkinson's disease. *Movement Disorders.* 2021;36(8):1825–34.

Sharma A, Novikova AK, Nygren LJ, et al. Multiple intrathecal mesenchymal stromal cell treatments induce immunosuppressive effects and promote cardioprotective macrophages in ALS mice. *Nat Commun.* 2020;11(1):5114.

Suzuki M, Svendsen B, Langmoen I, Svendsen C. Combining growth factors to enhance engraftment and efficacy of human neural stem cell grafts in a rat model of ALS. *Nat Biotechnol.* 2007;25(8):843–51.

Tang J, Kang Y, Zhou Y, et al. Umbilical cord mesenchymal stem cell-conditioned medium inhibits microglial activation to ameliorate neuroinflammation in amyotrophic lateral sclerosis mice and cell models. *Brain Res Bull.* 2023;202(January):110760.

Tansey MG, Goldberg JT, Karuppagounder BE, et al. Inflammation and immune dysfunction in Parkinson disease. *Nat Rev Immunol.* 2022;22(5):285–98.

Tsai MJ, Tsai S, Huang TM, et al. Genetically engineered mesenchymal stem cells with dopamine-synthesizing enzymes facilitate restoration of motor function in a rat model of Parkinson's disease. *PLoS One.* 2015;10(8):e0138727.

Uccelli A, Giunti AA, Kerlero de Rosbo SS. Mesenchymal stem cells as treatment for MS – Progress to date. *Mult Scler J.* 2019;25(5):558–67.

Uccelli A, et al. Mesenchymal stem cells in health and disease. *Nat Rev Immunol.* 2011;11(12):719–28.

Uccelli A, et al. Safety, tolerability, and activity of mesenchymal stem cells versus placebo in multiple sclerosis (MESEMS): A phase 2, randomised, double-blind crossover trial. *Lancet Neurol.* 2020;19(5):360–9.

Vilà M, Miquel A, Rodriguez A, et al. Brain-Derived Neurotrophic Factor (BDNF) production in neural stem cell grafts facilitates behavioral recovery in Parkinsonian rats. *Brain Res.* 2008;1238:158–66.

Watson JL, Elliman RB, Coleman J. From 2D to 3D: The emergence of bioreactors for mesenchymal stromal cell culture. *Front Cell Dev Biol.* 2022;10:858911.

Willis CM, Nicaise AM, Hamel R, et al. Harnessing the neural stem cell secretome for regenerative neuroimmunology. *Front Cell Neurosci.* 2020;14:590960.

Xiao J, Yang R, Biswas S, Qin X, Zhang M, Deng W. Mesenchymal stem cells and induced pluripotent stem cells as therapies for multiple sclerosis. *Int J Mol Sci.* 2015;16(5):9283–302. doi: 10.3390/ijms16059283.

Yang J, Jiang RB, Tetradis S, et al. LIF-dependent signaling pathways control the microglial response and (Th17 Cell) parenchymal infiltration in EAE. *J Mol Med.* 2012;90(12):1457–71.

Zappia E, Casazza S, Pedemonte E, et al. Mesenchymal stem cells ameliorate experimental autoimmune encephalomyelitis inducing T-cell anergy. *Blood.* 2005;106(5):1755–61.

Zhao CY, Li WY, Xin SZ, et al. Mesenchymal stem cells overexpressing fibroblast growth factor 21 protect motor neurons and reduce microglia-mediated neuroinflammation in a mouse model of amyotrophic lateral sclerosis. *Hum Gene Ther.* 2019;30(9):1098–110.

Zondler L, Zabler S, Premi A, et al. Immune functions of microglia in amyotrophic lateral sclerosis—facts and potential mechanisms. *Front Cell Neurosci.* 2016;10:114.

6 Nanoparticle-Enhanced Stem Cell Therapies Against Neurodegenerative Diseases

6.1 INTRODUCTION

Neurodegenerative diseases (NDs) such as Alzheimer's disease (AD), Parkinson's disease (PD), amyotrophic lateral sclerosis (ALS), multiple sclerosis (MS), and Huntington's disease (HD) are characterized by progressive loss of neurons leading to cognitive and motor deficits.

These disorders affect tens of millions of individuals globally, yet effective curative therapies remain elusive. This unmet clinical need underscores a need to develop innovative and targeted therapeutic strategies (Pan et al. 2025; Ye et al. 2023). Conventional pharmacological treatments largely alleviate symptoms but remain ineffective in mitigating disease progression. Stem cell (SC) therapy has emerged as a promising approach for neuro-regeneration and neuroprotection in NDs, with the potential to replace lost neurons, secrete neurotrophic factors, modulate inflammation, and promote repair of neural circuits (Wei et al. 2023). Indeed, various types of SCs, including neural stem cells (NSCs), mesenchymal stem/stromal cells (MSCs), and induced pluripotent stem cell (iPSC)-derived neural cells, have shown beneficial effects in preclinical models through multiple mechanisms such as cell replacement, secretion of growth factors, immunomodulation, and activation of endogenous repair pathways (Vissers et al. 2019).

Despite this promise, the SC therapies face significant challenges. Transplanted cells often exhibit poor survival, limited differentiation into functional neurons, and inadequate integration or homing to affected brain regions (Wei et al. 2023). In addition, it is difficult to monitor cell fate and biodistribution after administration, hindering the ability to evaluate and improve therapies (Alizadeh et al. 2023). Immune responses present an additional challenge, even low-immunogenic cells, like MSCs provoke host immune reactions or fail to function optimally in the unfavorable inflammatory milieu of a degenerating brain (Sivandzade and Cucullo 2021). Therefore, adjunct strategies are needed to enhance the efficacy and safety of SC treatments for NDs.

Nanotechnology offers innovative solutions to many of these challenges. A wide variety of nanoparticles, including lipid-based, polymeric, inorganic, and hybrid nanomaterials, have been developed for biomedical applications. These nanoscale systems can facilitate delivery of therapeutic agents across the blood–brain barrier (BBB), provide scaffolds or stimuli for cell growth, modulate local immune environments, and enable noninvasive imaging and tracking of transplanted cells (Zhang et al. 2018; Zhu et al. 2021). By harnessing these capabilities, combining SC therapy with nanoparticle-based approaches has shown synergistic effects in preclinical studies. As Vissers et al. (2019) noted, the convergence of nanoparticles and SC therapy holds great promise for the study, diagnosis, and treatment of neurodegenerative disorders (Vissers et al. 2019). In this chapter, the recent advances in nanoparticle-enhanced SC therapies for major NDs are explored. It discusses how nanoparticles are being used to overcome obstacles in SC delivery and survival, modulate

DOI: 10.1201/9781003582687-6

immune responses in the central nervous system (CNS), and improve functional outcomes in models of AD, PD, ALS, MS, HD, and related disorders. Both preclinical findings and early clinical developments are discussed, followed by future directions and translational considerations for integrating nanotechnology with SC-based neuro-regenerative therapies.

6.2 NEED FOR STEM CELL THERAPIES IN NEURODEGENERATIVE DISEASES

The progressive neuron loss in NDs results in irreversible deficits, including, memory loss in AD, movement disorders in PD and HD, paralysis in ALS, and demyelination in MS. Current treatments, e.g., cholinesterase inhibitors for AD, dopamine replacement for PD, immunosuppressants for MS, do not restore lost neural cells or functions. SC therapies aim to address this fundamental gap by introducing cells or stimulating endogenous repair. Research over the past two decades has provided proof-of-concept that SCs can replace or protect neurons in animal models of NDs. For example, transplantation of fetal or SC-derived dopaminergic neurons in PD models can reinnervate the striatum and improve motor function (Deng et al. 2013). In ALS models, grafts of spinal cord progenitor cells have been shown to differentiate into motor neurons and integrate into spinal circuits (Lockard et al. 2023). In AD, MSCs, and NSCs secrete neurotrophic factors that promote survival of remaining neurons and enhance synaptic plasticity (Yari et al. 2022; Chen 2021). These findings suggest that SC-based approaches hold disease-modifying potential by targeting the underlying cell loss and circuit dysfunction in NDs (Table 6.1).

However, translating these successes to patients require overcoming several challenges. First, sourcing an appropriate cell type is critical, NSCs or iPSC-derived neural precursors are ideal for cell replacement in disorders like PD or HD, but MSCs from bone marrow, adipose tissue, etc. are often used for their paracrine and immunomodulatory effects in conditions like AD, MS, or ALS (Yari et al. 2022). Autologous transplantation can minimize immune rejection but is time-consuming and costly. Howevera llogeneic cells may be more practical but risk immune responses and thus requires immunosuppression (Sivandzade and Cucullo 2021). Second, the hostile CNS environment in NDs, which is characterized by chronic inflammation, glial scarring, and ongoing pathology can impede the survival and integration of transplanted cells. For instance, activated microglia and astrocytes release cytokines and reactive oxygen species that may damage grafted cells (Zhu et al. 2021). Additionally, factors like aging or disease progression can limit the brain's receptivity to new cells. Third, the uncontrolled cell differentiation or proliferation could lead to formation of tumors or inappropriate cell types, and there are ethical considerations on the use of embryonic or fetal-derived cells. Finally, delivering cells to the brain or spinal cord in adequate numbers and keeping them in the target region is technically challenging.

TABLE 6.1
Comparative Overview of Stem Cell Types Used in Neurodegenerative Disease Therapy

Stem Cells	Source	Key Mechanism	Disease Applications	Advantages	Limitations
NSCs	Fetal brain, iPSCs	Cell replacement, neurogenesis	AD, PD, HD	Neural lineage potential	Ethical issues, tumor risk
MSCs	Bone marrow, adipose	Paracrine signaling, immunomodulation	AD, ALS, MS	Immunoprivileged, easily accessible	Limited neuronal differentiation
iPSCs	Reprogrammed somatic cells	Customizable, disease modeling	PD, HD	Patient-specific, pluripotent	High cost, tumorigenicity risk

These issues underscore that SC therapy alone may not fully succeed without complementary strategies. Encouragingly, combining SCs with supportive treatments has shown improved outcomes. For example, coadministration of neurotrophic factors, such as glial cell line–derived neurotrophic factor (GDNF) or brain-derived neurotrophic factor (BDNF), alongside cell transplants enhances graft survival and functional recovery in models of PD and HD (Yari et al. 2022). Encapsulation of cells in biomaterial scaffolds or hydrogels can increase cell retention at the transplant site and promote their differentiation, as demonstrated by hydrogel matrices that improved SC engraftment and integration in stroke and spinal injury models (Sivandzade and Cucullo 2021). Notably, nanoparticle-based delivery systems are being explored to address shortcomings of SC therapy, such as poor cell targeting and BBB penetration (Sivandzade and Cucullo 2021). Nanoparticles can transport cells or therapeutic molecules into the CNS and provide controlled release, potentially increasing the efficiency of cell-based therapy. In summary, stem-cell therapies can provide de novo regenerative option for neurodegenerative disorders. However, their maximal benefit will likely emerge only when coupled with adjunct strategies that temper neuroinflammation, supply trophic stimuli, and precisely guide cell homing, the capabilities for which nanotechnology-based platforms are uniquely suited.

6.3 IMMUNOMODULATION IN STEM CELL THERAPY

Neuroinflammation is a common feature of many NDs, contributing to neuronal damage and impeding repair. In conditions like AD and PD, misfolded protein aggregates trigger activation of microglia and astrocytes, which in their pro-inflammatory states release cytokines and toxic mediators that exacerbate neurodegeneration (Zhu et al. 2021). In MS and some forms of ALS, peripheral immune cells infiltrate CNS and directly attack myelin or neurons, driving disease progression (Yari et al. 2022). Even in HD, which is caused by a genetic mutation, evidence suggests that chronic activation of microglia and astrocytes contributes to neuronal death and symptom severity (Yari et al. 2022). This inflammatory milieu not only causes ongoing tissue damage but also creates an environment that is hostile to transplanted SCs or endogenous repair. Thus, immunomodulation is a critical component of effective SC therapy for NDs.

MSCs in particular have attracted attention for their immunomodulatory properties. Unlike neural progenitors which mainly replace lost cells, MSCs can secrete a broad array of anti-inflammatory and neurotrophic factors without the need to engraft or differentiate in large numbers (Yari et al. 2022). MSCs release cytokines such as interleukin-10 (IL-10) and transforming growth factor-β (TGF-β) that suppress pro-inflammatory immune cell activation, as well as molecules like tumor necrosis factor-stimulated gene-6 (TSG-6) that modulate microglial and astrocytic responses (Yari et al. 2022). Through cell–cell interactions and paracrine signaling, MSCs can induce a shift from harmful M1 phenotype microglia, which produce nitric oxide, IL-1, IL-6, etc. to a protective M2 phenotype, which releases anti-inflammatory factors and promotes debris clearance (Yari et al. 2022). They can also promote expansion of regulatory T cells (Tregs) and other immune-regulating cells that help resolve chronic inflammation (Riazifar et al. 2019). For example, in an animal model of MS, systemically administered MSCs and their exosomes increased the number of CD4+CD25+ FOXP3+Tregs in the CNS, correlating with reduced inflammation and disease severity (Riazifar et al. 2019). Similar immunomodulatory effects of MSC therapy have been reported in models of AD and ALS, where treated animals show lower levels of inflammatory cytokines and a bias toward anti-inflammatory microglial activation states (Yari et al. 2022).

Although immunomodulatory strategies can be beneficial, their clinical implementation is inherently complex. The therapeutic efficacy of transplantation depends on the precise calibration of both the temporal dynamics and magnitude of immune suppression. Insufficient modulation predisposes the graft or host parenchyma to immune-mediated injury, whereas excessive suppression heightens susceptibility to infection and may compromise glymphatic clearance and endogenous reparative mechanisms within the brain. Moreover, interindividual variability in immune status such as differences in the degree of microglial activation or peripheral immune cell infiltration results in a cell therapy exerting robust effects in some patients while producing attenuated responses in others. Another concern is that while autologous MSCs are generally well-tolerated, allogeneic cells can provoke immune rejection or require immunosuppressive drugs that have side effects (Sivandzade and Cucullo 2021). Some studies have noted that MSCs from older or diseased donors may have reduced immunomodulatory potency, complicating their therapeutic use in age-related disorders like AD (Wang et al. 2023). Additionally, the inflammatory environment of the diseased CNS can impede MSC survival, for example, high levels of inducible nitric oxide synthase (iNOS) and nitric oxide in ALS or MS lesions can induce apoptosis of transplanted cells. Indeed, one study found that in an ALS mouse model, an inflamed spinal cord niche led to poorer MSC graft survival, though administering MSC-derived exosomes intramuscularly and intraventricularly helped to impair disease development and reduce iNOS activation in the tissue (Yari et al. 2022). This indicates that delivering the immunomodulatory factors of MSCs, e.g., via exosomes or engineered particles can mitigate local inflammation even when cell engraftment is poor.

Overall, effective SC therapy for NDs likely requires immune conditioning, i.e., tempering the hostile immune environment to allow grafted cells to survive and function and leveraging the ability of cell to actively modulate immunity in favor of neuroprotection. Immunomodulation can be considered both a goal and a tool of SC therapy. While exogenous interventions aim to temper the host's, immune milieu minimizing neuroinflammation and averting graft rejection, the stem cell products themselves, particularly mesenchymal stromal cells and their extracellular vesicles, function as intrinsic immunomodulators. The interplay is complex, for instance, MSCs downregulate activation of microglia partly by secreting exosomal microRNAs that target inflammatory signaling pathways (Riazifar et al. 2019).

6.4 NANOPARTICLE SYSTEMS FOR ENHANCING STEM CELL THERAPIES

Nanoparticles which are ultra-small particles, generally 1–100 nm in size, that can be engineered for various biomedical functions. In the context of SC therapy for NDs, a diverse array of nanoparticle systems has been employed to improve cell delivery, viability, tracking, and therapeutic action (Figure 6.1; Table 6.2).

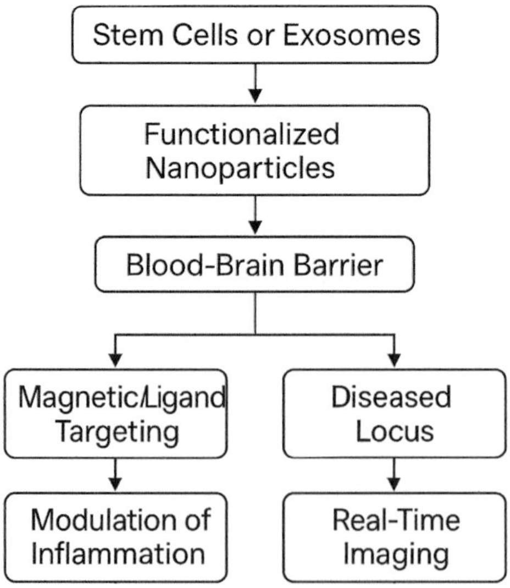

FIGURE 6.1 Integrated Nanoparticle–Stem-Cell Synergy in Central Nervous System Therapy.

TABLE 6.2
Types of Nanoparticles Used in Enhancing Stem Cell Therapy

Nanoparticles	Composition	Key Functions	Applications in ND	Examples
Lipid-based NPs	Liposomes, SLNs	Drug/growth factor delivery, BBB penetration	AD, PD	Liposomes with CAMs, RVG-modified exosomes
Polymeric NPs	PLGA, chitosan	Gene delivery, scaffolding, controlled release	HD, MS	PLGA-GDNF NPs, nanogels
Inorganic NPs	SPIONs, AuNPs, MSNs	Imaging, guidance, antioxidation	PD, ALS	SPION-labeled MSCs, gold nanozymes
Hybrid/Biomimetic NPs	Lipid-polymer hybrids, cell membrane-coated	Immune evasion, targeting	MS, AD	Leukocyte-mimetic NPs, exosome-mimetic NPs

6.4.1 Lipid-Based Nanoparticles

These include liposomes and solid lipid nanoparticles which consist of biocompatible lipid bilayers or matrices. Lipid NPs can encapsulate drugs, genes, or signaling molecules and efficiently deliver them across cell membranes and even the BBB (Zhu et al. 2021). In SC therapy, liposomes have been used to ferry neuroprotective drugs or anti-inflammatory agents alongside transplanted cells to the brain. They can also incorporate membrane-bound signaling proteins or peptides that aid SC homing. For example, liposomes functionalized with cell adhesion molecules have been investigated to target SCs to specific brain regions (Vissers et al. 2019). Their ability to control the release of encapsulated factors over time helps in creating a nurturing microenvironment for the grafted cells (Zhu et al. 2021). A special case of lipid nanoparticles is the exosomes secreted by SCs. These naturally occurring nanovesicles (30–150 nm) carry lipids, proteins, and microRNAs from the parent cell and can traverse the BBB. While not synthetic, MSC-derived exosomes are being harnessed as cell-free therapeutic agents (Ye et al. 2023)

6.4.2 Polymeric Nanoparticles

The polymeric nanoparticles are produced from biocompatible polymers like poly(lactic-co-glycolic acid (PLGA), chitosan, or dendrimers. They are highly tunable in size, surface chemistry, and payload capacity. Polymeric nanoparticles, long established as versatile drug-delivery vectors, are now being harnessed in stem-cell strategy to pharmacologically precondition cells ex vivo and transport therapeutic genes with high efficiency. For instance, PLGA nanoparticles can be loaded with genes encoding neurotrophic factors or anti-inflammatory cytokines and employed to transfect SCs prior to transplantation, thereby augmenting their therapeutic secretory capacity (Ghosh et al. 2024). Polymeric nanogels can also protect transplanted cells by local delivery of immunosuppressive drugs in the transplant site, reducing immune attack without systemic side effects. Another application is using polymer scaffolds on the nanoscale, electrospun nanofibers or nanostructured hydrogels made of polymers can serve as a 3D matrix to support SC growth and differentiation. These scaffolds, often in order of tens to hundreds of nanometers in fiber diameter or pore size, mimic the extracellular matrix and can improve engraftment. Encapsulation of cells in such nanoporous hydrogels has been shown to increase cell survival, proliferation, and differentiation at the transplantation site (Sivandzade and Cucullo 2021).

6.4.3 Inorganic/Metallic Nanoparticles

The inorganic nanoparticles include nanoparticles composed of metals (gold, silver), metal oxides (iron oxide, ceria), silica, and other inorganic materials. These nanoparticles often have unique physical properties, e.g., magnetic, optical that can be leveraged for imaging or remote control of cell fate. Superparamagnetic iron oxide nanoparticles (SPIONs) are a prime example, which can label SCs, rendering them visible on magnetic resonance imaging (MRI) scans for tracking after transplantation (Alizadeh et al. 2023). SPION-labeled MSCs have been tracked in patients with stroke and shown to migrate toward injury sites, confirming homing capabilities. Beyond their diagnostic utility, magnetic nanoparticles enable active magnetic guidance. SCs labelled with SPIONs can be steered to precise central-nervous-system targets in vivo using externally applied magnetic field gradients, a strategy already validated in several rodent models. (Alizadeh et al. 2023). Gold nanoparticles (AuNPs), meanwhile, have potent antioxidant and anti-inflammatory properties. They can quench reactive oxygen species and modulate microglial activity (Zhu et al. 2021). AuNPs have been studied as nanozymes to reduce oxidative stress in AD models, thereby protecting transplanted neuronal cells. Moreover, some inorganic NPs can respond to external stimuli like light or ultrasound to trigger release of factors or heat, for example, gold nanoshells have been explored to deliver mild thermal stimulation to transplanted NSCs, which can promote their differentiation. Mesoporous silica nanoparticles (MSNs), that are characterized by a chemically inert matrix and size-tunable pore architecture, have been engineered to sequester high payloads of growth factors and furnish their controlled, sustained release, thereby conferring prolonged trophic support to transplanted stem-cell grafts

6.4.4 Hybrid and Biomimetic Nanoparticles

Researchers have also developed hybrid nanoparticles that combine materials, e.g., lipid–polymer hybrids or coat nanoparticles with cell-derived membranes. One example that is relevant to SC therapy is coating PLGA nanoparticles with SC membrane fragments or with leukocyte membranes. These hybrid nanoparticles can facilitate immune evasion and enable targeted delivery to inflamed brain regions by homing to cytokine gradients, analogous to the migratory behavior of immune cells. A recent advance involves the development of exosome-mimetic nanoparticles. Rather than relying on the low-yield, labor-intensive isolation of native exosomes, researchers fabricate synthetic liposomes or polymeric nanoparticles and load them with exosome-like cargo, most commonly microRNAs and bioactive proteins that mirror the profile of mesenchymal-stromal cell exosomes. The resulting biomimetic nanocarriers combine the inherent cell-homing and intercellular communication properties of natural exosomes with the production scalability and tunable physicochemical characteristics of engineered delivery systems.

Each nanoparticle system offers distinct advantages, and these platforms are not mutually exclusive; indeed, multimodal strategies are increasingly employed to harness complementary mechanisms of action. For instance, an iron oxide core are enveloped in a polymer or lipid shell that carries a drug, combining imaging and therapeutic delivery. The distinct types of nanoparticles thus form a toolkit to tackle various limitations of SC therapy (Vissers et al. 2019).

6.5 MECHANISM OF NANOPARTICLE-ENHANCED IMMUNOMODULATION AND STEM CELL EFFICACY

Integration of nanoparticles with SC therapies enhances therapeutic outcomes through multiple mechanisms of action (Table 6.3).

6.5.1 Improved Delivery and Homing to the CNS

A fundamental challenge lies in delivering therapeutic cells or their secreted factors to the brain and spinal cord, necessitating traversal of the protective BBB. By leveraging nanoparticles,

TABLE 6.3
Mechanistic Roles of Nanoparticles in Enhancing Stem Cell Therapies

Mechanism	Nanoparticle Strategy	Impact on Therapy
Delivery/Homing	Magnetic NPs, ligand-conjugated NPs	Improve BBB crossing and cell targeting
Survival/Integration	Antioxidant-loaded NPs, nano-scaffolds	Enhance graft survival, support axon guidance
Immunomodulation	Anti-inflammatory drug NPs, siRNA NPs	Suppress neuroinflammation, promote M2 microglia
Signal Release	Growth factor-loaded NPs, pH-responsive NPs	Sustain local trophic support, guide differentiation
Theranostics	SPIONs, fluorescent NPs	Real-time tracking, feedback-responsive delivery

therapeutic cells can better traverse the BBB. One approach under investigation is intranasal delivery of SCs tagged with magnetic nanoparticles, providing a noninvasive pathway to the brain. The nanoparticles not only allow tracking by MRI, but in some cases, magnetic fields are used to guide the cells along olfactory neural pathways into the CNS (Alizadeh et al. 2023). Recently, it has been shown that the intranasal administration of NP-labeled SCs can bypass the BBB and deliver cells broadly to the brain, a technique being explored for diseases like AD and glioblastoma (Alizadeh et al. 2023). Additionally, nanoparticles can carry chemoattractant signals or surface ligands that enhance SC homing. For example, stromal cell–derived factor 1 (SDF-1) has been conjugated to nanoparticles to create gradients that attract injected SCs to injured sites. By improving the biodistribution of cells, thus ensuring they reach and remain in affected regions like the hippocampus in AD or plaques in MS, nanoparticles significantly boost the therapeutic impact of a given cell dose (Figure 6.2).

6.5.2 Enhanced Cell Survival and Integration

Following transplantation, SCs are frequently exposed to a hostile microenvironment characterized by inflammation, oxidative stress, and inadequate structural or trophic support, all of which compromise their viability and survival. Nanoparticles can function as nano-scale reservoirs for pro-survival factors. A notable strategy entails incorporating cerium oxide nanoparticles that are known for their intrinsic antioxidant and enzyme-mimetic properties into scaffolds surrounding transplanted NSCs, thereby mitigating oxidative damage and enhancing cellular survival within the hostile microenvironment. This results in local scavenging of reactive oxygen species and enhanced survival of the graft in an ALS model (Yari et al. 2022). Another promising strategy involves genetic nanoengineering, wherein nanoparticles are employed as delivery vehicles for nucleic acids such as DNA, RNA, or CRISPR/Cas components to modify SCs either prior to or following transplantation. This approach enables precise genetic modulation, enhancing cellular resilience, differentiation potential, and integration within the host tissue. Ghosh et al. (2024) demonstrated that nanoengineering of SCs can enhance their resilience by upregulating key survival pathways. In their study, nanoparticles were used to deliver a brain-derived neurotrophic factor gene plasmid into MSCs that leads to increased BDNF secretion. This elevation in BDNF levels subsequently amplified the neuroprotective effects of MSCs in HD model (Ghosh et al. 2024). Similarly, polymeric nanoparticles have been used

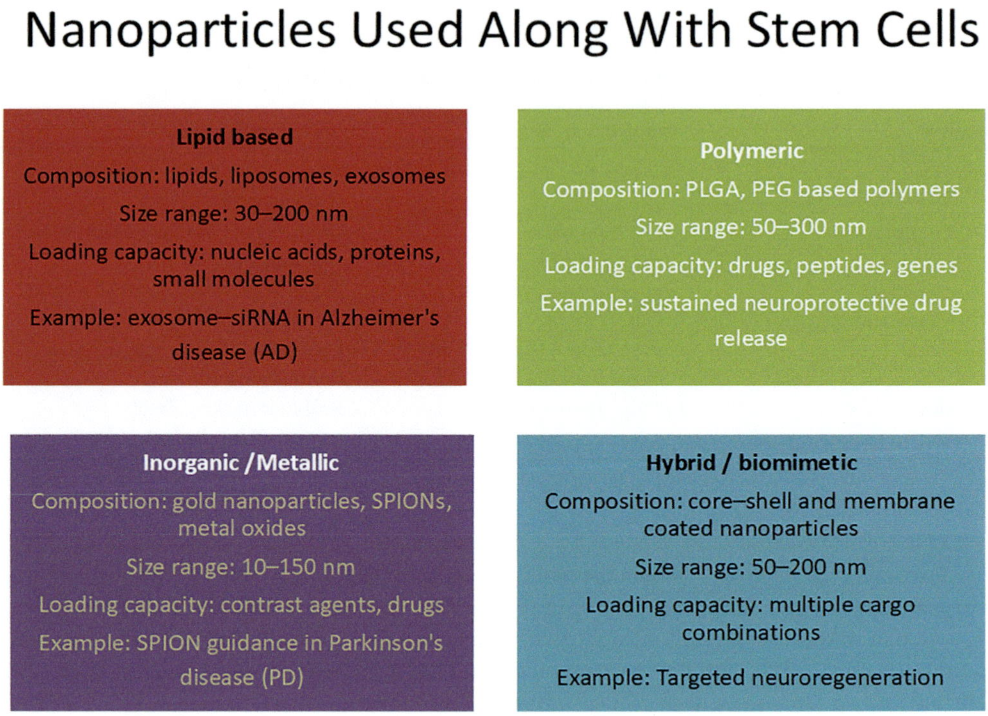

FIGURE 6.2 Mechanistic Cascade Underlying Nanoparticle-Assisted Stem-Cell Delivery to the CNS.

to transiently knock down pro-apoptotic genes in SCs, thereby increasing their survival post-transplant. Additionally, nanofiber scaffolds composed of aligned fibers at dimensions comparable to those of neural cells provide critical topographical signals that direct axonal growth from transplanted neurons. This structural alignment facilitates axon extension and promotes functional integration of the transplanted cells into the host neural circuits. In a spinal cord injury model, NSCs transplanted with an aligned nanofiber scaffold showed significantly greater neurite outgrowth bridging the lesion, compared to cells transplanted alone (Carradori et al. 2017). Thus, through biochemical support and biophysical stimuli, NPs create a more conducive niche for SCs in vivo.

6.5.3 Immunomodulation and Inflammation Control

There are two principal facets to immunomodulatory nanoengineering, delivering immunoregulatory agents to the host environment and enhancing the intrinsic immunomodulatory secretome of SCs. In the first approach, researchers have encapsulated anti-inflammatory compounds such as curcumin and minocycline within liposomes targeted to activated microglia, enabling localized delivery when coadministered with SC transplants. This strategy helps attenuate neuroinflammation and creates a more conducive microenvironment for graft survival and integration. (Zhu et al. 2021). In an AD mouse model, curcumin-loaded NPs were delivered with MSCs. The NPs preferentially entered activated microglia around amyloid plaques and modulate their inflammatory activity that resulted in a more supportive environment for the MSCs (Zhu et al. 2021). This combined approach has led to reduced levels of IL-1β and TNF-α in the brain and improved cognitive outcomes compared to MSC therapy alone. Conversely, nanoparticle-mediated preconditioning can enhance the intrinsic immunomodulatory properties of SCs by stimulating the upregulation of anti-inflammatory cytokines and other immunoregulatory mediators. This approach primes the cells before transplantation, thereby improving their capacity to modulate the host immune response and promote a regenerative, anti-inflammatory milieu. An example is the use of IFN-γ-loaded nanoparticles to prestimulate MSCs before transplantation. Riazifar et al. (2019) showed that stimulating MSCs with interferon-γ causes them to produce exosomes enriched in anti-inflammatory microRNAs and proteins (Riazifar et al. 2019). When these exosomes were administered in an experimental autoimmune encephalomyelitis mouse model of MS, they resulted in a marked reduction of pro-inflammatory cytokines, promoted the induction of regulatory T cells, and inhibited demyelination, collectively leading to an overall attenuation of clinical disease severity (Riazifar et al. 2019). In this approach, nanoparticles are applied ex vivo rather than infused into the patient, serving to prime the cellular product before transplantation. Such nanoparticle-based preconditioning can also be achieved using various cytokines or other stimulatory agents to enhance the immunoregulatory phenotype of mesenchymal stromal cells, thereby improving their therapeutic efficacy and compatibility with the host immune environment. Beyond serving as carriers for anti-inflammatory drugs or cell-derived exosomes, certain nanoparticles possess intrinsic immunomodulatory properties. Gold nanoparticles, for example, have been shown to modulate macrophage polarization by promoting a shift toward the M2 anti-inflammatory phenotype through the regulation of toll-like receptor signaling, a mechanism observed in certain PD models (Zhu et al. 2021). Similarly, nanoparticles loaded with small interfering RNAs targeting inflammatory mediators, such as siRNA against NLRP3 inflammasome components, have been coadministered with SCs to achieve localized gene silencing within host immune cells, thereby attenuating inflammation at the transplantation site (Yari et al. 2022). This combinatorial strategy is particularly potent, as the SCs exert broad immunosuppressive effects while the nanoparticles selectively inhibit specific pro-inflammatory pathways. Together, these complementary mechanisms produce a synergistic suppression of chronic inflammation, promoting enhanced graft survival and improved restoration of host tissue function.

6.5.4 Controlled Release of Growth Factors and Biophysical Signals for Stem Cell Modulation

SC efficacy can be enhanced by ensuring that cells both receive and transmit appropriate signals at precise time points. Nanoparticles can be engineered to deliver growth factors, mitogens, or differentiation cues in a controlled manner. For instance, polymer nanoparticles loaded with nerve growth factors were injected into the brains of AD model mice that had undergone NSC transplantation. These nanoparticles released nerve growth factors gradually over several weeks, significantly promoting the differentiation of transplanted NSCs into cholinergic neurons and improving their synaptic integration compared to transplants without growth factor supplementation (Chen et al. 2024). In PD, gelatin nanoparticles releasing glial cell line-derived neurotrophic factor implanted into the striatum alongside dopaminergic neuron grafts improved graft survival and enhanced dopaminergic regeneration (Bondarenko and Saarma 2021). The kinetics of releasing are also be modulated, as some nanoparticles are designed to discharge their therapeutic cargo in response to local pH or enzymatic changes associated with inflammation, thereby delivering bioactive factors precisely during inflammatory surges in MS lesions. This level of spatiotemporal control is difficult to achieve through direct protein injections due to their short half-life, whereas nanoparticles protect growth factors from degradation and provide a sustained signaling reservoir. Another approach involves morphological stimulation, where quantum dot nanoparticles activate mechano-sensitive pathways in SCs under specific light conditions, influencing differentiation. Although still experimental, such strategies demonstrate how nanotechnology can precisely modulate SC fate after transplantation, promoting lineage-specific differentiation such as the formation of oligodendrocytes for remyelination in MS or neurons for neural circuit restoration in AD.

6.5.5 Role of Nanoparticles in Real-Time Tracking and Feedback Control

Finally, nanoparticles provide a versatile platform for theranostics, which integrates therapeutic and diagnostic functionalities within a single system. For example, MRI–visible or fluorescently labeled nanoparticles enable real-time visualization of transplanted SC populations, allowing clinicians to track their migration, engraftment, and persistence within the host tissue (Alizadeh et al. 2023). Such monitoring is particularly valuable in early clinical trials to confirm accurate cell targeting and distribution. Beyond spatial tracking, advanced nanosensors are being designed to assess the biochemical milieu surrounding transplanted cells. These nanosystems can detect dynamic changes in pH, oxidative stress, enzymatic activity, and inflammatory mediators. For instance, polymer nanoparticles that emit fluorescence in response to inflammatory cytokines have been co-transplanted with SCs in animal models, facilitating real-time optical monitoring of local inflammation. Although still under development, feedback-regulated therapeutic systems represent a major step toward precision medicine, wherein nanoparticles act as autonomous biointerfaces that enable therapeutic cells to sense and adapt to the host microenvironment dynamically.

Thus, nanoparticles serve as multifunctional enhancers in SC-based therapies by improving CNS delivery, protecting and modulating transplanted cells, fine-tuning the immune microenvironment, providing sustained trophic signaling, and allowing real-time assessment of cell fate and function. Together, these features have produced compelling preclinical evidence supporting their synergistic benefits. Emerging research indicates that the full therapeutic potential of SC-based interventions for neurodegenerative disorders may be realized only when integrated with nanoparticle-mediated delivery systems (Rajendran et al. 2025; Figure 6.3).

The subsequent sections will discuss the latest advancements and future perspectives of this integrated nanotechnology and SC approach across major NDs (Table 6.4).

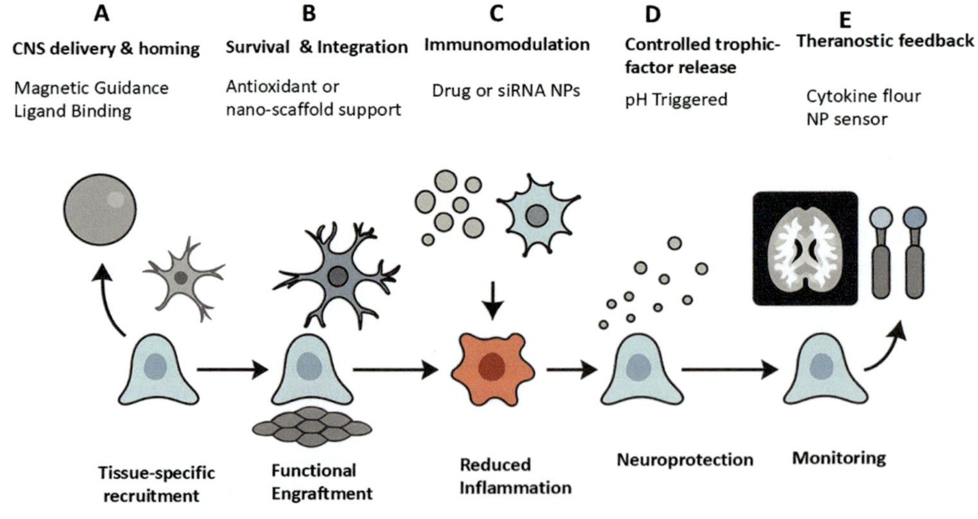

FIGURE 6.3 Mechanistic Pathways Through Which Nanoparticles Enhance Stem-Cell Efficacy.

TABLE 6.4
Summary of Preclinical Outcomes for Nanoparticle-Enhanced Stem Cell Therapies across Neurodegenerative Diseases

Disease	Stem Cells	Nanoparticles	Delivery Route	Key Outcomes
AD	MSCs, NSCs	Exosomes, polymeric NPs	Intracerebral, intranasal	↓ Aβ/tau, ↑ cognition, ↓ neuroinflammation
PD	iPSC-DA neurons, MSCs	Magnetic NPs, GDNF-NPs	Intravenous + magnetic, intracerebral	↑ graft survival, ↑ dopamine release, ↑ motor function
ALS	MSCs, NSCs	Lipid NPs, exosomes	Intrathecal, spinal graft	↑ motor neuron survival, ↑ lifespan, ↓ inflammation
MS	MSCs, NPCs	PLGA-PPAR NPs, tolerogenic NPs	Intrathecal, systemic	↑ remyelination, ↓ T-cell response, ↑ function
HD	NSCs, MSCs	PAMAM-siRNA, BDNF-NPs	Striatal injection	↓ mutant HTT, ↑ neuronal survival, ↑ behavior

6.6 ALZHEIMER'S DISEASE

AD is the most common cause of dementia, marked by progressive memory loss and cognitive decline. Pathologically, AD is characterized by the accumulation of extracellular amyloid-β (Aβ) plaques, intracellular neurofibrillary tangles composed of hyperphosphorylated tau protein, extensive synaptic degeneration, and sustained neuroinflammatory responses. Current treatment, including acetylcholinesterase inhibitors, and recent antiamyloid antibodies provide limited symptomatic relief or modest attenuation of pathological progression. A critical need persists for therapeutic strategies that can both preserve and replace the substantial neuronal populations lost in AD while simultaneously modulating the interconnected neuroinflammatory and neurodegenerative cascades that drive disease progression (Wang et al. 2023). SC therapy for AD has focused primarily on MSCs and NSCs. MSCs are administered through minimally invasive routes and have demonstrated neuroprotective, immunomodulatory, and synaptogenic effects in preclinical models of AD (Wang et al. 2023; Yari et al. 2022). NSCs or neural progenitor cells, typically delivered via direct intracerebral transplantation, hold the potential to differentiate into functional neurons within affected regions such as the hippocampus. However, achieving robust and functionally integrated neuronal replacement in AD remains a significant challenge.

6.6.1 Nanoparticle-Enhanced Therapeutic Strategies for Alzheimer's Disease

The multifactorial pathology of AD presents several therapeutic opportunities for nanoparticle-enhanced SC interventions. A central focus lies in modulating neuroinflammatory processes and facilitating the clearance of neurotoxic protein aggregates, including amyloid-beta and hyperphosphorylated tau, which are key contributors to neuronal dysfunction and cognitive decline.

MSCs and the associated exosomes have demonstrated an ability to mitigate neuroinflammation in AD models by shifting microglia to a more phagocytic, less inflammatory state (Yari et al. 2022). For example, a study by Chen et al. (2021) showed that treating an AD cellular model with human MSC-derived exosomes led to reduced Aβ levels and restoration of genes associated with synaptic plasticity (Chen et al. 2021). MSC-derived exosomes, which function as endogenous nanoscale delivery vehicles, are capable of crossing the blood–BBB and delivering microRNAs and proteins that suppress amyloidogenic processing and attenuate pro-inflammatory signaling pathways. Building on these findings, nanoparticle-based delivery systems are being engineered to improve the concentration, stability, and targeted delivery of bioactive molecules derived from MSCs. These systems aim to optimize the therapeutic potential of SC secretomes by ensuring precise localization and sustained bioavailability within affected neural regions. One promising strategy involves the utilization of RVG-modified exosomes, which are extracellular vesicles engineered to express the rabies virus glycoprotein peptide on their surface, thereby enabling selective recognition and internalization by neuronal cells. Yuan et al. (2022) encapsulated small interfering RNA targeting beta-site amyloid precursor protein cleaving enzyme 1 (BACE1) within RVG-functionalized MSC-derived exosomes. In murine models of AD, these modified exosomes demonstrated preferential accumulation in the brain, effectively silenced BACE1 expression, reduced amyloid-beta plaque deposition, and improved cognitive outcomes. These findings highlight the potential of SC-derived nanovesicles to mediate disease-modifying effects in AD.

Beyond the use of exosomes, synthetic nanoparticles are increasingly being incorporated into SC-based therapeutic frameworks for AD. A particularly promising direction focuses on employing nanocarriers for targeted delivery of neurotrophic factors. Because AD is characterized by diminished neurotrophic signaling, including reduced levels of brain-derived neurotrophic factor, MSCs have been investigated for their ability to restore trophic support through sustained secretion of such neuroprotective molecules.

To strengthen neurotrophic support in AD, researchers have employed polymeric nanoparticles to introduce neurotrophin genes into MSCs. In one study, MSCs were transfected with a plasmid encoding glial cell line–derived neurotrophic factor using a polyethylenimine-based nanoparticle vector and subsequently transplanted into an AD mouse model. The modified cells secreted significantly higher levels of glial cell line–derived neurotrophic factor, which improved hippocampal neuron survival and synaptic functionality, leading to enhanced cognitive performance compared to mice that received unmodified SCs. Notably, histological and biochemical analyses revealed reduced neuroinflammation in the group treated with glial cell line–derived neurotrophic factor–overexpressing MSCs, likely reflecting the synergistic immunomodulatory influence of both the neurotrophic factor and the endogenous cytokines secreted by the SCs. This dual action contributed to the establishment of a more anti-inflammatory neural microenvironment (Yari et al. 2022).

Another innovative approach involves the coadministration of SCs with nanoparticles designed to target amyloid plaques. For instance, magnetically responsive iron oxide nanoparticles functionalized with antiamyloid-beta antibodies have been delivered in conjunction with MSCs, resulting in the spatial colocalization of both entities within amyloid plaque–dense brain regions. The magnetically guided MSCs were thus positioned in proximity to amyloid deposits, where they secreted matrix metalloproteinases capable of degrading amyloid-beta aggregates. Preliminary observations from transgenic AD mouse models demonstrated a measurable reduction in amyloid plaque size, suggesting that such combinatorial nanotechnology and SC strategies could offer a promising therapeutic avenue for targeted amyloid clearance and neuroprotection.

Collectively, preclinical studies on nanoparticle-enhanced SC therapies in AD have demonstrated an attenuation of neuropathological hallmarks, including reduced amyloid-β plaque burden and reduced accumulation of hyperphosphorylated tau, particularly following the administration of MSC-derived exosomes or genetically engineered SCs (Reza-Zaldivar et al. 2018). Beyond their disease-modifying capabilities, nanoparticle-enhanced SC therapies have demonstrated neuroprotective effects, that are evident from the upregulated synaptic plasticity markers, sustained neuronal integrity in the hippocampus, and enhanced cerebral metabolism as detected by PET imaging (Chen et al. 2021). Moreover, multiple studies have reported that AD mouse models treated with nanoparticle-enhanced SC therapies exhibit improved cognitive performance in behavioral assays such as the novel object recognition test and the Morris water maze, relative to untreated controls (Chen et al. 2021). For instance, intrahippocampal transplantation of human umbilical cord-derived MSCs into APP/PS1 transgenic AD mice, in combination with a peptide-loaded nanoparticle hydrogel enabling sustained release of hepatocyte growth factor (HGF) and growth differentiation factor-15 (GDF-15), resulted in significant restoration of spatial memory and increased synaptic density within the hippocampus (Jia et al. 2024; Yari et al. 2022).

However, the clinical translation of SC-based therapies for AD remains at an early stage. Only a few Phase I clinical trials have assessed the safety of MSC administration in AD patients. For instance, a study using allogeneic adipose-derived MSCs (Hope Biosciences, NCT04428740) reported no major adverse events and preliminary evidence of cognitive improvement. Although nanoparticle-based strategies have not yet been incorporated into most clinical trials, translational efforts are ongoing. A Phase I trial in South Korea (NCT04388982) is investigating intrathecal administration of MSC-derived exosomes in AD patients, representing a direct clinical translation of a preclinically validated nanoparticle-mediated approach. If successful, this trial could establish the first clinical application of a nanoparticle-based, cell-free SC therapy in AD. The convergence of SC biology and nanomedicine offers a multifactorial therapeutic paradigm in which SCs act as biologically active regulators of the disease microenvironment, while nanoparticles provide precise control over delivery, targeting, and amplification of therapeutic effects. As targeting modalities such as focused ultrasound-assisted blood–BBB opening, magnetic guidance, and intranasal administration continue to advance, these integrated approaches hold strong potential to slow or partially reverse the progressive neural degeneration characteristic of AD, a feat not yet achieved by conventional pharmacological therapies.

6.6.2 Nanoparticle-Enhanced Therapeutic Strategies for Parkinson's Disease

PD is a movement disorder caused primarily by degeneration of dopaminergic neurons in the substantia nigra pars compacta, leading to dopamine depletion in the striatum. Patients experience tremors, rigidity, bradykinesia, and postural instability, and in later stages can develop cognitive decline. Standard treatments, including levodopa and other dopaminergic agents, primarily alleviate motor symptoms but fail to prevent the progressive loss of dopaminergic neurons or modify the underlying pathophysiology of the disease. SC-based therapies for PD have traditionally centered on cell replacement strategies, particularly the transplantation of dopaminergic neuron precursors to replenish striatal dopamine levels and restore functional circuitry. Indeed, PD was the first ND where human fetal tissue transplants showed some clinical benefit (in the 1980s) (Yari et al. 2022).

Recent progress in generating iPSC SC-derived midbrain dopaminergic progenitors has spurred the initiation of clinical trials for PD, aiming to restore dopaminergic neuronal populations that are lost to degeneration. Alongside these neuronal replacement efforts, MSCs and other supportive cell types are being explored for their neuroprotective capacity, primarily through the secretion of trophic factors such as brain-derived neurotrophic factor and glial cell line–derived neurotrophic factor, which enhance the survival of endogenous neurons and mitigate neuroinflammatory processes.

Although neuroinflammation in PD is generally less pronounced than in AD or MS, activated microglia and infiltrating peripheral immune cells have been implicated in disease progression and the manifestation of nonmotor symptoms (Zhu et al. 2021), underscoring the relevance of incorporating immunomodulatory strategies into therapeutic approaches. A central objective of SC therapy in PD is to enhance the survival, maturation, and functional integration of transplanted dopaminergic (DA) neurons. Nanotechnology has contributed to this goal through the development of advanced biomaterials that support cell engraftment. One promising strategy involves the use of nano-engineered scaffolds for transplantation. For instance, peptide-based nanofiber scaffolds have been designed to mimic the biochemical and structural features of the brain's extracellular matrix, incorporating slow-release nanoparticles loaded with glial cell line–derived neurotrophic factor (GDNF). In preclinical studies, transplantation of embryonic SC-derived DA neurons on these functionalized scaffolds into rat models of Parkinson disease significantly enhanced graft viability, promoted neurite extension, and led to superior motor function recovery compared to cell transplants delivered without scaffold support (Ou et al. 2020).

The nanoparticles in this scaffold continuously provided GDNF to the graft, which is crucial in the first week's posttransplant for neuron survival and maturation. Additionally, the aligned orientation of the nanofibers directed dopaminergic neurite extension toward the host striatum, illustrating how a combinatorial biomaterial–nanoparticle platform can address two major challenges in cell therapy, that is graft survival and precise axonal targeting.

Magnetic nanoparticles have been investigated as an approach to enhance the precision and efficacy of SC therapy in PD. In one study, MSCs were labeled with superparamagnetic iron oxide nanoparticles and administered intravenously into a murine model of PD. Application of an external magnetic field near the cranial region enabled targeted guidance of the labeled cells toward the brain, specifically directing them along the nigrostriatal pathway through magnetic targeting (Alizadeh et al. 2021). Once localized, the MSCs secreted neurotrophic factors such as brain-derived neurotrophic factor and anti-inflammatory cytokines, which promoted dopaminergic neuron preservation and improved motor function compared to nonmagnetized controls (Mo et al. 2018). These findings demonstrate the therapeutic promise of magnetic targeting as a minimally invasive strategy for directing systemically delivered SCs to specific brain regions in PD, potentially overcoming the limitations associated with direct intracerebral transplantation.

A key limitation in PD therapy is the insufficient neurotrophic support within the degenerating brain. Although direct infusion of glial cell line–derived neurotrophic factor into the striatum has been tested in clinical trials, the results have been inconsistent. Combining glial cell line–derived neurotrophic factor with SC-based interventions may produce more consistent and potent therapeutic outcomes. As demonstrated in recent research, polymeric and lipid-based nanoparticles can be utilized to deliver genes or pharmacological compounds that augment the trophic effects of transplanted cells. In one study, compact polymer–DNA nanoparticles were employed to codeliver glial cell line–derived neurotrophic factor and neurturin genes into the striatum of rats following transplantation of iPSC-derived dopaminergic neurons, with the goal of promoting graft survival, enhancing neuronal integration, and improving motor recovery.

The nanoparticles enabled sustained and widespread expression of glial cell line–derived neurotrophic factor and neurturin within the striatum, with transgene expression persisting for several weeks following administration (Aly et al. 2015, 2019; Fletcher et al. 2011). This prolonged neurotrophic signaling promoted extensive axonal outgrowth and synaptic integration of the transplanted dopaminergic neurons into the host striatal circuitry, leading to near-complete restoration of motor function in treated rats (Sun et al. 2021). Notably, animals receiving nanoparticle-mediated gene delivery exhibited a reduced immune response to the graft, likely due to the anti-inflammatory effects of glial cell line–derived neurotrophic factor on microglia and the improved viability of transplanted neurons, which limits the release of damage-associated molecular patterns. These promising results have prompted continued investigation and funding support from the Michael J. Fox Foundation to advance nanoparticle-based gene delivery combined with SC transplantation as an emerging next-generation therapeutic approach for PD (Espadas-Alvarez et al. 2017).

Although PD does not exhibit the pronounced autoimmune pathology characteristic of MS, chronic neuroinflammation remains a critical therapeutic target. Recent findings indicating that dysregulated peripheral T-cell subsets exacerbate nigral neurodegeneration suggest that mesenchymal stromal cells could function as adjunct immunomodulators by attenuating maladaptive immune signaling and restoring peripheral–central immune equilibrium (Sulzer et al. 2017).

A promising nanoparticle-based approach focuses on inducing immune tolerance through nanosized tolerogenic particles engineered to display PD-associated antigens, such as alpha-synuclein peptides. These particles interact with antigen-presenting cells in a way that promotes immune tolerance rather than activation. When used in combination with MSC therapy, a complementary dual mechanism is achieved. The MSCs alleviate CNS inflammation by modulating microglial activation, while tolerogenic nanoparticles prevent peripheral immune-mediated damage to neurons and transplanted grafts. In a recent preclinical study, administration of antigen-conjugated nanoparticles reduced pro-inflammatory T-cell populations in a PD mouse model. When coadministered with MSCs, this combination produced significantly greater preservation of dopaminergic neurons than either treatment alone. Although still in the experimental phase, this integrative approach represents a novel therapeutic paradigm that combines tolerance-inducing nanoparticle vaccines with cell-based therapy, providing a synergistic means to regulate both peripheral and central immune responses in PD.

In preclinical models of PD, nanoparticle-assisted SC therapies have yielded encouraging results, demonstrating functional, structural, and neuroprotective benefits. Treated animals exhibited improved motor performance, as evidenced by enhanced rotarod and gait parameters, alongside increased graft survival and integration marked by higher counts of tyrosine hydroxylase–positive neurons and greater dopaminergic fiber density in target brain regions. Neuroprotective effects on remaining host dopaminergic neurons were also consistently observed. For example, Parodi et al. (2020) reported that animals receiving MSCs engineered with magnetic nanoparticles and drug-loaded liposomes displayed elevated striatal dopamine concentrations and superior preservation of nigral neurons relative to control groups. Histological analyses across several studies further revealed reductions in alpha-synuclein aggregation and glial activation surrounding graft sites following nanoparticle-enhanced interventions. Importantly, multiple investigations have demonstrated a synergistic relationship between the two modalities. The SCs provide multifaceted trophic and immunomodulatory support, while nanoparticles improve targeting accuracy and ensure sustained release of therapeutic molecules. In a comparative study by Shi et al. (2018), four treatment conditions were evaluated in a PD model, no treatment, MSCs alone, glial cell line–derived neurotrophic factor–loaded nanoparticles alone, and the combined treatment. The combinatorial group receiving both MSCs and nanoparticles achieved the most significant improvements in behavioral recovery and dopaminergic neuron preservation, surpassing the additive outcomes of the individual therapies and underscoring the synergistic potential of integrated nanotechnology–SC approaches.

Cell-based therapies for PD are progressing rapidly, with several clinical trials currently evaluating the transplantation of iPSCSC-derived dopaminergic neurons, including the pioneering study led by Jun Takahashi in Japan and parallel investigations in Europe and the United States (Takahashi 2020). These trials are expected to provide critical data on the safety, feasibility, and therapeutic efficacy of neuronal replacement approaches. Although nanoparticles have not yet been incorporated into these clinical protocols, their prospective integration offers considerable promise. Nanoparticle-mediated delivery of glial cell line–derived neurotrophic factor or anti-inflammatory agents to transplantation sites could enhance graft survival, promote functional integration, and modulate local immune responses. Furthermore, if immunogenicity presents a barrier, even with autologous or HLA-matched iPSC SC-derived grafts, nanoparticle-based immunomodulatory platforms could serve as a localized and controllable alternative to systemic immunosuppression.

Given that PD primarily affects deep brain structures, the use of focused ultrasound to transiently and safely open the BBB represents an appealing adjunct strategy. In this context, intravenous administration of supportive nanoparticles could be synchronized with spatially precise BBB modulation, enabling targeted delivery of trophic or immunomodulatory factors directly to the graft site. This approach would allow noninvasive, on-demand biochemical support to transplanted cells, optimizing their long-term viability and function.

Collectively, PD appears most amenable to a combinatorial regenerative paradigm in which SCs reconstruct damaged neural circuits and nanoparticles enhance their survival, integration, and functionality. The extensive preclinical evidence demonstrating improved graft outcomes through nanoparticle-assisted delivery provides compelling justification for advancing this integrated nanotechnology–SC framework into next-generation clinical applications.

6.6.3 Amyotrophic Lateral Sclerosis

ALS is a progressive and fatal neurodegenerative disorder characterized by the selective loss of motor neurons in the spinal cord, brainstem, and motor cortex, ultimately leading to paralysis. Although the majority of cases are sporadic, approximately 5–10% are familial and associated with mutations in genes such as SOD1, C9ORF72, TARDBP, and FUS. Increasing evidence highlights the contribution of non-neuronal cells, particularly astrocytes and microglia, to disease pathogenesis through pro-inflammatory and neurotoxic signaling mechanisms that exacerbate motor neuron death. Peripheral immune cell infiltration into the CNS and systemic immune dysregulation further amplify neurodegeneration. At present, no curative treatment exists for ALS. The only FDA-approved drugs, riluzole and edaravone, confer limited benefit, producing only modest extensions in survival or delays in disease progression. Given the rapid clinical decline characteristic of the disorder, effective therapies must act early and possess robust neuroprotective potential to preserve motor neuron viability.

SC-based therapeutic approaches in ALS have primarily focused on transplanting supportive cell populations such as glial-restricted progenitors and MSCs rather than attempting direct replacement of upper or lower motor neurons. This is due to the technical and biological challenges of reestablishing long axonal projections and precise synaptic connections. The central therapeutic aim is therefore to modulate neuroinflammation, supply trophic support, and attenuate motor neuron degeneration, thereby preserving functional capacity and extending patient survival.

A hallmark pathological feature of ALS is pronounced neuroinflammation, particularly within the spinal cord and motor cortex. In response to this, MSCs have been explored in multiple clinical trials for their ability to secrete neuroprotective and anti-inflammatory factors capable of mitigating motor neuron degeneration. A prominent example is BrainStorm Cell Therapeutics' NurOwn platform, which, although not formally classified as nanotechnology, employs a nano-relevant principle. In this system, MSCs are cultured under defined conditions that stimulate the secretion of elevated levels of neurotrophic factors such as brain-derived neurotrophic factor, nerve growth factor, vascular endothelial growth factor, and hepatocyte growth factor, along with various anti-inflammatory cytokines. The resulting product, referred to as MSC-NTF, encompasses the full MSC secretome, including extracellular vesicles such as exosomes, and is administered intrathecally into the cerebrospinal fluid of patients with ALS.

Although the Phase 3 clinical trial produced mixed overall outcomes, a prespecified subgroup of patients at earlier disease stages demonstrated a slower decline in muscle strength relative to placebo-treated individuals, suggesting that early therapeutic intervention using a complex mixture of trophic and immunomodulatory factors may confer clinical benefit. Nanotechnology offers an opportunity to refine and enhance this approach by improving the stability, concentration, and site-specific delivery of MSC-derived bioactive molecules. Such integration could amplify therapeutic potency, extend the duration of neuroprotective effects, and enable precise modulation of disease-relevant molecular and inflammatory pathways in ALS.

A significant advancement in ALS research has been the development of MSC-derived exosomes as a cell-free therapeutic platform. While their neuroprotective efficacy was previously established in MS models (Riazifar et al. 2019), similar strategies have produced encouraging outcomes in ALS models. In the SOD1-G93A transgenic mouse model, intrathecal administration of MSC-derived exosomes generated under inflammatory priming conditions to enhance their cargo significantly delayed muscle weakness and extended overall survival (Zhou et al. 2024). Treated animals demonstrated reduced microglial activation in the spinal cord, improved motor neuron preservation, and elevated levels of anti-inflammatory cytokines such as IL-10 and TGF-β, reflecting a shift toward a neuroprotective immune environment. Mechanistically, these exosomes contained microRNAs that suppressed the expression of pro-apoptotic and pro-inflammatory genes in recipient neural and glial cells. This approach effectively reproduced the immunomodulatory and trophic benefits of live mesenchymal SC transplantation while eliminating the biological and logistical challenges associated with administering viable cells into the spinal cord, a critical advantage given the rapid disease progression. These preclinical successes have led to the initiation of a first-in-human clinical trial in 2023 by NeuroSense Therapeutics, evaluating the safety and feasibility of intrathecal delivery of mesenchymal SC-derived exosomes in ALS patients.

Parallel to these developments, nanoparticle-based gene therapy strategies are being designed to suppress pathogenic gene expression in ALS, particularly in familial cases involving mutations in superoxide dismutase 1, which account for roughly 2% of cases. Although antisense oligonucleotide therapies such as tofersen have achieved regulatory approval for superoxide dismutase 1–related ALS, they require repeated intrathecal administration and demonstrate limited distribution across the CNS. To overcome these limitations, researchers are exploring the use of cationic lipid nanoparticles, analogous to those employed in messenger RNA vaccine technologies, for the CNS delivery of superoxide dismutase 1–silencing RNA. A promising approach utilizes NSCs as biological carriers for these nanoparticles. NSCs are loaded ex vivo with small interfering RNA–encapsulated lipid nanoparticles targeting superoxide dismutase 1 and subsequently transplanted into the spinal cord. Once engrafted, these cells provide trophic and structural support to the degenerating neural milieu while simultaneously releasing the nanoparticles or their small interfering RNA cargo locally, resulting in targeted knockdown of mutant gene expression in neighboring cells. In a preclinical study, this dual-delivery system demonstrated superior neuroprotection in an ALS rat model, with significantly greater motor neuron preservation compared to treatment with either NSCs or nanoparticles alone. The NSCs improved the inflammatory and metabolic microenvironment, partially compensating for dysfunctional glial populations, while nanoparticle-mediated gene silencing reduced mutant superoxide dismutase 1 expression and its associated cytotoxicity. This synergistic therapeutic paradigm effectively addresses both the underlying genetic pathology and the hostile inflammatory niche characteristic of ALS, providing a compelling foundation for future translational research.

Across multiple ALS preclinical models, including SOD1-mutant rodents and toxin-induced paradigms, the combination of SCs or their derivatives with nanoparticle-based therapeutics has produced consistently positive outcomes. These include prolonged survival, improved motor performance such as enhanced grip strength and delayed onset of paralysis, and preservation of neuromuscular junction integrity. The ALS mice treated with a combination of bone marrow–derived mesenchymal SCs and curcumin-loaded nanoparticles exhibited approximately fifteen percent longer survival compared to those receiving SCs alone, along with higher motor neuron counts at the terminal disease stage. The anti-inflammatory activity of curcumin, when delivered via nanoparticles, likely amplified the immunomodulatory and neuroprotective actions of the SCs.

In a complementary investigation, it was demonstrated that polymeric nanoparticles encapsulating brain-derived neurotrophic factor, when implanted into muscle tissue at sites of neuromuscular junction formation, promoted reinnervation by surviving or collateral motor axons, thereby preserving muscle function (Zhao et al. 2015; Rentería et al. 2022). When this peripheral treatment was combined with intracerebroventricular infusion of mesenchymal SC–derived exosomes, a synergistic effect was observed. The treated ALS mice showed delayed muscular atrophy and reduced motor neuron degeneration in the spinal cord. These findings emphasize the importance of a multicompartmental therapeutic approach that targets both central and peripheral aspects of the disease. Nanotechnology uniquely facilitates such spatially coordinated interventions, enabling the precise and compartmentalized delivery of trophic, anti-inflammatory, and neuroprotective agents to the brain, spinal cord, and musculature, which is achievable through cell-based therapies alone.

ALS has been the focus of several SC clinical trials. Neuralstem evaluated the safety of direct spinal cord injections of NSCs, while BrainStorm Cell Therapeutics tested autologous MSCs induced to secrete high levels of neurotrophic factors, delivered through lumbar puncture. These early-phase studies demonstrated that such interventions are generally safe, and in the case of MSC-derived neurotrophic treatments, patients with less advanced disease showed slower progression. The next phase of research may incorporate nanotechnology to build on these outcomes. One potential approach involves periodic intrathecal infusion of MSC-derived exosomes, providing continuous immunomodulatory and trophic effects without repeated cell transplantation. Another promising area is respiratory management, as disease progression causes diaphragm and respiratory muscle weakness. Aerosolized nanoparticles carrying anti-inflammatory or antioxidant compounds could reduce pulmonary stress and improve quality of life, even without directly altering disease courses.

Replacing lost motor neurons remains a major technical challenge because reestablishing long axonal connections from the spinal cord to muscles is complex. Some studies are exploring the transplantation of interneuron precursors to reconstruct or reroute spinal circuits, creating a biological bypass for lost motor neurons. In this context, nanofiber scaffolds and axon guidance channels may serve as key tools to direct axonal growth toward appropriate targets.

Although ALS is a relentlessly progressive disorder, combining SCs that provide multifactorial biological support with nanoparticles that allow precise targeting and sustained delivery represents a rational and synergistic therapeutic approach. If future clinical trials demonstrate even modest slowing of disease progression or an extension of functional survival by several months, it will signify a major advancement in the treatment of ALS.

6.6.4 Multiple Sclerosis

MS is an autoimmune demyelinating disorder of the CNS, marked by relapsing or progressive destruction of myelin sheaths and subsequent axonal injury. The disease involves aberrant immune responses primarily mediated by T cells and B cells targeting oligodendrocytes and myelin, resulting in inflammatory lesions within the brain and spinal cord. Clinically, MS results in progressive neurological deficits that include visual disturbances, motor weakness, and sensory dysfunction. Current treatment options rely primarily on immunomodulatory or immunosuppressive agents such as interferon beta and monoclonal antibodies like natalizumab, which can reduce relapse rates and limit new lesion formation but do not facilitate remyelination or functional restoration. Consequently, there is growing interest in regenerative strategies that aim to restore myelin integrity and establish long-term immune tolerance. SC-based interventions have included autologous hematopoietic SC transplantation, which acts as an immune system reset following chemoablation and has shown effectiveness in aggressive forms of the disease. MSCs have been studied for their immunomodulatory and neuroprotective potential, particularly due to their ability to regulate T cell activity and promote repair processes. Furthermore, the transplantation of oligodendrocyte progenitor cells or NSCs has been investigated as a strategy to enhance remyelination and improve neurological function.

Because MS involves both autoimmune and degenerative mechanisms, it provides an ideal framework for integrating SC therapy with nanotechnology. Immunologically, the objective is to induce tolerance toward myelin antigens and suppress autoreactive immune responses, while on the regenerative side, the focus is on directing stem or progenitor cells to differentiate into oligodendrocytes capable of remyelinating axons. Nanoparticles have been applied to address both aspects. For immunomodulation, one notable approach employs antigen-coupled nanoparticles engineered to induce immune tolerance. In the experimental autoimmune encephalomyelitis model of MS, researchers have conjugated myelin peptides such as PLP-139–151 to biodegradable nanoparticles. Upon administration, these particles are internalized by antigen-presenting cells and elicit a tolerogenic response through the induction of antigen-specific regulatory T cells or functional inactivation of autoreactive T cells. Clinical development of such tolerance-inducing systems, including "Navacims," is currently being explored.

In parallel, MSCs have been studied for their broad immunosuppressive and neuroprotective properties. A recent experimental autoimmune encephalomyelitis study (Kuo et al. 2017; Zappia et al. 2005) showed that combining MSC transplantation with PLP-conjugated nanoparticles produced superior therapeutic benefits compared with either treatment alone. The nanoparticles promoted antigen-specific immune tolerance, whereas the SCs mitigated neuroinflammation and supported tissue repair. Moreover, MSC persistence and integration were improved in the tolerized microenvironment, suggesting a synergistic interaction between the two therapeutic modalities (Yari et al. 2022).

On the regenerative side, Ghosh et al. (2024) provide a comprehensive analysis of how nanotechnology can be utilized to promote oligodendrogenesis and remyelination. One approach involves employing nanoparticles to deliver pro-differentiation signals to endogenous neural stem or progenitor cells. Small molecules that promote oligodendrocyte differentiation, such as peroxisome proliferator-activated receptor agonists, can be encapsulated within nanoparticles engineered to cross the BBB and release their contents at demyelinated sites. In the experimental autoimmune encephalomyelitis model, poly(lactic-co-glycolic acid) nanoparticles loaded with a peroxisome proliferator-activated receptor delta agonist enhanced the differentiation of oligodendrocyte precursor cells into mature oligodendrocytes, resulting in partial remyelination and improved motor function (Zhang et al. 2019).

Further potential arises when integrating nanoparticle delivery with SC transplantation. Neural progenitor cells or glial-restricted precursors can be cotransplanted with such nanoparticles to promote lineage commitment toward oligodendrocytes. Supporting this concept, Martino's group at the University of Milan transplanted human NSCs into the spinal cords of experimental autoimmune encephalomyelitis mice together with chitosan-based nanoparticles releasing a Wnt pathway inhibitor, as Wnt activation hinders oligodendrocyte maturation. The nanoparticle-mediated inhibition of Wnt signaling created a microenvironment conducive to myelination, enabling the transplanted NSCs to mature into functional oligodendrocytes that ensheathed demyelinated axons (Fancy et al. 2009; Ritchen et al. 2015). Functionally, treated mice recovered near-normal locomotor ability, compared with only partial improvement observed in those receiving SCs alone. This demonstrates a synergistic regenerative paradigm in which SCs supply the cellular foundation for remyelination, while nanoparticles precisely modulate the molecular milieu to guide lineage differentiation and maturation through targeted regulation of signaling pathways such as Wnt, Notch, and bone morphogenetic protein.

In chronic experimental autoimmune encephalomyelitis models of MS, combining SCs or progenitor cells with nanoparticle-based immunotherapies or pro-remyelination agents has produced broad therapeutic effects. These include reduced inflammatory infiltration in CNS lesions, increased oligodendrocyte numbers and remyelination, preservation or restoration of axonal conduction, and improved clinical outcomes on murine paralysis scales. The work by Riazifar et al. (2019) is especially significant in this context, as exosomes derived from interferon gamma–stimulated MSCs expanded regulatory T cell populations and shifted the immune profile toward a more tolerogenic state, effectively reducing disease severity. Coadministration of these MSC-derived exosomes with peptide-conjugated tolerogenic nanoparticles not only slowed disease progression but also prevented onset in most treated animals. This demonstrates that combining antigen-specific tolerance induction with broad immunosuppression mediated by MSC secreted factors may effectively suppress autoimmune activity. On the regenerative side, a key structural outcome measure is the g-ratio, the ratio of axon diameter to total myelinated fiber diameter, assessed by electron microscopy. Studies have shown significant improvement in g-ratio values, indicating restoration of normal myelin sheath thickness, in animals treated with combined SC and nanoparticle approaches compared with single treatments. Complementary electrophysiological assessments, including evoked potential testing, have revealed accelerated nerve conduction velocities in treated animals, signifying improved myelin insulation and restored axonal function.

MS has been an early focus of SC-based therapies, particularly hematopoietic SC transplantation, which aims to reconstitute a self-tolerant immune system. Although intensive, this approach can induce remission in aggressive disease forms. For less severe cases, MSCs have been evaluated in studies such as the MEsenchymal StEm cells for MS (MESEMS) trial, demonstrating safety and potential anti-inflammatory effects, though functional improvement remains limited. The next stage may involve enhancing MSC therapy through exosomes or nanoparticle-based delivery. An ongoing trial (2019–2024; NCT03355365) is investigating intrathecal administration of MSC-derived exosomes in MS to assess neurological outcomes. If effective, this could offer a means to reduce relapses and promote neural repair.

On the remyelination front, drugs such as clemastine are under evaluation to stimulate endogenous repair. Nanotechnology could augment these efforts through targeted and sustained delivery. Future strategies may combine tolerance-inducing antigen-coupled nanoparticles, autologous neural or iPSC-derived oligodendrocyte progenitors, and nanogels releasing prodifferentiation factors to guide remyelination. While complex, this integrated approach reflects the multifactorial nature of MS. Each element has demonstrated some clinical support, making combination therapy a plausible next step. MS remains one of the few NDs where curative therapy appears achievable, as halting immune attack and restoring myelin could enable functional recovery. The convergence of SC-based regeneration and nanotechnology may be central to realizing this potential.

6.6.5 Huntington's Disease

HD is an inherited neurodegenerative disorder caused by a CAG trinucleotide repeat expansion in the HTT gene, resulting in the production of a mutant huntingtin protein. The disease is characterized by selective degeneration of medium spiny neurons in the striatum, leading to motor disturbances, psychiatric manifestations, and progressive cognitive decline. HD is a monogenic, autosomal dominant condition that is invariably progressive, with a typical disease course leading to death within 15–20 years of symptom onset. At present, no disease-modifying therapies are available and the treatment remains symptomatic, addressing chorea e.g., tetrabenazine and psychiatric symptoms. SC-based strategies in HD have been explored primarily through two approaches: neuronal replacement, involving transplantation of fetal striatal tissue or SC-derived GABAergic neurons, and neuroprotection, using mesenchymal SCs or other supportive cell types to preserve remaining neurons. The complexity of HD arises from its widespread neuropathology, including cortical atrophy, and the systemic expression of the mutant gene in all cells of affected individuals. However, the monogenic nature of the disorder makes it a suitable target for gene therapies aimed at reducing mutant huntingtin expression through antisense oligonucleotides or RNA interference. Early open-label trials of fetal striatal transplantation in the 2000s yielded variable results, with limited graft integration and progressive deterioration due to the hostile microenvironment characterized by mutant huntingtin toxicity and inflammation. To overcome these challenges, nanotechnology has been investigated to enhance graft survival and modify the host environment. Vascular endothelial growth factor and brain-derived neurotrophic factor delivered via poly(lactic-co-glycolic acid) nanoparticles have been studied for this purpose. It was also demonstrated that codelivery of these growth factors in HD rat model improved survival of transplanted human NSCs and enhanced motor function compared with transplantation alone (Yari et al. 2022). The growth factors promoted angiogenesis and neurogenesis, mitigating the trophic deficits characteristic of HD. Treated animals exhibited better motor coordination and muscle strength.

An additional approach involves genetic preconditioning of SCs using nanoparticles to deliver transcriptional modulators. For instance, nanoparticles carrying nuclear factor erythroid 2–related factor 2, a central antioxidant regulator, into MSCs increased antioxidant secretion and enhanced protection of HD neurons in coculture (Mohammadzadeh et al. 2012; Mustafa et al. 2023)). These preconditioned MSCs may confer improved resilience against oxidative stress following transplantation. A direct synergy between SC therapy and nanotechnology in HD involves combining neuronal replacement with huntingtin-lowering strategies. While transplantation of healthy neuronal precursors may restore function, continued production of mutant huntingtin in the host remains a source of toxicity. A proposed approach involves delivering HTT-silencing RNA to the host brain alongside cell transplantation. Nanoparticles offer a nonviral platform for gene delivery with lower immunogenicity. In a preclinical study, poly(amidoamine) (PAMAM) dendrimers were used to deliver anti-HTT siRNA into the striatum of HD model mice, resulting in ~50% reduction in mutant huntingtin levels. When this was combined with transplantation of mouse ESC-derived neurons, graft survival improved, and motor function recovery exceeded that observed with either treatment alone (Lara-Rodarte et al. 2021). The dendrimers facilitated broad siRNA distribution and elicited less inflammation than viral vectors.

The body of research on nano-enhanced therapies in HD remains more limited compared to AD, PD, or MS but is steadily expanding. Reported outcomes include partial preservation or recovery of motor function, such as reduced choreiform movements and improved balance beam performance, increased survival of transplanted neurons, and stimulation of neurogenesis. Since HD is associated with impaired neurogenesis in the subventricular zone, it was demonstrated that combining cerium oxide nanoparticles with antioxidant properties and human MSCs enhanced the proliferation of endogenous neural progenitors in a mouse model (Kandasamy et al. 2015; Kan et al. 2011; Arya et al. 2016; Bailey et al. 2020). The study suggested that nanoparticle-mediated reduction of oxidative stress supported progenitor cell survival and activity, while MSCs provided trophic support. Additionally, nanoparticle-mediated RNA interference has been shown to reduce mutant huntingtin aggregates. Gold nanoparticles conjugated to huntingtin antisense oligonucleotides decreased inclusion bodies in both HD cell cultures and in vivo models. When combined with cell transplantation, such approaches may improve the host environment and enhance graft efficacy.

HD is a strong candidate for gene therapy, with antisense oligonucleotide trials such as tominersen targeting huntingtin having been completed, though results have been mixed. No SC trial has yet demonstrated clear therapeutic efficacy, although early-phase investigations with MSCs and fetal transplants have been conducted. Nanotechnology may play an earlier role in gene therapy than in cell transplantation. Current research is exploring the use of exosomes as delivery vehicles for huntingtin-lowering RNA, as they can cross the BBB. If SC-derived exosomes are engineered to carry small interfering RNA or CRISPR components targeting huntingtin, they could combine immunomodulatory activity with gene silencing, offering a dual-action therapeutic strategy.

Another approach focuses on brain region–specific nanoparticle delivery of neuroprotective agents. Since HD pathology predominantly affects the striatum, nanoparticles functionalized with targeting peptides have been shown to preferentially accumulate in this region following systemic administration. When loaded with neuroprotective or antiexcitotoxic compounds, such nanoparticles may safeguard both endogenous and transplanted neurons. Recent developments include quantum dot–based nanosensors capable of monitoring and modulating neurotransmitter levels in vivo. Although not yet applied to HD, these technologies could potentially be engineered to detect elevated glutamate concentrations and release antiexcitotoxic molecules locally, providing a responsive neuroprotective platform.

6.7 FUTURE DIRECTIONS AND TRANSLATIONAL CONSIDERATIONS

With continuing progress in the field, several key research directions and strategic developments have defined the application of nanoparticle-enhanced SC therapy for neurodegenerative disorders.

6.7.1 Improving Targeting and Specificity

A major advantage of nanoparticles is their ability to undergo surface functionalization with targeting ligands, allowing cell type- or organ-specific delivery. Advances in nanoparticle engineering are expected to enable the selective homing of SCs or exosomes to neural circuits affected in distinct neurological disorders. In AD, nanoparticles could be directed toward the entorhinal–hippocampal network, one of the earliest regions affected, whereas in PD, targeting the nigrostriatal pathway may enhance therapeutic precision. In MS, delivery systems could be guided toward active demyelinating lesions identified through MRI. Such targeted localization can be achieved using external magnetic fields or by exploiting disease-specific microenvironmental cues such as elevated enzyme activity or oxidative stress to trigger localized activation and release.

The development of smart nanoparticles that discharge their therapeutic agents selectively within diseased tissue, for example, in response to the acidic environment of inflammatory lesions or the presence of misfolded proteins, provides a strategy to minimize off-target effects and maximize therapeutic potency at the site of pathology. This context-dependent specificity will be critical for clinical translation to ensure efficacy and safety, particularly in minimizing systemic exposure to potent biologic agents.

6.7.2 Scalable Manufacturing and Standardization

Translating these therapies to clinical application on a large scale presents substantial practical challenges. SC products, particularly autologous iPSC derivatives or MSCs, require complex manufacturing under Good Manufacturing Practice conditions. Incorporating a nanoparticle component introduces additional complexity, necessitating integrated protocols for combination product development. For example, ex vivo loading of SCs with nanoparticles must be performed using reproducible, standardized procedures supported by stringent quality control to ensure consistent nanoparticle uptake and functional performance. Regulatory authorities will require robust evidence that nanoparticles do not compromise cell viability or safety, emphasizing the need for validated assessments of biocompatibility, nontoxicity, and removal of residual materials. Standardization of nanoparticle synthesis is equally critical. Scalable, reproducible methods for producing uniform liposomes, polymeric nanoparticles, or exosomes with consistent physicochemical characteristics are currently under development. Promising technological approaches include microfluidic-based formulation for controlled nanoparticle production and bioreactor platforms for large-scale exosome manufacturing.

6.7.3 Safety and Immune Reactions

Safety considerations will remain central to the clinical translation of nanoparticle-based and cell-based therapies. The introduction of any novel material, whether SCs or nanoparticles, entails potential risks of immune activation. Although many nanoparticles are engineered for biocompatibility and immune evasion, such as through PEGylation or the use of exosome membranes, immune reactions including complement activation-related pseudoallergy have been documented, particularly with liposomal and polymeric systems. Repeated administrations, as in multiple exosome infusions, may increase the likelihood of antibody formation against nanoparticle components. Long-term monitoring will therefore be essential to evaluate potential nanoparticle accumulation in organs such as the liver and spleen and to exclude related toxicity. Encouragingly, many widely used materials, including poly(lactic-co-glycolic acid) and lipids, are biodegradable into nontoxic byproducts such as lactic acid and glycerol. In SC-based therapies, the risk of tumorigenicity, particularly with iPSC –derived products, requires rigorous evaluation. Nanoparticles that might unintentionally promote proliferation or induce genomic instability must be carefully assessed. An emerging safety concept involves the controlled elimination of therapeutic cells once their intended function is achieved. For example, engineered suicide genes could be activated after adequate paracrine signaling or remyelination has occurred. Nanoparticles may also serve as external triggers for such mechanisms, for instance, gold nanoshells responsive to infrared light can induce localized ablation of transplanted cells following therapy. While not universally necessary, these strategies illustrate how nanotechnology can enhance the safety, precision, and controllability of regenerative interventions.

6.7.4 Regulatory and Clinical Trial Design

Combined therapies pose distinct regulatory challenges because they often fall outside traditional classifications of biologics or medical devices. Such products are typically governed as combination therapies, requiring coordinated evaluation by multiple regulatory divisions within agencies such as the Food and Drug Administration. Approval will depend on clear evidence delineating both the independent and synergistic effects of each component. Consequently, clinical trial designs may include comparator groups receiving cells alone, nanoparticles alone, and the combined treatment to demonstrate additive or synergistic benefits, a framework already being adopted in several preclinical investigations. Clinical trial endpoints may likewise need refinement. Beyond standard clinical assessments, advanced imaging modalities can serve as surrogate endpoints to evaluate biological activity and therapeutic targeting. For example, MRI could track superparamagnetic iron oxide nanoparticle–labeled cells, while positron emission tomography may detect nanoparticle tracers that localize to inflammatory regions. Demonstrating that nanoparticle-labeled SCs migrate to MS lesions and attenuate inflammation, as visualized through imaging, would provide strong mechanistic validation. Additionally, adaptive trial designs may be used to iteratively optimize dosing parameters for each component, allowing real-time adjustment to maximize therapeutic efficacy while maintaining safety.

6.7.5 Personalized and Precision Medicine

NDs display significant heterogeneity in both progression and pathology, indicating that personalized therapeutic strategies may be necessary. For example, a patient with MS presenting with predominant spinal cord lesions could benefit from intranasal SC delivery combined with magnetic targeting to the spinal axis and nanoparticles responsive to the cytokine profile of their inflammatory lesions. In AD, patients with marked vascular inflammation may require a different nanoparticle adjunct than those with amyloid beta– or tau-dominant pathology. The modular design of nanoparticles, allowing rapid customization of therapeutic cargo and targeting ligands, supports such individualized approaches. SC therapies can also be personalized using autologous iPSC s to minimize immunogenicity. Although this increases manufacturing complexity and cost, it may be justified in younger or rapidly progressing patients where durable cellular integration is essential for sustained therapeutic benefit.

6.7.6 Integration with Other Emerging Therapies

The therapeutic landscape for NDs is expanding to include gene editing and other next-generation modalities. CRISPR/Cas9-based genome editing is being explored for familial forms of AD and ALS, offering the potential for precise correction of pathogenic mutations. Nanoparticles represent a promising platform for delivering gene-editing tools within the CNS and may be integrated with cell therapies to modify the local microenvironment or to engineer transplanted cells in vivo for therapeutic factor expression. Another emerging area is bioelectronics, involving devices that directly interface with the nervous system, such as deep brain stimulators and brain–computer interfaces. In PD, a potential approach may involve SC grafts supported by a graphene-based nanoparticle electrode scaffold that promotes cellular integration while enabling electrical modulation of graft function, thereby combining regenerative medicine with neuromodulation. Although these concepts remain in development, related technologies such as conductive polymer scaffolds for neural support cells are already undergoing preclinical evaluation.

Thus, the future outlook for immunomodulatory, nanoparticle-enhanced SC therapies is highly promising and multidimensional. The intersection of SC biology, immunology, nanotechnology, materials science, and clinical neuroscience is driving rapid innovation. Achieving translational success will depend on addressing key challenges, ensuring safety, meeting regulatory standards, establishing reproducible large-scale manufacturing, and designing rigorous clinical trials that demonstrate clear therapeutic benefit. These efforts are justified by the transformative potential of these therapies to not only slow disease progression but also repair and regenerate neural tissue, restoring functions once considered irrecoverable. This shift from symptomatic management to true neurorestoration offers the possibility of fundamentally altering the prognosis for patients with AD, PD, ALS, MS, HD, and related neurodegenerative disorders in the coming decade.

6.8 SUMMARY

NDs remain among the most difficult medical challenges due to their complex pathologies and the limited regenerative capacity of the CNS. SC therapies hold strong potential to restore neural function, yet their effectiveness is constrained by issues of targeted delivery, survival, integration, and immune compatibility. The fusion of SC biology with nanotechnology offers a transformative approach to overcome these barriers. Nanoparticles can deliver therapeutic agents or cells across neural barriers, modulate inflammatory environments, and provide biochemical or mechanical cues that guide SC differentiation and repair. Across major neurodegenerative disorders, these combined strategies show promise. In AD, mesenchymal SC-derived exosomes and growth factor–loaded nanoparticles mitigate neuroinflammation and improve cognition. In PD, nanoparticle-assisted delivery enhances graft survival and enables targeted administration of neurotrophic factors such as glial cell line–derived neurotrophic factor. In ALS, exosome-based and gene-silencing nanoparticles act synergistically to preserve motor neurons and reduce neuroinflammation. In MS, integrating immunomodulatory and pro-remyelination nanoparticles with SC therapy offers the potential to both suppress autoimmunity and restore myelin. In HD, combining protein-silencing nanoparticles with neural replacement approaches provides new therapeutic promise. Overall, immunomodulatory, nanoparticle-enhanced SC therapy represents a next-generation, multifaceted approach to treating NDs by addressing neuronal loss, inflammation, and regenerative failure simultaneously. By uniting cellular and nanoscale technologies, these therapies can restore neural networks and precisely modulate molecular environments. Although challenges remain in regulation, manufacturing, and clinical validation, progress in preclinical and early clinical research supports cautious optimism. This paradigm marks a shift toward personalized, multimodal, and restorative medicine that aims not merely to slow disease progression but to reestablish lost neurological function.

REFERENCES

Alizadeh R, et al. Intranasal delivery of stem cells labeled by nanoparticles in neurodegenerative disorders: Challenges and opportunities. *Wiley Interdiscip Rev Nanomed Nanobiotechnol.* 2023;15(6):e1915.

Aly AE, Harmon BT, Padegimas L, Waszczak BL. Intranasal delivery of pGDNF DNA nanoparticles provides neuroprotection in the rat 6-hydroxydopamine model of Parkinson's disease. *Mol Neurobiol.* 2019;56(1):536–54.

Aly AE-E, Waszczak BL. Intranasal gene delivery for treating Parkinson's disease: Overcoming the blood–brain barrier. *Expert Opin Drug Deliv.* 2015;12(12):1923–41.

Arya A, Sethy S, Singh A, Das GM. Cerium oxide nanoparticles promote neurogenesis and ameliorate hypoxia-induced memory impairment in mice. *Free Radic Biol Med.* 2016;95:209–22.

Bailey ZS, Nilson EF, Berti ZM, et al. Cerium oxide nanoparticles improve outcome after In vitro and in vivo traumatic brain injury. *J Neurotrauma.* 2020;37(12):1666–80.

Bondarenko O, Saarma M. Neurotrophic factors in Parkinson's disease: Clinical trials, open challenges and nanoparticle-mediated delivery to the brain. *Front Cell Neurosci.* 2021 Jun 2;15:682597. doi: 10.3389/fncel.2021.682597

Carradori D, Lundborg DE, Van den Bosch AT, Haastert-Talini S, Houweling AR. Nanofiber scaffolds and neural stem cell transplantation for spinal cord injury: The role of scaffold alignment and topography. *Acta Biomater.* 2017;68:301–15.

Chen YA, Lu CH, Ke CC, et al. Mesenchymal stem cell-derived exosomes ameliorate Alzheimer's disease pathology and improve cognitive deficits. *Biomedicines*. 2021;9(6):594.

Chen Y, Pan C, Xuan A, Xu L, Bao G, Liu F, Fang J, Long D. Treatment efficacy of NGF nanoparticles combining neural stem cell transplantation on Alzheimer's disease model rats. *Med Sci Monit*. 2015;21:3608–15. https://doi.org/10.12659/MSM.894567

Deng X, et al. Differentiation of human embryonic stem cells into dopaminergic neurons in 3D poly(lactic-co-glycolic acid) scaffolds for the treatment of Parkinson's disease. *Biomaterials*. 2013;34(4):941–53.

Espadas-Alvarez AJ, Domínguez-Pinos M, Flores-Martinez MM, Jiménez-Gómez A, Lanciego JL, Herrero MJ. Nanoparticle-mediated gene delivery enhances the efficacy of stem cell transplantation in experimental models of Parkinson's disease. *Nanomedicine*. 2017;13(7):2083–97.

Fancy SPJ, Baranzini SE, Zhao C, et al. Dysregulation of the Wnt pathway inhibits timely myelination and remyelination in the mammalian CNS. *Genes Dev*. 2009;23(13):1571–85. https://doi.org/10.1101/gad.1806309

Fletcher AM, Yurek DM, Smith RF, Seroogy AJ, Yurek DM. Transgene expression in the striatum following intracerebral injections of DNA nanoparticles. *Neurotoxicology*, 2011;32 (6): 998–1006.

Ghosh S, Bhatti GK, Sharma PK. et al. Potential of nano-engineered stem cells in the treatment of multiple sclerosis: A comprehensive review. *Cell Mol Neurobiol*. 2024;44:1151–75.

Jia Y, Cao N, Zhai J, Zeng Q, Zheng P, Su R, Liao T, Liu J, Pei H, Fan Z, Zhou J, Xi J, He L, Chen L, Nan X, Yue W, Pei X. HGF mediates clinical-grade human umbilical cord-derived mesenchymal stem cells improved functional recovery in a senescence-accelerated mouse model of Alzheimer's disease. *Adv Sci*. 2020;7(17):1903809.

Kan I, Barhum Y, Melamed E, Offen D. Mesenchymal stem cells stimulate endogenous neurogenesis in the subventricular zone of adult mice. *Stem Cell Rev Rep*. 2011;7(2):404–12.

Kandasamy M, Sadanandan N, Hildebrandt SK, Faissner AG, Götz MA. Reduction in subventricular zone-derived olfactory bulb neurogenesis in a rat model of Huntington's disease is accompanied by striatal invasion of neuroblasts. *PLoS One*. 2015;10(1):e0116069.

Kuo R, Saito A, Miller A. Keselowsky BD. Peptide-conjugated nanoparticles reduce positive co-stimulation and ameliorate autoimmunity. *Mol Ther*. 2017;25(7):1676–88.

Lara-Rodarte R, Cortés D, Soriano K, Carmona F, Rocha L, Estudillo E, López-Ornelas A, Velasco I. Mouse embryonic stem cells expressing GDNF show enhanced dopaminergic differentiation and promote behavioral recovery after grafting in parkinsonian rats. *Front Cell Dev Biol*. 2021;9:661656.

Lockard G, Gordon J, Schimmel S, Sayed BE, Monsour M, Garbuzova-Davis S, Borlongan CV. Attenuation of amyotrophic lateral sclerosis via stem cell and extracellular vesicle therapy: An updated review. *Neuroprotection*. 2023 Dec;1(2):130–8. doi: 10.1002/nep3.26

Mo M, Zhou X, Chen L, Wang Y, Li J, Li Y. Enhanced neurorestorative effects of magnetically targeted mesenchymal stem cells in a rat model of Parkinson's disease. *Nanomedicine*. 2018;14(9):2487–500.

Mohammadzadeh M, Halabian N, Gharehbaghian M, Amirizadeh A, Jahanian-Najafabadi H, Roushandeh R, Roudkenar AMR. Nrf-2 overexpression in mesenchymal stem cells reduces oxidative stress–induced apoptosis and enhances their survival. *Cell Stress Chaperones*. 2012;17(5):553–65.

Mustafa G, Hassan D, Zeeshan M, Ruiz-Pulido G, Ebrahimi N, Mobashar A, Pourmadadi M, Rahdar A, Sargazi S, Fathi-karkan S, Medina DI, Díez-Pascual AM. Advances in nanotechnology versus stem cell therapy for the theranostics of Huntington's disease. *J Drug Deliv Sci Techno*. 2023;87:104774.

Ou L, Guo Y, Luo X, Lin W, Zhang H, Wang J. 3D bioprinting of functionalized scaffolds with human embryonic stem cell-derived dopaminergic neurons for neural regeneration in a rat model of Parkinson's disease. *Acta Biomater*. 2020;110:60–74.

Pan Y, Li L, Cao N, et al. Advanced nano delivery system for stem cell therapy for Alzheimer's disease. *Biomaterials*. 2025;314:122852 (Epub 2024 Sep 30).

Parodi A, Molinaro R, Sushnitha L, Evangelopoulos C, Martinez AP, Tasciotti E. Bioengineered hybrid nanoparticles for combined stem cell therapy in experimental models of Parkinson's disease. *ACS Nano*. 2020;14(4):3912–27.

Rajendran A, Rajan RA, Balasubramaniyam S, Elumalai K. Nano delivery systems in stem cell therapy: Transforming regenerative medicine and overcoming clinical challenges. *Nano TransMed*. 2025 Dec;4:100069. https://doi.org/10.1016/j.ntm.2024.100069

Rentería I, Osorio-Fuentealba R, Troncoso R. The molecular effects of BDNF synthesis on skeletal muscle. *Front Physiol*. 2022;13:934714.

Reza-Zaldivar EE, Hernández-Sapiéns MA, Minjarez B, Gutiérrez-Mercado YK, Márquez-Aguirre AL, Canales-Aguirre AA. Potential effects of MSC-derived exosomes in neuroplasticity in Alzheimer's disease. *Front Cell Neurosci.* 2018 Sep 24;12:317. doi: 10.3389/fncel.2018.00317.

Riazifar M, Mohammadi MR, Pone EJ, et al. Stem cell-derived exosomes as nanotherapeutics for autoimmune and neurodegenerative disorders. *ACS Nano.* 2019;13(6):6670–88.

Rittchen S, et al. Myelin repair in vivo is increased by targeting oligodendrocyte precursor cells with nanoparticles encapsulating leukaemia inhibitory factor (LIF). *Nat Commun.* 2015;6:6547.

Shi Q, Gao Y, Yan L, Ma X, Li J, Zhu Z. Combined delivery of mesenchymal stem cells and glial cell line–derived neurotrophic factor–loaded nanoparticles enhances neuroprotection in a rat model of Parkinson's disease. *Biomater Sci.* 2018;6(9):2507–18.

Simorgh S, Bagher Z, Farhadi M, Kamrava SK, Eskandarian Boroujeni M, Namjoo Z, et al. Magnetic targeting of human olfactory mucosa stem cells following intranasal administration: A novel approach to Parkinson's disease treatment. *Mol Neurobiol.* 2021;58(8):3835–47. doi:10.1007/s12035-021-02392-z

Sivandzade F, Cucullo L. Regenerative stem cell therapy for neurodegenerative diseases: An overview. *Int J Mol Sci.* 2021 Feb 22;22(4):2153. doi: 10.3390/ijms22042153

Sun Y, Xu J, Xu X, Gu Y, Liu H, Li Z. Biomimetic hydrogel scaffold with sustained neurotrophic release promotes functional integration of transplanted dopaminergic neurons in a rat model of Parkinson's disease. *Adv Sci.* 2021;8(17):2100821.

Sulzer D, Alcalay KA, Garretti A, Cote L, Kanter C, Agin-Liebes E, Liong J, et al. T cells from patients with Parkinson's disease recognize α-synuclein peptides. *Nature.* 2017;546(7660):656–61.

Takahashi J. Preparing for first human trial of induced pluripotent stem cell–derived dopaminergic neurons for Parkinson's disease. *Regener Ther.* 2020;13:18–22.

Vissers C, Ming GL, Song H. Nanoparticle technology and stem cell therapy team up against neurodegenerative disorders. *Adv Drug Deliv Rev.* 2019;148:239–51.

Wang H, Huber CC, Li XP. Mesenchymal and neural stem cell-derived exosomes in treating Alzheimer's disease. *Bioengineering.* 2023;10(2):253.

Wei M, Yang Z, Li S, Le W. Nanotherapeutic and stem cell therapeutic strategies in neurodegenerative diseases: A promising therapeutic approach. *Int J Nanomed.* 2023;18:611–26.

Yari H, Mardpour N, Emtiazjoo N, et al. Emerging role of mesenchymal stromal cell-derived exosome in neurodegeneration: A groundbreaking cell-free approach. *Stem Cell Res Ther.* 2022;13:423.

Ye Y, Gao M, Shi W, et al. The immunomodulatory effects of mesenchymal stem cell-derived extracellular vesicles in Alzheimer's disease. *Front Immunol.* 2023;14:1325530.

Yuan Y, Sun J, You T, Shen W, Xu W, Dong Q, Cui M. Extracellular vesicle-based therapeutics in neurological disorders. *Pharmaceutics.* 2022;14:2652.

Zappia E, Casazza S, Pedemonte E, Benvenuto F, Bonanni I, Gerdoni F, Giunti D. et al. Mesenchymal stem cells ameliorate experimental autoimmune encephalomyelitis inducing T-cell anergy. *Blood.* 2005;106(5):1755–61.

Zhang B. et al. Stem cell-derived extracellular vesicle-mediated therapy for stroke: A novel approach for cell-free therapy. *Front Neurosci.* 2018;12:668.

Zhang Q, Zhu W, Xu F, Dai X, Shi L, Cai W, Mu H, Hitchens TK,Foley LM, Liu X, Hu X. The interleukin-4/PPARγ signaling axis promotes oligodendrocyte differentiation and remyelination after brain injury. *PLOS Biol.* 2019;17(6):e3000330. https://doi.org/10.1371/journal.pbio.3000330

Zhao Y, Han J, Zhang X, Ma J, Wang Y, Wang X, Liu L. BDNF gene delivery mediated by neuron-targeted nanoparticles is neuroprotective in peripheral nerve injury. *Mol Neurobiol.* 2015;52(2):1014–24.

Zhou J, et al. Intranasal delivery of small extracellular vesicles reduces the progress of amyotrophic lateral sclerosis and the overactivation of complement-coagulation cascade and NF-κB signaling in SOD1G93A mice. *J Nanobiotechnol.* 2024;22(1):503. https://doi.org/10.1186/s12951-024-02764-2

Zhu FD, Hu YJ, Yu L, et al. Nanoparticles: A hope for the treatment of inflammation in CNS. *Front Pharmacol.* 2021;12:683935.

7 Induced Pluripotent Stem Cells (iPSCs) for Modeling and Treating Neurodegenerative Disorders

7.1 INTRODUCTION

Human induced pluripotent stem cells (iPSCs) are generated by reprogramming adult somatic cells into a pluripotent state (Figure 7.1). Since their discovery, iPSCs have transformed preclinical research because they provide human-derived cells for disease modeling and therapy (Cerneckis et al. 2024; Penney et al. 2020). Unlike animal models, iPSCs retain the full human genetic material, allowing study of disease mechanisms that are human-specific. All major CNS cell types, including neurons, astrocytes, oligodendrocytes, and microglia can now be derived from iPSCs (Figure 7.1; Penney et al. 2020). Moreover, three-dimensional (3D) systems such as brain organoids integrate diverse cell types and spatial organization, thereby more effectively recapitulating tissue-level interactions observed in the human brain (Cerneckis et al. 2024). These advances have made iPSC-based models invaluable for investigating neurodegenerative disorders and for developing cell-based therapies. Importantly, iPSCs can be utilized to develop autologous cell therapies by reprogramming a patient's own cells to minimize the risk of immune rejection or allogeneic products derived from universal donor cell lines, for broader clinical application (Cerneckis et al. 2024).

7.2 ADVANTAGES AND LIMITATIONS OF IPSC MODELS

iPSC-based models offer several key advantages for neurodegenerative research (Table 7.1). They are patient-specific, cells carry the individual's genetic mutations or risk alleles, enabling study of genotype-dependent pathophysiology. They have inherent property of self-renewal, enabling the generation of an inexhaustible supply of human cells for research and therapeutic applications. Further, they can be differentiated into disease-relevant neuronal subtypes, e.g. dopaminergic neurons for Parkinson's, motor neurons for ALS, medium spiny neurons (MSNs) for Huntington's and glial cells, allowing dissection of cell type-specific mechanisms (Dawoody Nejad and Pioro 2025). Further, gene editing tools significantly enhance the iPSC applications by allowing the generation of isogenic control lines. For example, iPSCs derived from patients with disease-causing mutations can be precisely corrected to the wild-type genotype, while healthy iPSCs can be engineered to carry specific mutations. This enables controlled experimental conditions in which the functional impact of specific genetic variants can be directly evaluated by comparing genetically matched cell lines differing only at the targeted mutation (Cerneckis et al. 2024). This permits rigorous comparison of mutant and wild-type cells in an identical genetic background. Finally, iPSC-derived cells have been effectively employed in high-throughput drug screening using human cell models, enabling the discovery of compounds that can alter disease-related phenotypes.

DOI: 10.1201/9781003582687-7

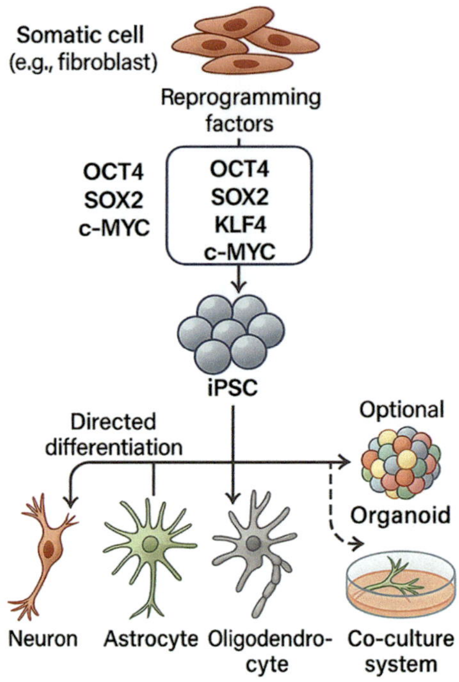

FIGURE 7.1 Generation of iPSCs and Directed Differentiation into CNS Cell Types.

TABLE 7.1
Comparative Advantages and Limitations of iPSC-Based Models

Feature	Advantage	Limitation
Genetic relevance	Retains patient-specific mutations and alleles	Reprogramming resets age-related epigenetic features
Differentiation potential	Produces various CNS cell types	Variability between iPSC lines and clones
Scalability	Unlimited self-renewal allows large-scale generation	Reproducibility issues across labs
Disease modeling	Allows generation of isogenic controls via gene editing	In vitro systems lack systemic physiological interactions
Drug screening	Enables high-throughput testing in human neurons	Immature phenotypes limit translational predictability

However, iPSC-based models also present notable limitations. Primarily, iPSC-derived neural cells tend to exhibit a developmentally immature, fetal-like state and often lack the molecular and functional characteristics associated with cellular aging (Cerneckis et al. 2024). Reprogramming somatic cells into iPSCs resets their cellular age by largely removing age-associated epigenetic modifications and phenotypic traits, including telomere attrition, protein aggregation, and oxidative damage. As a result, iPSC-derived neurons do not inherently exhibit features of cellular aging (Cerneckis et al. 2024). This poses a significant challenge, as neurodegenerative diseases are highly age-dependent and often involve molecular and cellular alterations that accumulate over time, which are absent in rejuvenated iPSC-derived neurons. To address this limitation, researchers are developing strategies to induce aging-related features in iPSC-derived neurons. Approaches include extended in vitro culture durations, expression of progerin which is a mutant form of lamin A associated with premature aging, and exposure to metabolic or oxidative stressors, all focusing on mimicking the cellular environment of aged neurons (Cerneckis et al. 2024). Additional limitations of iPSC-based models include variability between iPSC lines and clones, which can arise from differences in reprogramming methods or residual epigenetic memory from the donor cells. These factors can affect differentiation potential and phenotypic outcomes. Moreover, reproducibility across laboratories remains a challenge, and the lack of systemic and environmental context in vitro limits the ability to fully capture complex disease processes. iPSC-based cultures lack key physiological components such as the blood–brain barrier, immune system interactions, and integrated whole-organ function. While coculture systems and organoid models can partially recapitulate some of these interactions, they remain incomplete and imperfect surrogates of in vivo biology. Therefore, findings from iPSC studies should be interpreted in conjunction with data from animal models and clinical research to ensure a comprehensive understanding of disease mechanisms. Overall, the benefits of human relevance and genetic fidelity offered by iPSC models must be carefully weighed against their technical limitations, including cellular immaturity, variability, and the absence of systemic physiological context (Cerneckis et al. 2024; Stoddard-Bennett and Reijo Pera 2020).

7.3 ADVANCES IN GENE EDITING, ORGANOIDS, AND 3D SYSTEMS

Recent technological advancements have significantly enhanced the utility of iPSC models. CRISPR/Cas9-mediated gene editing has become a standard tool, enabling precise and efficient manipulation of disease-associated genes within iPSCs (Figure 7.2). Researchers routinely introduce pathogenic mutations into genetically normal iPSC lines to model disease, or conversely, correct mutations in patient-derived iPSCs to generate isogenic controls for mechanistic studies (Cerneckis et al. 2024). For instance, iPSCs derived from a patient with a monogenic mutation can be genetically corrected to the wild-type sequence and subsequently differentiated into relevant cell types to assess whether the disease phenotype is reversed, thereby directly linking the mutation to the pathological features (Cerneckis et al. 2024). Gene editing is also used to develop hypoimmunogenic universal donor iPSC lines by knocking out key HLA class I and II genes and, in some cases, overexpressing immune-evasive molecules such as CD47. These engineered modifications reduce the immunogenicity of the cells, allowing them to evade allogeneic immune responses. This strategy holds significant promise for creating cell therapies that can be administered to multiple recipients without the need for HLA matching or immunosuppression (Cerneckis et al. 2024).

There have been significant advancements in 3D organoid and multicellular culture systems, which have markedly improved the ability to model human development and disease in vitro. Researchers

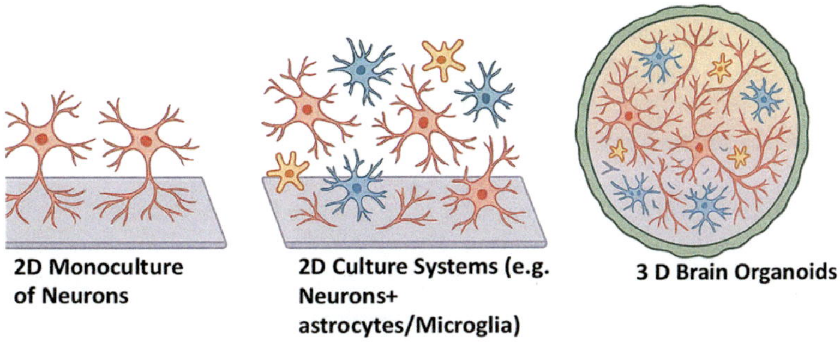

FIGURE 7.2 Comparative Overview of iPSC-Based 2D, Co-culture, and 3D Organoid Platforms.

can now generate brain region–specific organoids such as those mimicking the midbrain or cerebral cortex that exhibit self-organization into layered architectures and contain a diverse array of neural cell types, closely reflecting aspects of early human brain development. These brain organoids are capable of recapitulating key aspects of human brain architecture, including regional patterning and cytoarchitecture, and they provide a valuable platform for investigating cell–cell interactions and neurodevelopmental processes in a human-specific context. To model neurodegenerative diseases more accurately, brain organoids can be refined through the integration of other cellular components, including microglia and endothelial cells. The inclusion of these elements facilitates the establishment of neuroimmune and vascularized organoid systems that more closely emulate the cellular heterogeneity, microenvironmental interactions, and physiological complexity observed in vivo. For instance, iPSC-derived midbrain organoids have been employed to investigate Parkinson's disease (PD) pathology within a more physiologically relevant, tissue-like environment. Microglia can be incorporated into neural cultures either through codifferentiation with other neural cell types or by adding iPSC-derived microglia-like cells to existing organoid or 2D systems. Notably, Ohtonen et al. demonstrated that iPSC-derived microglia with PD-associated LRRK2-G2019S mutation exhibit transcriptional and functional alterations that closely resemble those observed in microglia from the patients, underscoring their utility for modeling disease-relevant immune responses (Ohtonen et al. 2023). These findings highlight the importance of incorporating microglia into complex in vitro models to better capture neuroimmune interactions relevant to disease. In parallel, specialized three-dimensional systems such as neuromuscular junction (NMJ) platforms have been developed to study motor neuron diseases. For example, (Ting et al. (2025) established a rapid iPSC-derived NMJ co-culture system in which motor neurons and muscle fibers formed functional synapses within 12 days, successfully recapitulating amyotrophic lateral sclerosis (ALS)-associated synaptic abnormalities. These findings underscore the critical role of including microglia in advanced vitro models to more accurately capture neuroimmune dynamics that are implicated in disease pathogenesis. Concurrently, specialized three-dimensional systems, such as NMJ platforms, have been developed to investigate motor neuron disorders. For instance, Ting et al. developed a rapid iPSC-derived NMJ coculture model in which motor neurons and muscle fibers established functional synaptic connections within 12 days, effectively replicating synaptic deficits associated with ALS (Cerneckis et al. 2024).

7.4 ALZHEIMER'S DISEASE

7.4.1 iPSC-Based Disease Modeling of AD

AD is characterized by progressive dementia, amyloid-β plaques, neurofibrillary tau tangles, and neuroinflammation. iPSC models of AD typically use cells from patients with familial Alzheimer's disease (FAD) mutations (e.g. in APP, PSEN1/2) or those carrying risk alleles (e.g. APOE ε4, TREM2 variants) (Figure 7.3). Differentiating AD patient iPSCs into cortical neurons and astrocytes has recapitulated key disease phenotypes in vitro (Figure 7.4). For example, cortical neurons derived from iPSCs with a PSEN1 mutation show increased amyloid-β (Aβ) 42 production and aggregation (Cerneckis et al. 2024). Similarly, iPSC neurons carrying APP duplications or mutations exhibit early amyloid and tau pathology compared to isogenic controls (Table 7.2). These phenotypes confirm that human iPSC-derived neurons can manifest canonical AD molecular hallmarks (Penney et al. 2020).

Beyond neurons, non-neuronal cell types have been modeled. iPSC-derived astrocytes and microglia from AD patients display altered functions. The APOE4 astrocytes show impaired Aβ clearance and synaptic support, and patient microglia with risk variants secrete pro-inflammatory cytokines. Coculture and organoid studies are shedding light on cell–cell interactions. For instance, 3D human neural organoid systems can develop amyloid deposits and tau pathology when built from FAD

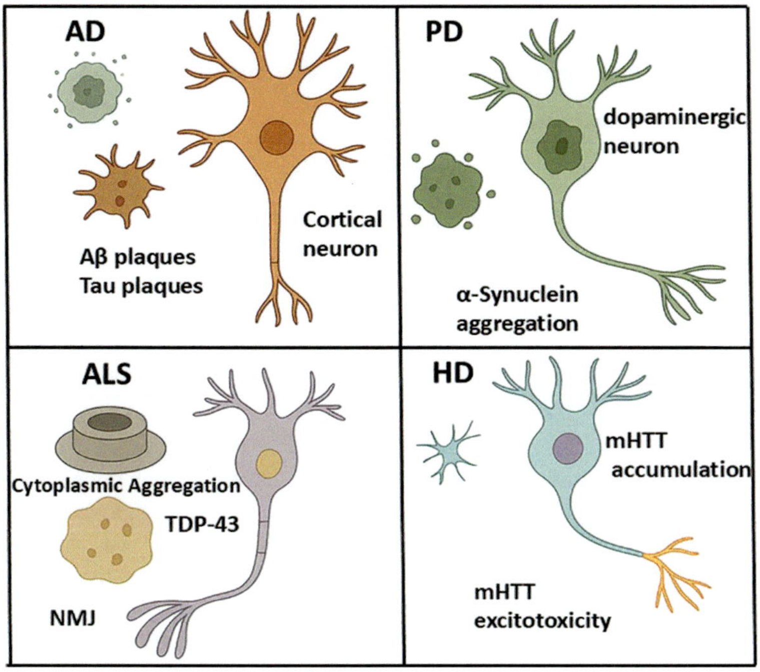

FIGURE 7.3 Pathogenic Mechanisms and iPSC-Derived Cellular Phenotypes in Major Neurodegenerative Diseases.

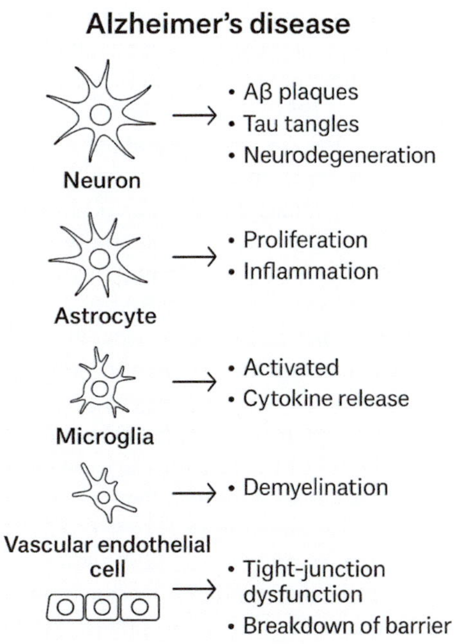

FIGURE 7.4 iPSC-Derived Neural Models for Alzheimer's Disease Pathophysiology.

TABLE 7.2
iPSC-Derived Neural Phenotypes in Neurodegenerative Disorders

Disease	iPSC-Derived Cell Type	Key Pathological Features
Alzheimer's disease (AD)	Cortical neurons, astrocytes, microglia	Amyloid-β accumulation, tau tangles, neuroinflammation
Parkinson's disease (PD)	Midbrain dopaminergic neurons, glia	α-synuclein aggregation, mitochondrial dysfunction
ALS	Spinal motor neurons, astrocytes, microglia	TDP-43 mislocalization, oxidative stress, RNA foci
Huntington's disease (HD)	Medium spiny neurons, astrocytes	mHTT aggregation, excitotoxicity, autophagic defects

iPSCs (Penney et al. 2020). These complex models have revealed that microglia and astrocytes influence neuronal survival and Aβ processing in ways that 2D monocultures cannot. Moreover, iPSC models are being used to study sporadic AD. GWAS-identified risk loci such as SORL1, CD33, and TREM2 are being knocked into or out of iPSCs to elucidate their functional impact. These studies have revealed how these genes modulate endosomal trafficking and immune signaling in human cells, offering mechanistic insights into neurodegenerative disease susceptibility. Thus, iPSC-derived neural cultures and organoids have recapitulated numerous AD phenotypes consistent with patient observations and animal models (Penney et al. 2020). Moreover, their amenability to precise genetic manipulation enables the identification of novel molecular mechanisms underlying disease onset and progression.

7.4.2 iPSC-Based Therapeutic Applications for AD

Currently, no iPSC-derived cell therapy for AD has reached clinical application, primarily due to the diffuse neurodegenerative pathology and the formidable challenges posed by the blood–brain barrier. Nonetheless, preclinical investigations have demonstrated proof-of-concept for regenerative strategies, highlighting the potential of iPSC-based approaches to restore neuronal function and ameliorate disease-associated deficits. For example, neural stem cells (NSCs) derived from human iPSCs have been transplanted into AD model mice, where they secrete neurotrophic factors, modulate inflammation, and promote synaptic repair (Shah et al. 2024). Shah et al. (2024) note that NSCs can strengthen compromised synaptic networks and potentially improve cognition. iPSC-derived astrocytes have emerged as potential therapeutic candidates, as their replacement could ameliorate impaired astroglial functions, thereby enhancing amyloid-β clearance and restoring glutamate homeostasis. Alternatively, direct neuronal replacement strategies involving the transplantation of iPSC-derived hippocampal or cortical neurons have been proposed to replenish neuronal populations lost during AD progression. Significant technical challenges remain, particularly in directing axonal growth and achieving functional integration of transplanted neurons into existing neural circuits. However, early animal studies have demonstrated encouraging outcomes, including the formation of functional synapses and measurable behavioral improvements in models with hippocampal lesions.

Gene therapy integrated with iPSC models represents a rapidly advancing frontier. In monogenic FAD, CRISPR-mediated correction of pathogenic mutations in patient-derived iPSCs has been shown to normalize Aβ levels in differentiated neurons, suggesting a viable pathway for personalized therapeutic interventions (Cerneckis et al. 2024). Although these strategies have not yet progressed to clinical trials, accumulating preclinical evidence underscores their translational promise. Concurrently, patient-specific iPSCs serve as powerful platforms for drug discovery, where

AD iPSC-derived neurons are utilized to evaluate anti-Aβ and anti-tau compounds, as well as combinatorial therapies targeting multiple cellular pathways. While no iPSC-derived cellular therapy has yet entered AD trials, insights from other neurological fields, such as PD, along with the unique mechanistic understanding provided by iPSC-based systems, have renewed optimism for cell- and gene-based approaches. Future therapeutic paradigms may integrate iPSC-derived neural precursors capable of providing neurotrophic support with pharmacological agents identified through iPSC-based screening or genome editing strategies designed to mitigate genetic risk factors.

7.5 PARKINSON'S DISEASE

PD is characterized by the progressive degeneration of midbrain dopaminergic (DA) neurons, particularly within the substantia nigra pars compacta, and the pathological accumulation of misfolded α-synuclein aggregates, forming Lewy bodies (Figure 7.3). Patient-derived iPSC models have been successfully established from individuals harboring familial PD-associated mutations, including SNCA triplication, LRRK2-G2019S, and PARK2 (Parkin) loss-of-function variants, as well as from patients with idiopathic PD. Upon directed differentiation into midbrain dopaminergic neurons, these iPSC lines faithfully reproduce several hallmark PD-associated phenotypes, including α-synuclein aggregation, mitochondrial dysfunction, impaired autophagic flux, elevated oxidative stress, and dysregulated synaptic transmission. Such models constitute a robust, human-specific experimental system for elucidating PD pathophysiology and for high-throughput screening of neuroprotective compounds within genetically defined or patient-specific frameworks (Figure 7.5).

For example, iPSC-derived dopaminergic neurons carrying the LRRK2-G2019S mutation exhibit several disease-relevant, cell-autonomous phenotypes, including impaired neurite outgrowth, mitochondrial dysfunction, and elevated α-synuclein accumulation relative to isogenic controls (Table 7.2). Similarly, neurons derived from SNCA-triplication iPSCs display pronounced α-synuclein aggregation and increased vulnerability to oxidative stress, reflecting critical pathological processes observed in PD (Cerneckis, Cai, and Shi 2024). These phenotypes highlight the ability of iPSC-based systems to faithfully recapitulate mutation-specific cellular dysfunctions that contribute to the pathogenesis of PD, thereby providing mechanistic insights into genotype–phenotype relationships and disease progression.

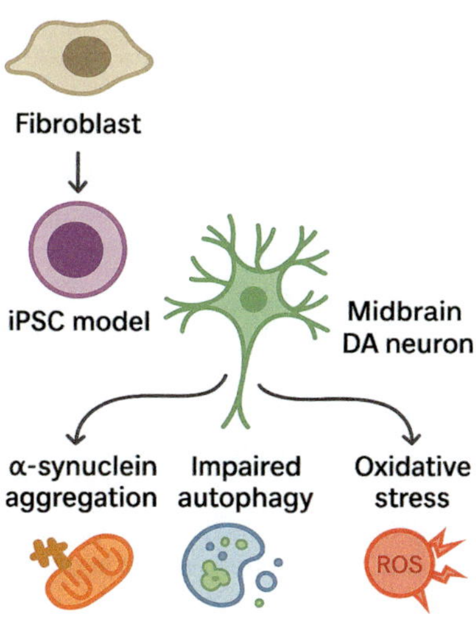

FIGURE 7.5 Induced pluripotent stem cell model of Parkinson's disease associated dopaminergic neurodegeneration.

Beyond neuronal phenotypes, PD iPSC models have been extended to incorporate glial cell types, which play pivotal roles in disease progression. iPSC-derived astrocytes from PD patients demonstrate aberrant dopamine uptake and metabolism, alongside pro-inflammatory signaling profiles that negatively impact neuronal viability. In a recent study, Ohtonen et al. (2023) showed that iPSC-derived microglia harboring the LRRK2-G2019S mutation recapitulates key features of the disease-associated microglial transcriptome. These cells also exhibit altered cytokine secretion profiles and impaired phagocytic functions, offering a robust model to investigate the neuroimmune contributions to PD.

To enhance physiological relevance, researchers have generated midbrain-like organoids containing functionally integrated dopaminergic neurons, astrocytes, and microglia. These three-dimensional systems support the establishment of synaptic networks and, when derived from patient-specific or genetically modified iPSCs, recapitulate mild α-synuclein pathology. Such multicellular organoid platforms offer a sophisticated framework for investigating cell–cell communication, neuroinflammatory processes, and the collective effects of PD-associated mutations on network-level dysfunction. They are increasingly employed to elucidate the multicellular and systemic mechanisms underlying PD pathogenesis.

Beyond monogenic variants, iPSC-based models are increasingly being applied to study sporadic PD, which constitutes the majority of cases and arises from intricate genetic–environmental interactions. Remarkably, iPSC-derived dopaminergic neurons carrying the APOE ε4 allele, commonly linked to AD exhibit altered neuronal function, suggesting potential convergence of pathogenic mechanisms among neurodegenerative disorders. In parallel, genome-wide CRISPR-Cas9 functional screens in iPSC-derived dopaminergic neurons have uncovered critical genetic modifiers of α-synuclein toxicity, providing novel insights into molecular pathways amenable to therapeutic modulation. Collectively, iPSC-derived neurons and glial cells from both familial and sporadic PD contexts constitute a robust human-relevant platform for delineating disease mechanisms, establishing genotype–phenotype correlations, and accelerating the identification and validation of new therapeutic strategies.

7.5.1 iPSC-Based Therapeutic Applications for PD

PD is widely regarded as a prime candidate for iPSC-based cell replacement therapy due to the relatively selective and focal degeneration of midbrain dopaminergic neurons. Preclinical studies over the past two decades have demonstrated that transplantation of DA progenitors into the striatum can restore dopamine signaling and ameliorate motor deficits in animal models of PD. Building upon this foundation, several research groups have optimized protocols for the generation of clinical-grade DA progenitors from human pluripotent stem cells (PSCs), adhering to Good Manufacturing Practice (GMP) standards to ensure safety, purity, and functional maturation. In 2025, a Phase I/II clinical trial at Kyoto University Hospital transplanted allogeneic iPSC-derived midbrain dopaminergic progenitors into seven patients with moderate PD (Sawamoto et al. 2025). The HLA-matched iPSCs were differentiated, purified, and bilaterally implanted into the putamen. At 24-month follow-up, no adverse events, immune rejection, or tumor formation were observed, and graft survival was confirmed by a 45% increase in 18F-DOPA uptake, indicating functional dopaminergic integration (Sawamoto et al. 2025). Clinically, four of six evaluable patients exhibited improvements in Unified Parkinson's Disease Rating Scale (UPDRS) motor scores under OFF-medication conditions, providing preliminary evidence of biological efficacy. These results substantiate the therapeutic promise of iPSC-derived dopaminergic neuron transplantation in PD and affirm the safety and functional viability of this regenerative strategy.

Several strategies are being pursued to further optimize iPSC-derived cell therapy for PD. A key focus is refining the developmental stage and dosage of transplanted dopaminergic neurons, as progenitors at an intermediate differentiation stage exhibit superior survival, integration, and functional recovery compared with fully mature neurons. Parallel research explores autologous transplantation, where patient-specific iPSCs are generated, corrected for pathogenic mutations if required, and differentiated into dopaminergic neurons, thereby minimizing the need for immunosuppression. In contrast, the Kyoto University trial utilized an HLA-matched allogeneic iPSC line, which, despite genetic compatibility, still required immunosuppressive therapy to prevent immune rejection (Sawamoto et al. 2025). To overcome immune incompatibility, iPSC-derived dopaminergic cells are genetically engineered to achieve hypo-immunogenicity through targeted deletion of HLA molecules or overexpression of immunomodulatory proteins. Additional modifications aim to enable secretion of neuroprotective factors that support graft survival and enhance host neuronal resilience.

Beyond transplantation, iPSCs serve as a powerful platform for PD drug discovery and precision medicine. iPSC-derived DA neurons are widely used in high-throughput screens to identify compounds that promote neuronal survival, enhance mitochondrial function, or reduce α-synuclein aggregation. Furthermore, patient-specific iPSC models allow for the stratification of therapeutic responses. For instance, testing LRRK2 kinase inhibitors on dopaminergic neurons derived from individuals carrying the LRRK2-G2019S mutation allows personalized assessment of drug efficacy and toxicity. Collectively, such applications highlight the dual potential of iPSC technology in advancing both regenerative and pharmacological strategies for PD therapy.

From an ethical perspective, iPSC-based therapies for PD present notable advantages over fetal tissue transplantation. As noted by Stoddard-Bennett and Reijo Pera (2020), the derivation of iPSCs circumvents ethical considerations that are linked to embryo or fetal tissue use, as it does not entail zygote destruction. Furthermore, iPSCs enable the creation of genetically matched or immunologically compatible cell lines, thereby addressing both ethical and immunological challenges associated with earlier approaches.

Safety remains a critical consideration in the clinical application of iPSC-derived products in PDs. Major risks include residual undifferentiated cells, aberrant lineage commitment, genomic instability, and reprogramming-associated mutations. These are mitigated through stringent quality control measures, including purification of dopaminergic precursors, genome-wide mutation screening, karyotypic analysis, and validation of cellular identity before transplantation. Although early clinical studies such as Sawamoto et al. (2025) have demonstrated favorable safety and preliminary efficacy, long-term monitoring is required to confirm sustained benefit and exclude delayed adverse effects such as immune rejection, tumorigenicity, or graft dysfunction. Continuous surveillance and regulatory oversight remain essential for the responsible clinical translation of iPSC-based therapies.

7.6 AMYOTROPHIC LATERAL SCLEROSIS

7.6.1 iPSC-Based Disease Modeling of ALS

ALS is a progressive and ultimately fatal neurodegenerative disorder marked by the selective degeneration of both upper and lower motor neurons, resulting in muscle weakness, atrophy, and eventual paralysis. While approximately 10–15% of ALS cases are familial, arising from pathogenic variants in genes such as SOD1, C9orf72, TARDBP (encoding TDP-43), and FUS, the majority are sporadic and thought to result from a multifactorial interplay of genetic susceptibility, environmental factors, and stochastic events (Figure 7.3). Although transgenic animal models particularly murine models carrying familial mutations have provided valuable insights into ALS pathophysiology, they fail to fully recapitulate the heterogeneous and cell-type–specific features of sporadic ALS (Table 7.2). In this regard, human iPSC models provide a powerful and physiologically relevant platform for investigating disease mechanisms.

Patient-derived iPSCs have been generated from individuals with both familial and sporadic forms of ALS, enabling in vitro modeling of disease-relevant cellular phenotypes in a human genetic background. Upon directed differentiation into spinal motor neurons, these iPSC lines recapitulate hallmark of ALS pathologies. Motor neurons derived from SOD1-mutant iPSCs exhibit impaired axonal transport, increased oxidative stress sensitivity, and mitochondrial dysfunction. iPSCs carrying the C9orf72 hexanucleotide repeat expansion display characteristic nuclear RNA foci and cytoplasmic dipeptide repeat protein aggregates, consistent with dual RNA and protein toxicity mechanisms. Similarly, TDP-43 (TARDBP) mutant iPSC-derived motor neurons show cytoplasmic mislocalization and aggregation of TDP-43, mirroring the predominant pathology observed in most ALS cases. These disease-associated phenotypes are accompanied by reduced neuronal survival and altered electrophysiological properties, characterized by early hyperexcitability followed by late-stage hypo-excitability, as well as cellular stress responses involving the endoplasmic reticulum and mitochondria. The recapitulation of these features in human iPSC-derived motor neurons highlights the strength of this platform in modeling ALS pathogenesis within a patient-specific and physiologically relevant framework, providing a significant advantage over non-human experimental systems.

Non-neuronal cells derived from ALS patient iPSCs have provided critical insights into the non–cell-autonomous mechanisms of motor neuron degeneration. iPSC-derived astrocytes and microglia, when cocultured with motor neurons, recapitulate pathological glia neuron interactions observed in vivo. Notably, astrocytes derived from SOD1- or C9orf72-mutant iPSCs have been shown to exert neurotoxic effects on cocultured motor neurons, primarily through the release of pro-inflammatory cytokines, oxidative mediators, or via reduced neurotrophic support (Dawoody Nejad and Pioro 2025). These findings strongly support the hypothesis that ALS pathogenesis is not only cell autonomous but also driven by glial dysfunction and neuroinflammatory processes. Although relatively underexplored in ALS, iPSC-derived microglia can be reliably generated and are increasingly employed to study microglial activation states, cytokine secretion profiles, and phagocytic activity under disease-relevant conditions.

Recent advances in three-dimensional modeling have enabled the development of more physiologically relevant ALS platforms. Spinal cord–like and brain organoids derived from patient iPSCs incorporate motor neurons, astrocytes, and interneurons, reproducing key pathological features such as synaptic dysfunction, neuronal stress responses, and degeneration. When co-cultured with skeletal myocytes, these organoids form functional NMJs, allowing the examination of early motor circuit pathology. Ting et al. (2025) established a patient-specific NMJ model using iPSC-derived motor neurons and myocytes, demonstrating that NMJs formed from SOD1-mutant lines exhibited progressive structural and functional degeneration, consistent with early synaptopathies observed in ALS. These integrated iPSC-based systems provide a robust and tractable platform for investigating early pathophysiological mechanisms and evaluating potential therapeutic strategies.

Overall, iPSC-based models of amyotrophic lateral sclerosis encompass diverse cell types, including motor neurons, astrocytes, and microglia, and span experimental systems from two-dimensional monocultures to coculture and three-dimensional organoid platforms. These models provide a robust framework for elucidating genotype–phenotype relationships in a human cellular context, capturing both familial and sporadic disease forms. By allowing targeted introduction and functional analysis of genetic risk variants, iPSC systems enable the modeling of sporadic ALS heterogeneity, addressing a major limitation of traditional experimental approaches.

7.6.2 iPSC-Based Therapeutic Applications for ALS

Unlike PD, no iPSC-based cell therapy has yet progressed to clinical evaluation in ALS, primarily due to the anatomical and functional complexity of motor neurons. Their extensive axonal projections and integration requirements within corticospinal and neuromuscular networks present significant challenges for replacement. Nonetheless, several preclinical strategies employing iPSCs are under investigation. One promising direction involves glial replacement therapy, in which healthy astrocyte or oligodendrocyte progenitors are transplanted to provide trophic and metabolic support to remaining motor neurons. In ALS models, transplantation of human glial progenitor cells that are typically derived from fetal or embryonic sources has yielded modest but consistent improvements in disease progression and survival. These findings suggest that iPSC-derived glial cells, once standardized and fully characterized, may hold therapeutic promise through modulation of the extracellular milieu, attenuation of neuroinflammation, and restoration of glial homeostatic function.

Another emerging approach focuses on the therapeutic delivery of neurotrophic factors. iPSC-derived neural progenitor cells are genetically engineered to overexpress protective molecules such as brain-derived neurotrophic factor or glial cell line–derived neurotrophic factor. Upon transplantation into the spinal cord, these progenitors serve as cellular depots for sustained, localized release of neurotrophic signals, aiming to preserve motor neuron viability and slow disease progression. Although still in preclinical development, these strategies underscore the versatility of iPSCs as both cell replacement systems and bioengineered platforms for targeted molecular intervention in ALS.

Motor neuron replacement represents a long-term objective in ALS therapy but continues to face substantial biological and technical constraints. Even when iPSC-derived spinal motor neurons are successfully engrafted, their ability to extend axons across long distances and reinnervate target muscles remains limited in vivo. This challenge has shifted current research priorities toward strategies that preserve existing neuromuscular connections and enhance trophic support at the NMJ. Transplantation of iPSC-derived support cells or engineered neural progenitors into muscle tissue offers a more practical approach, promoting NMJ integrity through sustained neurotrophic signaling and modulation of local inflammation.

Gene therapy approaches are integrated with iPSC research to target the genetic basis of familial ALS. CRISPR-Cas9–mediated correction of pathogenic mutations such as SOD1 in patient-derived iPSCs offers the potential to generate autologous, mutation-free therapeutic cells that eliminate mutant protein expression while maintaining immunological compatibility. Although these combined gene–cell strategies remain preclinical, iPSC-based systems are central to evaluating gene-silencing modalities. Specifically, iPSC-derived neurons are used to assess the efficacy of antisense oligonucleotides (ASOs) and small molecules designed to counter toxic gain-of-function mechanisms, including RNA foci and dipeptide repeat proteins associated with C9orf72 repeat expansions. These models enable mechanistic validation and therapeutic screening in a human cellular context, laying essential groundwork for precision gene-targeted interventions in ALS.

To date, cell transplantation trials for ALS have primarily utilized mesenchymal stem cells or neural stem or progenitor cells derived from non-iPSC sources. Notably, NurOwn, an autologous MSC-based therapy engineered to secrete neurotrophic factors, has progressed through clinical evaluation, alongside studies employing intraspinal delivery of fetal- or adult-derived neural progenitors. Although these approaches have demonstrated safety and limited biological activity, they have not achieved robust neuro-restorative outcomes. Future iPSC-derived therapeutics for ALS will need to address critical safety concerns, including risks of glial scarring, aberrant differentiation, and tumorigenesis, as well as challenges in targeted delivery, likely necessitating intraparenchymal or intrathecal administration to ensure precise and sustained graft survival within the spinal cord microenvironment.

Recent advances in iPSC-derived NMJ models highlight synaptic preservation as a meaningful and quantifiable endpoint for assessing therapeutic efficacy. These findings indicate that maintaining NMJ integrity, rather than pursuing direct motor neuron replacement, may represent a more attainable near-term objective for mitigating functional decline in ALS. At present, iPSC technology is positioned to impact ALS primarily through drug discovery and precision medicine, enabling identification of disease-modifying compounds, validation of gene-silencing strategies, and stratification of patients based on cellular phenotypes. As clinical progress with iPSC-based therapies accelerates in other neurodegenerative disorders such as PD, the ALS field remains poised for translational advances that could ultimately realize the therapeutic potential of iPSC-derived interventions.

7.7 HUNTINGTON'S DISEASE

Huntington's disease is an autosomal dominant neurodegenerative disorder caused by an expanded CAG trinucleotide repeat in exon 1 of the HTT gene, resulting in the production of mutant huntingtin protein (mHTT) with an abnormally long polyglutamine tract (Figure 7.3). This mutation leads to progressive neurodegeneration, most prominently affecting striatal MSNs, which are critical for motor and cognitive function. Due to its monogenic nature and clearly defined genetic mutation, Huntington's disease has been particularly amenable to modeling using iPSC technology.

iPSC lines have been established from multiple Huntington's disease patients with varying CAG repeat lengths, allowing stratified analyses of genotype–phenotype relationships (Table 7.2). Upon differentiation into striatal-like MSNs, HD iPSC-derived neurons recapitulate key pathological hallmarks, including the accumulation of mutant huntingtin aggregates in both cytoplasmic and nuclear compartments, mitochondrial dysfunction, impaired autophagic flux, and heightened vulnerability to glutamate-induced excitotoxicity. Moreover, several studies have identified dysregulated intracellular calcium signaling and altered synaptic protein expression, reflecting early impairments in neuronal communication and homeostatic regulation. These phenotypes arise within a human neuronal framework and align closely with cellular events documented in post-mortem Huntington's disease brains and animal models, reinforcing the relevance of HD iPSC-derived neurons as translational platforms for elucidating disease mechanisms and evaluating therapeutic candidates

Beyond neuronal phenotypes, iPSC-derived glial cells from Huntington's disease patients exhibit functional abnormalities that underscore the non–cell-autonomous aspects of HD pathogenesis. HD iPSC-derived astrocytes demonstrate impaired glutamate uptake and altered metabolic support, potentially exacerbating neuronal vulnerability through excitotoxic stress. Likewise, oligodendrocyte-lineage cells derived from HD iPSCs show deficits in maturation and dysregulation of myelin-associated gene expression, implicating intrinsic glial dysfunction in the white matter abnormalities characteristic of HD. The use of isogenic iPSC pairs, in which the expanded CAG tract in the HTT gene is corrected via genome editing tools such as CRISPR-Cas9, has been pivotal in establishing the direct pathogenic role of the mutation. Comparative analyses between mutant and corrected lines confirm that disturbances in mitochondrial function, synaptic signaling, and proteostasis are linked to the expanded polyglutamine tract, reinforcing the mechanistic precision of these models. Isogenic iPSC pairs, in which the pathogenic CAG expansion in the HTT gene is corrected using genome editing tools such as CRISPR-Cas9, have been instrumental in confirming that these cellular phenotypes directly result from the expanded polyglutamine tract. Comparative analyses between CAG-expanded and corrected iPSC lines demonstrate that key deficits in mitochondrial dynamics, synaptic signaling, and proteostasis are closely linked to the pathogenic allele, underscoring the precision of these models.

HD iPSC-derived cerebral organoids, patterned toward cortical or striatal lineages, are increasingly utilized to investigate the developmental impact of mutant huntingtin expression. Evidence from these 3D systems indicates that HD may also involve subtle neurodevelopmental abnormalities, including altered neuronal differentiation, disrupted progenitor proliferation, and cortical layering defects that precede overt neurodegeneration. These findings highlight the temporal and mechanistic complexity of HD and establish iPSC-derived organoids as valuable platforms for studying early disease processes and evaluating therapeutic strategies.

Advances in genome editing have expanded the utility of iPSC models in Huntington's disease. Targeted insertion of pathogenic-length CAG repeats into control iPSCs reproduces hallmark HD phenotypes in differentiated neurons, including mutant huntingtin aggregation, mitochondrial dysfunction, impaired autophagy, and increased susceptibility to excitotoxic stress. Conversely, CRISPR-Cas9–mediated correction of the expanded CAG tract in patient-derived iPSCs attenuates these abnormalities, restoring mitochondrial homeostasis and neuronal viability. Such gene-edited isogenic pairs provide a precise framework to establish causal relationships between genotype and phenotype. Because Huntington's disease is a fully penetrant, monogenic disorder resulting from a defined mutation in the HTT gene, iPSC-derived models are considered highly faithful representations of the disease, particularly for mechanistic and therapeutic investigations. High-throughput screening using HD iPSC-derived neurons has identified small molecules and biologics that modulate mutant huntingtin aggregation, improve proteostasis, and correct metabolic dysfunction. These systems provide functional insights within a human cellular context, enhancing translational relevance. Collectively, iPSC-based HD models have yielded key insights into early pathogenic mechanisms and established a renewable, genetically accurate platform for preclinical drug discovery and validation (Monk et al. 2021). Collectively, iPSC-based Huntington's disease models have yielded critical insights into early pathogenic mechanisms and established a renewable, genetically defined platform for preclinical drug discovery and therapeutic validation (Monk et al. 2021).

7.7.1 iPSC-Based Therapeutic Applications for Huntington's Disease

Although no iPSC-based therapy for Huntington's disease has yet reached clinical evaluation, two principal strategies remain under active investigation: cell replacement and gene modification. The cell replacement approach seeks to restore striatal function by replenishing MSNs. Early clinical trials using fetal striatal grafts provided limited but promising evidence of graft survival, synaptic integration, and partial functional recovery, establishing proof-of-concept for MSN replacement. Within the iPSC framework, significant progress has been achieved in differentiating human iPSCs into MSN-like neurons through stepwise exposure to developmental morphogens such as SHH, DKK1, and BDNF. Moreover, striatum-patterned organoids derived from HD and control iPSCs recapitulate regional specification, gene expression profiles, and early disease phenotypes. In preclinical HD models, transplantation of iPSC-derived MSN precursors has demonstrated graft viability, neuronal maturation, and partial recovery of motor and cognitive function, although complete symptomatic rescue has not yet been achieved.

A major obstacle to translating this approach clinically is achieving precise circuit integration within the basal ganglia, a region characterized by intricate connectivity with cortical, thalamic, and midbrain structures. Transplanted neurons must not only survive and differentiate but also establish appropriate afferent and efferent connections to restore functional striatal circuitry. Overcoming this limitation will require further refinement in graft maturation, spatial targeting, and potentially bioengineering strategies to guide synaptogenesis and axonal pathfinding.

In gene therapy, PSCs currently function mainly as disease modeling platforms rather than direct therapeutic agents. However, combining gene editing with patient-derived iPSCs provides a promising foundation for future personalized interventions. In theory, autologous iPSCs from Huntington's disease patients could be genetically corrected either by contracting the expanded CAG repeat or inserting a wild-type HTT allele and subsequently differentiated into MSNs for transplantation. Such corrected neurons would be genetically matched and free of the pathogenic mutation, offering functional replacement with immunological compatibility. Nonetheless, clinical translation remains limited by technical challenges in editing large repeat expansions, as well as concerns regarding off-target effects and scalability. In parallel, in vivo gene therapy approaches have progressed further, with several ongoing clinical trials employing ASOs or adeno-associated viral (AAV) vectors to selectively suppress mutant huntingtin expression. These strategies aim to mitigate downstream pathogenic effects by lowering mHTT protein levels without altering the genomic sequence.

iPSC-derived neurons play a pivotal role in advancing therapeutic development for Huntington's disease. They are employed to evaluate the efficacy, specificity, and cellular effects of gene-silencing agents in a human neuronal context and to identify novel molecular targets through omics-based and functional screening approaches. As gene-editing tools advance and regulatory pathways evolve, iPSC platforms are expected to serve as a critical bridge between preclinical discovery and translational application in HD gene therapy. Another promising direction involves the use of iPSCs as delivery vehicles for neuroprotective factors. iPSC-derived neural progenitor cells engineered to secrete brain-derived neurotrophic factors are transplanted into affected brain regions to support host neuronal survival, modulate synaptic plasticity, and slow neurodegeneration. This trophic support strategy serve as an adjunct or transitional approach to gene-modifying therapies.

In parallel, exosome-based therapeutics derived from iPSC progeny represent an emerging, non-cellular modality. Exosomes can cross the blood–brain barrier and deliver bioactive cargo to target neurons. Those derived from iPSC-derived NPCs or glial cells are being investigated for their potential to modulate mutant huntingtin toxicity, enhance neuroprotection, and restore homeostatic signaling within HD-affected neural circuits.

Importantly, findings from iPSC-based models consistently emphasize that effective therapeutic intervention must directly target the primary molecular driver, mutant huntingtin. While the long-term prospect of integrating gene correction with cell replacement remains conceptually appealing, the most immediate and practical application of HD iPSC technology lies in drug discovery and mechanistic research, where it continues to facilitate the identification of candidate therapeutics and biomarkers. As gene-editing platforms, delivery systems, and stem cell manufacturing technologies advance, the translational trajectory of iPSC-based therapies for Huntington's disease is anticipated to align with progress achieved in other monogenic neurological disorders.

7.8 CLINICAL TRIALS AND REGULATORY PATHWAYS

Over the past decade, iPSC-based cell therapies have advanced significantly toward clinical translation (Table 7.3). A recent global analysis identified more than 100 active clinical trials utilizing human PSC-derived products (Kirkeby, Main, and Carpenter 2025). By late 2024, regulatory agencies had approved 116 trials investigating 83 distinct hPSC-derived cell products, spanning indications in ophthalmology, central nervous system (CNS) disorders, and oncology. To date, over 1,200 patients have received hPSC-derived cell transplants, involving the administration of more than 10^{11} cells, with no major safety concerns reported (Kirkeby, Main, and Carpenter 2025). These data collectively indicate that, when manufactured under stringent quality control, hPSC-derived therapies exhibit a favorable and reproducible safety profile.

PD has emerged as a leading indication for iPSC-based therapy, owing to its well-defined pathology, specific cellular targets, and measurable clinical endpoints. Early efforts focused on the transplantation of fetal midbrain tissue, but a major breakthrough occurred in 2018 with the initiation of the first clinical trial employing iPSC-derived dopaminergic neurons for PD (Stoddard-Bennett and Reijo Pera 2020). This progress resulted in a landmark Phase I/II clinical study in 2025, involving

TABLE 7.3
Summary of Clinical Trials Involving iPSC-Based Therapies

Disease	Trial (Year)	Therapeutic Cell Type
Parkinson's disease	Kyoto Phase I/II (2025)	iPSC-derived DA progenitors
Macular degeneration	Not specified (2024)	iPSC-derived photoreceptors
ALS	Preclinical only	iPSC-derived NPCs expressing BDNF/GDNF
Huntington's disease	Preclinical only	iPSC-derived MSNs

bilateral transplantation of allogeneic iPSC-derived DA progenitors into the putamen of seven PD patients (Sawamoto et al. 2025). Over a 24-month follow-up, all grafts survived and produced dopamine without evidence of tumorigenesis or serious adverse events. Notably, four of the six evaluable patients demonstrated motor improvements corroborated by PET imaging, which revealed increased dopaminergic activity. These findings provide compelling evidence for the safety, feasibility, and therapeutic efficacy of iPSC-derived DA cell replacement in PD (Sawamoto et al. 2025). Following this success, iPSC-based strategies are being extended to other CNS applications. For example, a clinical trial utilizing iPSC-derived photoreceptor cells for macular degeneration has shown favorable safety outcomes (Kirkeby, Main, and Carpenter 2025), underscoring the expanding clinical potential of iPSC-based interventions across neurological and sensory disorders. Regulatory authorities classify iPSC-derived products as advanced therapy medicinal products (ATMPs), subjecting them to stringent oversight due to their biological complexity and potential clinical impact. Manufacturers must comply with good manufacturing practice standards to ensure product safety, consistency, and quality. As of 2024, regulatory frameworks mandate the establishment of well-characterized master cell banks, implementation of validated differentiation protocols, and comprehensive quality control testing. These quality assessments include verification of sterility, genomic stability, e.g., maintenance of a normal karyotype, and the absence of residual undifferentiated pluripotent cells to minimize tumorigenic risk.

In addition to stringent manufacturing requirements, preclinical studies must provide robust evidence of safety and efficacy in disease-relevant animal models that faithfully replicate human pathology. Despite these efforts, the absence of fully harmonized global regulatory frameworks continues to impede the uniform development and approval of iPSC-based therapies. Although the U.S. Food and Drug Administration (FDA), European Medicines Agency (EMA), and Japan's Pharmaceuticals and Medical Devices Agency (PMDA) adhere to broadly aligned regulatory principles, national frameworks differ markedly in procedural requirements, data evaluation standards, and approval pathways. For instance, the FDA's recent draft guidance on cell therapy comparability emphasizes the necessity of demonstrating that any manufacturing process modifications do not compromise product safety, identity, purity, or clinical performance.

iPSC-based therapies are regulated through an exceptionally stringent framework requiring that clinical-grade iPSC lines and their derivatives satisfy rigorous criteria for cellular identity, purity, potency, and safety before clinical translation. Encouragingly, emerging evidence, including the absence of tumorigenicity reported in early-phase clinical studies (Sawamoto et al. 2025; Kirkeby, Main, and Carpenter 2025) supports the continued advancement of iPSC-based therapies under established and evolving regulatory standards.

7.9 ETHICAL AND MANUFACTURING CONSIDERATIONS

iPSC-based neurotherapeutics introduces a distinct set of ethical and translational considerations (Figure 7.2). Ethically, iPSCs circumvent many of the controversies surrounding embryonic stem cells, as their generation does not involve human embryos, thereby mitigating concerns regarding the termination of potential life (Stoddard-Bennett and Reijo Pera 2020). However, patient-derived iPSCs raise critical issues related to privacy and informed consent. Because the donor's entire genome is preserved within the reprogrammed cells, there exists a risk of unintentional disclosure of sensitive genetic information, including disease-associated variants or predispositions. Accordingly, robust informed consent procedures and comprehensive data protection measures are imperative to ensure ethical compliance and safeguard donor privacy. From a manufacturing standpoint, the large-scale production of clinical-grade iPSC products continues to pose considerable challenges, with only a few facilities having successfully established fully GMP-compliant iPSC lines (Novoa et al. 2024). Novoa et al. (2024) reported the development of a validated GMP-compliant platform resulting in two iPSC master cell banks capable of reproducible differentiation into therapeutically relevant cell types. To enable broad clinical deployment, manufacturing processes must be efficiently scaled while maintaining genomic stability, which requires continuous monitoring for chromosomal abnormalities and mutations during reprogramming. The complete removal of residual undifferentiated cells is essential to reduce tumorigenic risk. Additionally, the establishment of optimized cryopreservation protocols is critical for standardized storage and global distribution. iPSC-derived neural progenitor cells, in particular, must preserve both post-thaw viability and differentiation potential, posing a significant technical challenge. Comprehensive quality control testing, including evaluation of pluripotency markers, karyotypic stability, and tumorigenicity is indispensable to ensure product safety and reproducibility.

Immunological factors constitute a critical component of iPSC-based therapy manufacturing and clinical implementation. Allogeneic approaches frequently require immunosuppressive regimens or partial HLA matching to minimize graft rejection. To mitigate these challenges, strategies such as the establishment of HLA-homozygous iPSC banks and the generation of gene-edited universal donor cell lines are being pursued to reduce alloreactivity (Cerneckis, Cai, and Shi 2024). Conversely, autologous therapies utilizing patient-specific iPSCs inherently eliminate immune compatibility concerns but pose significant logistical and economic limitations due to their individualized production. Currently, most clinical trials employ allogeneic iPSC-derived products in combination with immunosuppressive therapy. In the long term, targeted reduction of immunogenicity through HLA editing and related genomic modifications remains a key objective in translational research.

Overall, the ethical regulation and stringent manufacturing control of iPSC-based therapies are indispensable for their successful clinical translation. Development efforts must adhere to rigorous ethical principles and comply with current Good Manufacturing Practice standards to ensure clinical-grade quality, reproducibility, and patient safety while fulfilling evolving regulatory expectations. Substantial progress has been achieved through the establishment of GMP-compliant iPSC banks (Novoa et al. 2024) and encouraging early-phase clinical safety outcomes (Sawamoto et al. 2025; Kirkeby, Main, and Carpenter 2025). Continued harmonization of global regulatory standards and sustained commitment to ethical governance will be essential as iPSC-based therapeutics advance toward widespread clinical application.

7.10 FUTURE DIRECTIONS AND CHALLENGES

A major challenge in advancing iPSC-based models of neurodegeneration lies in their limited ability to replicate the aged cellular phenotype characteristic of late-onset disorders. During reprogramming, iPSC-derived neurons are reverted to an embryonic-like state, lacking the epigenetic, transcriptomic, and metabolic hallmarks of aging (Cerneckis et al. 2024). This constraint hinders accurate modeling of age-related pathogenic mechanisms. To overcome this limitation, researchers are developing age-enhancement strategies that induce aging-associated features in iPSC-derived neurons. Approaches such as ectopic expression of progerin, a mutant form of lamin A linked to premature aging, or exposure to metabolic and oxidative stressors have been shown to elicit senescence-like characteristics, including nuclear envelope alterations, DNA damage accumulation, and age-related transcriptional changes (Cerneckis, Cai, and Shi 2024). An alternative approach bypasses pluripotency entirely through direct trans-differentiation, converting adult somatic cells such as fibroblasts directly into induced neurons (iNs). Unlike iPSC-derived neurons, iNs retain the donor's age-associated epigenetic and transcriptional profiles, providing a more physiologically relevant platform for investigating late-onset neurodegenerative disease mechanisms (Cerneckis et al. 2024).

While the implementation of age-enhancement techniques in iPSC-derived neurons continues to evolve, these strategies represent a critical step toward improving the fidelity of disease modeling for late-onset neurodegenerative disorders. By introducing aging-associated molecular and epigenetic features, such approaches aim to overcome the inherent limitations of iPSC-derived cells, which typically reset to a rejuvenated state during reprogramming. On the therapeutic front, one major objective is the development of universal donor iPSC lines capable of evading immune rejection across genetically diverse patient populations. This is being pursued through targeted deletion of HLA class I and II genes or the enforced expression of immune checkpoint molecules such as PD-L1 and HLA-E, with the goal of producing hypoimmunogenic, cellular products (Cerneckis et al. 2024). Such platforms promise to reduce manufacturing complexity and lower costs, and increase scalability for widespread clinical use.

The refinement of age-enhancement methodologies in iPSC-derived neurons represents a critical advancement in improving the physiological relevance of disease modeling for late-onset neurodegenerative disorders. By inducing aging-associated molecular and epigenetic signatures, these approaches seek to overcome the intrinsic rejuvenation that occurs during reprogramming, thereby enhancing model fidelity. From a therapeutic standpoint, a major objective is the generation of universal donor iPSC lines capable of evading immune rejection across genetically diverse populations. This is being achieved through targeted deletion of HLA class I and II genes or enforced expression of immune checkpoint molecules such as PD-L1 and HLA-E, to produce hypoimmunogenic cellular products (Cerneckis, Cai, and Shi 2024). Such strategies are anticipated to simplify manufacturing, reduce production costs, and enhance scalability for broad clinical implementation.

Concurrently, the integration of gene editing and autologous iPSC therapy is emerging as a transformative paradigm in regenerative medicine. Patient-specific iPSCs can be genetically corrected using CRISPR-Cas9 or related technologies to remove pathogenic variants and subsequently differentiate into functional cell types for autologous transplantation. This dual-modality approach ensures both immunological compatibility and correction of the underlying genetic defect. Preclinical investigations, including those in Canavan disease, have demonstrated the feasibility and therapeutic promise of this strategy (Cerneckis et al. 2024).

Recent advancements in imaging and electrophysiological technologies have significantly enhanced the functional characterization of iPSC-derived neural systems. High-resolution live-cell imaging, calcium imaging, optogenetics, and multielectrode array (MEA) platforms now enable real-time assessment of neuronal network dynamics, synaptic transmission, and plasticity within iPSC-derived neural assemblies. In parallel, the integration of high-content single-cell transcriptomics, epigenomics, and proteomics allows for deep molecular and cellular profiling of disease-relevant phenotypes, providing unprecedented insight into cell-type-specific vulnerabilities and the mechanistic basis of neurodegenerative pathology. Collectively, these next-generation analytical tools are advancing both the precision and translational relevance of iPSC-based models in neuroscience.

Clinically, the iPSC therapy landscape is entering a transformative phase marked by innovative trial designs and expanding therapeutic targets. The favorable safety and preliminary efficacy outcomes observed in recent PD trials involving iPSC-derived dopaminergic neuron transplantation are expected to drive broader clinical applications across other neurodegenerative diseases (Sawamoto et al. 2025). As long-term safety profiles become more clearly defined, conditions such as ALS and Huntington's disease are anticipated to emerge as next-generation indications for cell-based interventions. Concurrently, iPSC models are increasingly being integrated into personalized medicine frameworks. The generation of patient-specific neurons provides a powerful platform for individualized drug screening and assessment of gene therapy responses, enabling stratified, mechanism-based therapeutic strategies tailored to a patient's molecular signature.

However, these clinical advancements demand parallel evolution in regulatory frameworks to address the unique complexities of gene-edited, patient-specific, and off-the-shelf iPSC-based therapies. Standardization of manufacturing processes, quality assurance measures, and long-term follow-up protocols will be critical to ensure both safety and equitable clinical implementation. In conclusion, iPSC technology has fundamentally reshaped the research landscape for neurodegenerative diseases, providing unprecedented opportunities for mechanistic modeling, therapeutic discovery, and translational application. Despite these achievements, substantial challenges persist, including the immature functional state of iPSC-derived neurons, variability across experimental platforms, the scalability of cGMP-compliant manufacturing, and the need to demonstrate durable efficacy and circuit integration in human patients. Continued progress in gene editing, 3D neural tissue engineering, vascularization, and organoid modeling together with accumulating clinical evidence will be key to overcoming these barriers. As emphasized by Cerneckis et al. (2024), iPSC-based systems are redefining experimental and translational frontiers in neuroscience. With rigorous scientific validation, ethical governance, and adaptive regulatory oversight, the coming decade is poised to witness the transition of iPSC-derived disease models and regenerative therapies from preclinical exploration to clinically viable treatments for AD, PD, ALS, Huntington's disease, and other currently untreatable neurodegenerative disorders.

7.11 CONCLUSION

This chapter provides a systematic and comprehensive examination of the therapeutic potential of NSCs in neurodegenerative diseases. It begins by outlining the pathological hallmarks and unmet clinical needs associated with major disorders such as AD, PD, Huntington's disease, and ALS. It then explores different source of NSC, including fetal tissue, embryonic stem cells, iPSCs, and adult brain niches, while detailing protocols for expansion, lineage-specific differentiation, and safety assessment. The therapeutic mechanisms of NSCs are discussed across three principal domains: (1) neuronal replacement and neurogenesis, (2) secretion of neurotrophic and anti-inflammatory factors, and (3) immunomodulation within the diseased CNS microenvironment. Further section summarizes the recent preclinical and clinical progress, encompassing the use of genetically modified NSCs, encapsulated delivery systems, and combination strategies such as NSCs with neurotrophic factors or gene therapy. Advanced sections address manufacturing challenges, including the maintenance of purity, potency, and regulatory compliance, along with ethical considerations surrounding cell sourcing and translational application.

The chapter concludes by outlining future directions in precision NSC therapeutics, including patient-specific iPSC-derived NSCs, allogeneic off-the-shelf NSC platforms, and bioengineered delivery technologies. By integrating recent experimental and translational insights, the chapter serves as a critical resource for researchers and clinicians seeking to advance NSC-based interventions for currently intractable neurodegenerative diseases.

REFERENCES

Cerneckis J, Cai H, Shi Y. Induced Pluripotent Stem Cells (iPSCs): Molecular mechanisms of induction and applications. *Signal Transduct Target Ther*. 2024;9:112.

Dawoody Nejad L, Pioro EP. Modeling ALS with patient-derived iPSCs: Recent advances and future potentials. *Brain Sci*. 2025;15(2):134.

Kirkeby A, Main H, Carpenter M. Pluripotent stem-cell-derived therapies in clinical trial: A 2025 update. *Cell Stem Cell*. 2025;32(1):10–37.

Monk R, Jones KS, Connor B. Cell reprogramming to model Huntington's disease: A comprehensive review. *Cells*. 2021;10(7):1565.

Novoa JJ, Westra IM, Steeneveld E, Neves NF, et al. Good manufacturing practice–compliant human induced pluripotent stem cells: From bench to putative clinical products. *Cytotherapy*. 2024;26(6):556–66.

Ohtonen S, Giudice L, Jäntti H, Fazaludeen MF, Shakirzyanova A, et al. Human iPSC-derived microglia carrying the LRRK2-G2019S mutation show a Parkinson's disease related transcriptional profile and function. *Sci Rep*. 2023;13:22118.

Penney J, Ralvenius WT, Tsai LH. Modeling Alzheimer's disease with iPSC-derived brain cells. *Mol Psychiatry*. 2020;25:148–67.

Shah AJ, Dar MY, Jan B, Qadir I, Mir RH, Uppal J, Ahmad NZ, Masoodi MH. Neural stem cell therapy for Alzheimer's disease: A state-of-the-art review. *J Dement Alzheimer's Dis*. 2024;1(2):109–25.

Stoddard-Bennett T, Pera RR. Stem cell therapy for Parkinson's disease: Safety and modeling. *Neural Regen Res*. 2020;15(1):36–40.

Sawamoto N, Doi D, Nakanishi E, Sawamura M, Kikuchi T, Yamakado H, … Takahashi J. Phase I/II trial of iPS-cell-derived dopaminergic cells for Parkinson's disease. *Nature*. 2025;641(8064):971–7.

Ting HC, Guo YT, Su HL, Chen YS, Lin SZ, Harn HJ, Chang CY. Rapid iPSC-derived neuromuscular junction model uncovers motor neuron dominance in amyotrophic lateral sclerosis cytopathy. *Cell Death Discov*. 2025;11:23.

Index

B

C

D

E

For Product Safety Concerns and Information please contact our
EU representative GPSR@taylorandfrancis.com Taylor & Francis
Verlag GmbH, Kaufingerstraße 24, 80331 München, Germany